Crop Postharvest and Storage

Crop Postharvest and Storage

Edited by **Fernando Plath**

SYRAWOOD
PUBLISHING HOUSE

New York

Published by Syrawood Publishing House,
750 Third Avenue, 9th Floor,
New York, NY 10017, USA
www.syrawoodpublishinghouse.com

Crop Postharvest and Storage
Edited by Fernando Plath

International Standard Book Number: 978-1-68286-057-1 (Hardback)

Printed in the United States of America.

Contents

Preface

The quality of a crop is determined not only by a healthy harvest, but also on how well it is processed after that which includes preserving, cleaning, sorting and packing of the crop. Post-harvest management plays a significant role in avoiding physical, chemical and biological damage to the crop and storing it for a long duration of time. The book explains economics of storage, post-harvest losses and techniques to maintain grain quality in storage. It covers laboratory as well as field studies to solve storage related problems and preservation of stored products. Students, researchers and industry experts will find a comprehensive insight into the latest practices, tools and techniques in crop postharvest and storage.

All of the data presented henceforth, was collaborated in the wake of recent advancements in the field. The aim of this book is to present the diversified developments from across the globe in a comprehensible manner. The opinions expressed in each chapter belong solely to the contributing authors. Their interpretations of the topics are the integral part of this book, which I have carefully compiled for a better understanding of the readers.

At the end, I would like to thank all those who dedicated their time and efforts for the successful completion of this book. I also wish to convey my gratitude towards my friends and family who supported me at every step.

Editor

Proximate composition of selected groundnut varieties and their susceptibility to *Trogoderma granarium* Everts attack

A. K. Musa[1]*, D. M Kalejaiye[1], L. E. Ismaila[1] and A. A. Oyerinde[2]

[1]Department of Crop Protection, University of Ilorin, P. M. B. 1515, Ilorin, Nigeria.
[2]Department of Biology, University of Abuja, Abuja, Nigeria.

The proximate composition including moisture content, crude fat, crude protein and ash were determined for six groundnut varieties. Moisture content was highest (8.9%) in SAMNUT 19 followed by SAMNUT 23(8.6%) and was lowest (6.6%) in EX-DAKAR while ash ranged from 3.0% in RRB to 7.4% in RMP-91. The crude protein was highest (31.3%) in EX - DAKAR and lowest (19.7%) in SAMNUT 19. RRB had the highest crude fat of 53.1% while EX-DAKAR had the lowest value of 32.7% showing significant difference ($p < 0.05$) between them. Groundnut variety RRB had the highest percentage kernel damage assessment and kernel weight loss during the storage period. However, these findings may offer scientific basis for *Trogoderma granarium* attack of groundnuts as manifested by nutritional contents of the kernels, percentage kernel damage, percentage weight loss and adult population of the dermestid.

Key words: *Arachis hypogaea*, nutrient composition, *Trogoderma granarium*.

INTRODUCTION

Groundnut, *Arachis hypogaea* L, is an important oil crop of Brazilian origin, is cultivated in tropical and warm temperate climates. The crop is grown usually as a component of a variety of crop mixtures including sorghum, millet, cowpea and maize (Misari et al., 1988). Groundnut is an important oil seed and cash crop accounting for more than one-third of the total oil seed production in India (Sahayaraj and Martin, 2003).

Groundnuts are not only rich in proteins which are easily digestible and consequently, a higher biological value, but are also rich in B-complex vitamins. It is an important item in several confectionery products, and in supplementary feeding programmes such as in weaning food formulations in combination with cereals and pulses in many developing countries. Various cultivars of groundnut tested in Andhra Pradesh, southern India have shown high contents of P and K, possibly due to varietal differences (Pillari et al., 1984). The principal use of groundnut however, is in the production of oil (Cummings, 1986; Elegbede, 1998).

However, production and preservation of this crop is faced by numerous challenges. Since the pods are located underground, the challenge is the absence of suitable implements for harvesting. The curing or drying and storage facilities at the farmers' level also contribute much to the deterioration of seed or kernel quality in storage (Marthur and Jorgensen, 1992). The seeds are prone to quality and quantity losses during storage due to very serious insect pest damage especially Khapra beetle, *Trogoderma granarium* Everts. The insect was discovered in stored guinea corn in Nigeria in 1948 and may have been present in stored groundnuts as early as 1944 (Pasek, 1998). In Nigeria, this insect pest seriously threatens the sustainability of groundnut preservation. Adults *T. granarium* are short-lived and do not feed, but their larvae voraciously feed and cause heavy contamination to the stored product through mass webbing and frass. It will feed on almost any dried plant or animal matter, including dog food, dried orange pulp, bread, and dried coconuts (Szito, 2006). This is particularly distressing for young children, who develop vomiting and diarrhoea, and refuse food (Anonymous, 2001). The aim of the experiment was to determine the

Table 1. Proximate composition of groundnut varieties.

Variety	Moisture	Crude Fat	Crude Protein	Ash
RMP-12	7.6[a]	47.0[a]	25.6[a]	4.6[a]
RMP-91	7.3[a]	33.7[b]	24.4[a]	7.4[a]
SAMNUT 19	8.9[a]	49.1[a]	19.7[a]	6.5[a]
SAMNUT 23	8.6[a]	47.1[a]	24.9[a]	4.8[a]
EX-DAKAR	6.6[a]	32.7[b]	31.3[a]	5.6[a]
RRB	8.0[a]	53.1[a]	24.5[a]	3.0[a]

Values are expressed as g/100 g samples. Values in the column with different superscript differ significantly at $p < 0.05$.

proximate composition of selected groundnut varieties and severity of attack of *T. granarium* larvae on the groundnuts in three months storage period.

MATERIALS AND METHODS

Insect culture

Fifty adults *T. granarium* (1:1) were obtained from existing culture in the Nigerian Stored Products Research Institute, Ilorin, Nigeria. The insects were maintained on a groundnut variety (RRB) kept in a Kilner jar (250 ml) fitted with wire-mesh cap to allow aeration and prevent entry of other insects. All live and dead insects were removed two weeks after infestation while emerging larvae were used for the experiment. The culture was raised under prevailing temperature (25 ± 3°C) and relative humidity (68 ± 3%) determined by thermohygrograph. This is an instrument used to keep proper records of temperature and relative humidity changes.

Collection and preparation of groundnut varieties

Six improved groundnut varieties used for this study include RMP-12, RMP-91, SAMNUT 19, SAMNUT 23, EX-DAKAR and RRB. They were collected from the Institute for Agricultural Research (IAR), Samaru, Nigeria. The groundnut varieties were air-dried for 4 weeks and then wrapped in separate black polythene bags sealed in a clean dry air-tight container and placed in a deep freezer (-4°C) for 14 days to disinfest the decorticated seeds. The seeds were conditioned on separate plastic trays placed on the laboratory bench for 7 days.

Analysis of groundnut samples

The standard methods of the Association of Official Analytical Chemists (AOAC, 1990) were used to determine moisture, ash, crude fat and crude protein content. Moisture content was obtained by heating three 5.0 g portions of the groundnut samples in an oven (Gallenkamp QC, England) at 110°C until a constant weight was obtained. Ash determination was obtained by the incineration of three 3.0 g samples in a muffle furnace at 600°C for 3 h when a light-grey ash was produced. Crude protein (CP) was obtained using three 3.0 g portions of the samples. The CP was calculated by a multiplying factor (%N × 6.25). The crude fat (CF) was determined by extraction procedure using three 5.0 g samples in a Soxhlet apparatus using petroleum spirit (bp 40 - 60°C) as the solvent.

Determination of damage assessment/weight loss

One hundred grams (100 g) each of the groundnut varieties were weighed in plastic containers (12 cm in diameter) and replicated four times. Five freshly emerged larvae *T. granarium* (about 2 days old) were introduced into each container and then covered with muslin cloth secured with elastic rubber band to allow aeration and prevent entry of other insects or arthropods. An uninfested control of each groundnut variety was included in the experimental set up, laid out in completely randomized design (CRD). The experiment was left undisturbed to allow damage assessment and weight loss calculation at monthly intervals for three months.

Damage assessment involved sieving the groundnut samples using a sieve with a mesh diameter 0.01 mm to remove dust, frass and insects. The samples were then reweighed. Determination of their comparative weights was calculated in terms of the whole sample to give percentage kernel damage.

The percentage kernel weight loss was calculated using the following formula:

$$\frac{\text{Wt. of control kernels} - \text{Wt of infested kernels}}{\text{Wt. of control kernels}} \times 100$$

The total population of adult *T. granarium* in the infested groundnut kernels was recorded after counting with a tally counter.

Data analysis

Data were subjected to analysis of variance and the different means were separated using Student-Newman-Keuls (SNK) test.

RESULTS

The percentage chemical composition of selected groundnut varieties is shown in Table 1. The percentage moisture content ranged from 6.6 - 8.9%. It was highest in SAMNUT 19 followed by SAMNUT 23 (8.6%) and was lowest in EX-DAKAR. The crude fat content of the varieties ranged between 32.7 - 53.1%. Groundnut varieties RRB and EX-DAKAR had the highest and least crude fat content, respectively. There were significant ($p < 0.05$) differences among the varieties for crude fat. The crude protein cont-

Table 2. Groundnut kernel damage assessment following infestation with larvae *T. granarium.*

Variety	Percentage damage assessment (DAI)		
	30	60	90
RMP-12	33.1c	43.7c	41.4b
RMP-91	35.4bc	57.3b	67.4b
SAMNUT 19	37.3abc	61.3b	71.2a
SAMNUT 23	37.5abc	62.3b	80.3a
EX-DAKAR	38.6ab	75.6b	87.0a
RRB	41.4a	91.8a	87.5a
CV (%)	6.5	11.9	14.3
SE(±)	7.6	12.5	16.8

Values in the column with different superscript differ significantly at p < 0.05; DAI = Day After Infestation.

Table 3. Groundnut kernel weight loss following infestation with larvae *T. granarium.*

Variety	Percentage kernel weight loss (DAI)		
	30	60	90
RMP-12	24.9abc	31.0a	51.8a
RMP-91	17.4c	27.5a	45.5a
SAMNUT 19	22.7bc	29.0a	49.7a
SAMNUT 23	25.3abc	31.5a	55.0a
EX-DAKAR	29.2ab	34.8a	56.5a
RRB	33.9a	4108a	57.6a
CV (%)	18.8	29.4	14.9
SE(±)	4.82	7.89	9.55

Values in the same column followed by a common letter(s) did not differ significantly at p = 0.05 (SNK).

ent ranged between 19.7 - 31.3% for SAMNUT 19 and EX-DAKAR, respectively. The ash content of the different varieties ranged between 3.0 - 7.4%. RMP-91 and RRB had the highest and lowest ash content, respectively. There was no significant difference (p > 0.05) among the varieties for moisture, crude protein and ash content.

The groundnut kernel damage assessment following *T. granarium* infestation is shown in Table 2. The varieties differed significantly (p < 0.05) in percentage kernel damage assessment. RRB had the highest percentage damage assessment of 41.4, 91.8 and 87.5% at 30, 60 and 90 days after infestation (DAI), respectively. RRB differed significantly from RMP-12 (33.1%) and RMP-91 (35.4%) at 30 DAI. RRB also differed significantly from other varieties at 60 DAI, but it was not significantly different (p > 0.05) from SAMNUT 19, SAMNUT 23 and EX-DAKAR during the storage period. RMP-12 had the lowest percentage kernel damage assessment during the study period. However, RMP-12 differed significantly from other varieties at 60 DAI but it was not significantly

different from RMP-91 at 90 DAI.

The groundnut kernel weight loss following infestation with larvae *T. granarium* is shown in Table 3. The groundnut varieties differed significantly (p < 0.05) at 30 DAI in percentage kernel weight loss. Groundnut variety RRB had the highest percentage kernel weight loss (33.9%) which differed significantly from RMP-91 and SAMNUT 19. Similarly, at 60 and 90 DAI, RRB variety had the highest percentage kernel weight loss; however, the value was not significantly different from other varieties. RMP-91 had the lowest percentage kernel weight loss, but it was not significantly different from SAMNUT 19 and SAMNUT 23 at 30 DAI.

Groundnut variety (SAMNUT 19) had the highest and EX-DAKAR the lowest build up of adult *T. granarium* population at 60 and 90 DAI (Table 4). SAMNUT 19 was not significantly different from RMP- 91. EX-DAKAR was not significantly different from RMP- 12 and SAMNUT 23 at 60 DAI. However, EX-DAKAR was significantly different from other varieties.

Table 4. Comparison of total population of adult *T. granarium* following artificial infestation in stored groundnut.

Variety	Mean adult population(DAI)		
	30	60	90
RMP-12	0	10.0bc	11.2b
RMP-91	0	48.0a	54.6a
SAMNUT 19	0	54.3a	58.4a
SAMNUT 23	0	12.3bc	13.6b
EX-DAKAR	0	4.3c	5.4c
RRB	0	15.0b	16.8b
CV (%)	-	29.6	26.4
SE(±)	-	3.18	3.15

Values in the same column followed by a common letter(s) did not differ significantly at p = 0.05 (SNK).

DISCUSSION

The factors affecting insect infestation of feedstuffs include temperature, moisture, source of insects, available food, air, condition of the feedstuff, presence of other organisms, and the efforts to exclude or kill the pests (Durham, 2008). The nutritive content of groundnut and certain physical properties of feeds such as free water and heat will also determine the vulnerability of such materials to insect attack, especially *T. granarium*. The moisture content ranged between 6.6 - 8.9% in the groundnut varieties studied. The 6.6% moisture content was lower than the safe level required for proper storage reported by Ofuya and Lale (2001). The dermestid beetle attacked the groundnut varieties irrespective of the level of moisture. The beetle can feed on products with as little as 2% moisture content (Pasek, 1998). Moisture content is of importance in storage because the lower the moisture content of food material, the higher the keeping quality (Ajayi and Adedire, 2007).

Feeding by *T. granarium* larvae reduces the weight of kernels in a very short storage period. Feeding activity of the insect pest results in an increase in moisture, crude fat and total protein content (Mason, 2002). In India, loss of weight in groundnut infested by *T. granarium* larvae ranged from 2.2 - 5.5% in wheat (French and Venette, 2005). Previous studies have indicated that severe infestations of grain by Khapra beetle may make it unpalatable or unmarketable. Infestation levels of 75% in wheat, maize, and sorghum grains resulted in significant decreases in crude fat, total carbohydrates, sugars, protein nitrogen, and true protein contents and increases in moisture, crude fibre and total protein (Jood and Kapoor, 1993; Jood et al., 1993, 1996).

The ash content (3 - 7.4%) of the groundnut varieties is to some extent indicative of mineral content (Jossyln, 1973) and it is expected that RMP-91 with highest ash content would have highest mineral content (Abu, 2005). The varieties contained varying degrees of ash content.

This is due to the fact that the crop derives its nutrients from the soil.

Crude protein in the groundnut varieties studied ranged from 19.7 - 31.3%. This range compares favourably with the 25 - 30% reported by Metcalfe and Elkins (1980). The results from this study confirmed the observation that groundnut is rich in protein content. This might be attributed to genetic constitution, climatic and varietal differences. The high protein content makes groundnut a good food supplement for man and livestock. The groundnut probably contain protein at a level comparable to that in cowpea (23 - 30%) (Ngoddy and Ihekoronye, 1985).

The varieties had varying degrees of damage by *T. granarium* larvae. The attack of groundnut may be ascribed to the nutritive content, soft kernel coat, and kernel size. The insect attack and infestation on this crop needs to be carefully studied especially on the genetic composition to determine inherent factors responsible for response of *T. granarium* to groundnut varieties. However, storage insect infestation has been reported to be severe on improved varieties than local varieties (Enobakhare and Law-Ogbomo, 2002). Varietal differences exerted significant influence on parameters examined as higher kernel damage and loss were consistently recorded during the period. Despite high preference shown by SAMNUT 19 for adult build-up, it had moderate kernel damage and loss than RRB. Stibick (2007) reported that *T. granarium* is a dirty feeder because it damages more grain than it consumes. If the dermestid is left undisturbed in stored grain it can cause significant weight loss. Weight loss can be between 5 - 30%, sometimes in extreme cases 70% (Dwivedi and Shekhawat, 2004). In this investigation, it was observed that the population of adults was much lower than larvae.

Under optimal conditions, it was found that Khapra beetle can sustain a population increase of between 8.3 times per month. For that reason, population can build-up rapidly in a short time under hot, dry conditions (Ofuya and Lale, 2001). The larvae feed voraciously, causing heavy contamination to the stored groundnut through mass webbing, exuviae and frass. In an experiment, Mansoor-ul-Hassan et al. (2006) observed a build-up of *T. granarium* at 60 days post treatment. Musa et al. (2009) suggested repeated applications of *Hyptis suaveolens* Poit. seed extracts against *T. granarium* to control the competitiveness of the insect with man for nutritive value of the groundnuts. It is evident that RRB had the highest percentage kernel weight damage and loss during the three months storage period. Infestation of grains by Khapra beetle may present a serious health implication. Cast skins may cause dermatitis in people handling heavily infested grains (Pasek, 1998). Sneezing and eye irritation may be experienced by people handling heavily infested groundnut kernels.

ACKNOWLEDGEMENT

We are grateful to the Institute for Agricultural Research, Ahmadu Bello University, Zaria for supplying the groundnut varieties.

REFERENCES

Abu EA (2005). Comparative study of the effect of natural fermentation on some biochemical and anti-nutrient composition of soyabean (*Glycine max* (L.) Merr) and the locust bean (*Parkia filicoidea*) J. Trop. Bio. Sci., 5(1):6-11.

Ajayi OE, Adedire CO (2007). Nutrient characteristics of the subterranean termite, *Macrotermes subhyalinus* (Rambur) (Isoptera: Termitidae) Nig. J. Ent., 24:42-47.

Anonymous (2001). Bed bugs. Insects that can cause health problem. Pest Notes, 3(8):20-21.

AOAC (1990). Association of Official Analytical Chemists. Official methods of analysis. 13th edn.(Ed. W. Horwitz) Washington DC, p. 1141.

Cummings DG (1986). Groundnut. The unpredictable legume: Production constraints and Research needs. Proceedings International Symposium, ICRISAT Sahelian Center, Niamey, Niger.

Durham S (2008). Grain moisture measurements may divert mould, insect infestation. United States Department of Agriculture. Agricultural Research Service.

Dwivedi SC, Shekhawat NB (2004). Repellent effect of some indigenous plant extracts against *Trogoderma granarium* Everts. Asian J. Exp. Sci., 18 (1&2): 47-51 Post- harvest Sc.1.

Elegbede JA (1998). Legumes. In: Nutritional quality of plant foods. 1st edn. Post-harvest Research Unit. Department of Biochemistry, University of Benin, Benin City, Nigeria, pp. 53-83.

Enobakhare DA, Law-Ogbomo KE (2002). Reduction of harvest loss caused by *Sitophilus zeamais* (Motsch.) in three varieties of maize treated with plant products, pp. 1-6.

French S, Venette RC (2005). Mini risk assessment. Khapra beetle, *Trogoderma granarium* Everts (Coleoptera: Dermestidae), p. 22.

Jood S, Kapoor AC, Singh R (1993). Available carbohydrates of cereal grains as affected by storage and insect infestation. Plant Foods Hum. Nutr., 43: 45-54

Jood S, Kapoor AC, Singh R (1996). Chemical composition of cereal grains as affected by storage and insect infestation. Trop. Agric. Trinidad, 73: 161-164.

Jossyln MA (1973). Methods in food analysis. Academic Press. Inc. New York.

Mansoor-ur-Hassan, Sagheer ME, Ulah- Ahmad F, Waki W (2006). Comparative efficacy of ethanol leaf extracts of *Amaranthus viridis* L. and *Salsola baryosma* (Scbdres) and cypermethrin against *Trogoderma granarium* Everts. Pak. J. Agric. Sci., 43 (1-2): 52-58.

Mason LJ (2002). Khapra beetle. Stored grain management. Country Journal Publishing Co., Decatur, Illinois, Grainnet.

Metcalfe DS, Elkins DM (1980). Crop Production: Principles and Practices. 4th Ed. Macmillan Publishing Co., Inc. New York, p. 774.

Marthur SB, Jorgensen J (1992). Seed pathology. Proceedings of the CTA seminar held at Copenhagen, Denmark on 20-25 June, 1988, p. 412.

Misari SM, Ibrahim JM, Demski JMW, Ansa AO, Kuhn CW, Casper R, Breye E (1988). Aphid transmission of the viruses causing chlorotic rosette and green rosette diseases of peanut in Nigeria. Plant Dis., 72: 250-253.

Musa AK, Dike MC, Onu I (2009). Evaluation of nitta (*Hyptis suaveolens* Poit) seed and leaf extracts and seed powder for the control of *Trogoderma granarium* Everts (Coleoptera: Dermestidae) in stored groundnut. Am.-Russian J. Agron., 2(3): 176-179.

Ngoddy PO, Ihekoronye AL (1985). Integrated food science and technology. International College Edn. Macmillan Publisher Limited pp. 250-365.

Ofuya TI, Lale NES (2001). Pests of stored cereals and pulses in Nigeria. Biology, Ecology and Control. Dave Collins Publications, Nigeria, p. 174.

Pasek JE (1998). Khapra Beetle.*Trogoderma granarium* Everts: Pest-initiated pest risk assessment, USDA APHIS, Raleigh, NC, p. 32.

Pillari RN, Ranganakulu G, Padma Raju A, Sankara Reddi GH (1984). Mineral composition of kernels and shells of four cultivars of groundnut. Andhra Pradesh J. (India), 31(4):351-352.

Sahayaraj K, Martin P (2003). Assessment of *Rhynocoris marginatus* (Fab.) (Hemiptera: Reduviidae) as augmented control in groundnut pests. J. Central Eur. Agric., 4(2):103-110.

Stibick J (2007). New Pest Response Guidelines: Khapra Beetle. USDA-APHIS-PPQ-Emergency and Domestic Programs, Riverdale, Maryland, pp. 1-11.

Szito A (2006). *Trogoderma granarium* Global Invasive Species. Database Invasive Species. Specialist Group (ISSG). IUCN Survival Commission.

2

Effects of storage conditions and pre-storage treatment on sprouting and weight loss of stored yam tubers

Z. D. Osunde* and B. A. Orhevba

Department of Agricultural Engineering, Federal University of Technology, Minna, Niger State, Nigeria.

The effect of intermittent forced air flow using a fan and pre-storage treatment prepared by soaking neem bark in water for 12 h and blending neem leaves with water on weight loss and sprouting of stored yam tuber were investigated. A total of 36 tubers of white yam, (*Dioscorea rotundata*) were stored for six months in barn with fan and barn without fan. The results showed that the temperature in the barn with fan was slightly lower than that of the barn without fan. The neem bark extract treated tubers had lower sprout weights (25 g/kg) compared to 45 g/kg for the control. This difference is statistically significant at (P ≤ 0.05). The neem leaf slurry treated tubers also had less sprout weights (33 g/kg) tuber. The tubers treated with neem leaf slurry had the least weight loss (21%) compared to 26% for the tuber treated with neem bark extract. Tubers stored in the barn with fan had the least sprout weight and least weight loss. From the result, it can be concluded that intermittent air flow on stored yam tubers reduces sprouting and weight loss and neem bark extract treatment have an effect in suppressing sprouting in stored yam tubers.

Key words: Yam, storage, neem bark, neem leaf slurry, air flow, weight loss, sprouting.

INTRODUCTION

Yams (*Dioscorea* spp.) are the most important food crops in West Africa, next to cereals, (Onwueme, 1978; Opara, 1999). In 2005, 48.7 million tones of yams were produced worldwide. West and Central Africa account for about 94% of world production, Nigeria being the major producer (IITA, 2007). The white yam (*Dioscorea rotundata*) which originated in West Africa is the most important species of yam cultivated for human nutrition in the region (Gerardin et al., 1998). In Nigeria yam is not only an important staple food, but also has ritual and socio-cultural significance and it is considered a man's crop. Before the introduction of cereals and grains, also important staple foods in West Africa, yams were the major source of carbohydrate (Osunde and Yisa, 2003).

The storage life of the yam tuber is ended at the termination of dormancy, when new sprouts develop. Sprouting in stored yam causes weight and quality loss (Osunde and Orhevba, 2009; Sahore et al., 2007). Good

storage should therefore maintain tubers in their most edible and marketable condition by preventing large moisture losses, spoilage by pathogens, attack by insects and animals, and sprout growth (Osunde, 2008). In order to obtain good quality tubers after storage (that is fresh, edible and marketable yams), the freshly harvested yams to be stored must be clean and undamaged. Also, excessive temperature must be avoided and good aeration provided. Causes of storage losses of yam tubers include: sprouting, transpiration, respiration, rot due to mould and bacteriosis, insects, mammals, nematodes (Osagie, 1992; Ravi et al., 1996a; Opara, 1999). Methods of yam storage vary from delayed harvesting or storage in simple piles or clamps to storage in buildings specially designed for the purpose and application of sophisticated modern techniques (Ravi et al., 1996a). Also, Osagie (1991), Nwakiti and Makurdi (1989) adequately described yam storage practices.

Neem tree derivatives have been used as pest control in rural areas of developing countries (Ganguli, 2002). All parts of the neem tree have medicinal properties and are used for many different medical preparations. Ganguli

*Corresponding author. E-mail: zinashdo@yahoo.com.

(2002) reported that neem is used for the treatment of scabies mite and also, it is effective in treating infestation of lice in humans. Ibrahim et al. (1987) reported that neem tree extract treatment have favorable effect on sprouting in stored yam, as it was able to suppress sprouting for 5 months in stored yam tubers (*D. rotundata*). The effect of air flow in stored yam has been studied by Mozie (1983). In this study, a significant difference was observed in the percentage sprouting rate and the rate of weight loss of yam tubers stored in the conventional barn when supplied with airflow intermittently, continuously or without airflow. He observed that intermittent airflow allowed significant less weight loss and less sprouting than continuous airflow and no airflow. Other treatments such as oil coating and chemical treatments have been used to reduce sprouting and weight loss in stored yam. Ohu et al. (2007) reported that yam tubers coated in palm oil experienced significantly less loss of weight, moisture and dry matter. It was also reported that the Chloro-Isoprophyl Phenyl Carbamate (CIPC) chemicals which is successfully used to suppress sprouting in stored potato has no effect in stored yam (Orhevba and Osunde, 2006). Malic hydrazide (1000 ppm) has been reported to prolong dormancy of yam tubers of *Dioscorea alata* and *D. rotundata* (Ramanujam and Nair, 1982). The present study is aimed at investigating and evaluating the effect of intermittent forced airflow and neem tree bark and neem leaf slurry treatments on sprouting and weight loss of stored yam tubers.

MATERIALS AND METHODS

The experiment was conducted in Minna, Niger State of Nigeria which is located on the Guinea Savannah Ecological zone. Two traditional yam barns were used for this experiment; they were erected in the open air, where sufficient shade and ventilation was available. The frame of the barn consisted of vertically erected wooden poles of 2 m height, set at a distance of 1 m to each other. These wooden poles were stabilized by attaching horizontal poles to them. The dimensions for each barn were 2.5, 3.5 and 2 m, width, length and height, respectively. Locally knitted thatch (made of dried plant stalks) were wound round the frame and the top, this served as the roof and the wall. There was a slight opening between the roof and wall to allow for optimum ventilation and reduction in ambient temperature inside the barn.

Two of such structures were built and a standing fan to aid airflow was placed in one of them. The fan, with a blade diameter of 40 cm and airflow rate of 0.86 m^3/ second was placed at a corner and allowed to supply air at 2 hourly intervals throughout the experiment period (8 to 10 am, 12 to 2 pm, 4 to 6 pm, 8 to 10 pm) while the second barn was without a fan and airflow was natural. The "medium" fan speed with 27.24 m/s speed and which rotated at 1800 (to enable even ventilation for all the tubers) was used for providing the forced intermittent airflow. A total of 36 tubers of white yam *D. rotundata* tubers "giwa" variety were stored in each barn, this were further sub-divided into three sub-groups of 12 tubers each. The initial weight of the tubers were measured and recorded based on treatments used. The neem bark extract was prepared by soaking 5 kg of neem bark in water for 12 h and twenty four tubers of yam (twelve for each barn) with a total weight of 28.5 kg were treated using the prepared neem bark extract. The neem leaf slurry was prepared by blending 1 kg of neem leaves with 2 L of water and was used for twenty four tubers of yam with a total weight of 26.8 kg.

In addition to this, a control sample of 12 tubers in each barn with no treatment was stored. The tubers were arranged on a raised wooden platform, which were placed on the floor of the barns to reduce bruising and to facilitate airflow, weighing and making observations. Temperature and relative humidity inside the barns were measured three times a week and four times a day (8:00 am, 12 noon, 4:00 and 8:00 pm). Temperature and humidity readings were taken using a Mebus 4.0 digital thermo-hygrometer. The tubers were weighed before storage and at monthly intervals throughout storage period. To determine the sprout vigor the sprouts were removed manually twice a month and weighed. Percentage weight loss was determined based on the initial tuber weight while sprouting vigor was determined by weighing the sprouts of the yam tubers. The experimental design employed for this work was 3 x 2 factorial designs with 3 replicates. The results were analyzed using ANOVA and the means analyzed using F-LSD at P≤ 0.05.

RESULTS

Temperature in the barns

The summary of the average monthly minimum and maximum temperature in the two barns is presented in Table 1. The temperature in the barn with fan fluctuated between 20.5 and 36°C with an average of 29°C while that in the barn without fan fluctuated between 23 and 38°C, with an average of 33°C over the storage period. The average temperature in the barn with fan was 4°C less than in the barn without fan. The maximum temperature was obtained at 4:00 pm while the minimum temperature was recorded at 8:00 am. Figure 1 shows the average daily temperature variation for the two barns. From the figure, the barn without fan had the highest temperature (36.5°C) at 4:00 pm while the barn with fan had a temperature of (33.5°C) at the same period. The barn with fan had the lowest temperature (27.08°C) at 8:00 am while that of barn without fan was 30.92°C during the same period.

Relative humidity in the barns

The summary of the average monthly minimum and maximum relative humidity in the two barns is as presented in Table 2. The relative humidity in the barn with fan ranged between 26.5 and 60.4% with an average of 38% while that in the barn without fan ranged between 23 and 55% with an average of 44% over the storage period. The average relative humidity in the barn with fan was about 6% higher than in the barn without fan. Figure 2 shows the average daily humidity variation for the two barns. From the figure, the barn without fan had the lowest relative humidity (33.67%) at 4:00 pm while the barn with fan had 38.83% during the same period.

Table 1. Average monthly minimum and maximum temperature inside the two barns.

Months	Barn with fan		Barn without fan	
	Minimum	Maximum	Minimum	Maximum
January	20.5	29.5	23	32
February	26.5	34.5	29.5	36.5
March	28	36	31	38
April	28.5	35.5	32	37
May	30	33.5	32.5	35.5
June	29	32	31.5	34

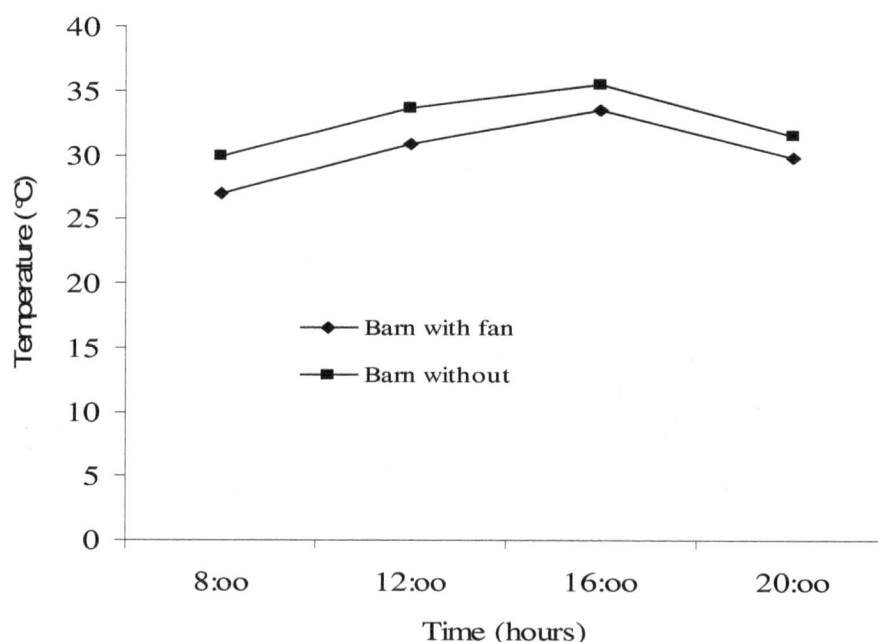

Figure 1. Average temperature in the two barns.

Table 2. Average monthly minimum and maximum relative humidity inside the two barns.

Months	Barn with fan		Barn without fan	
	Minimum	Maximum	Minimum	Maximum
January	26.5	32.4	23	29.5
February	28	32	25	30
March	32.4	51.5	26	49
April	35	57	32	55
May	45.8	57	42.5	55
June	53.4	60.4	48.5	55

Effect of neem bark extract and neem leaf slurry treatments on weight loss of stored yam tubers

Figure 3 shows the percentage weight loss of tubers treated with neem tree bark extract and neem leaf slurry stored in barn with fan while Figure 4 shows that of the barn without fan. From Figure 3 tubers treated with neem leaf slurry had the lowest weight loss throughout thestorage period. A similar observation was made in the barn without fan (Figure 4). The neem bark extract treated tubers and the control had higher weight loss compared to the neem leaf slurry treated tubers throughout the storage period. Summary of average weights loss of yam tubers treated with neem extract at the end of the storage period for the two storage methods is presented in Table 3. From the table, it is seen that tubers stored in the barn with fan has the lowest weight loss for all the treatment. It is also observed that in all

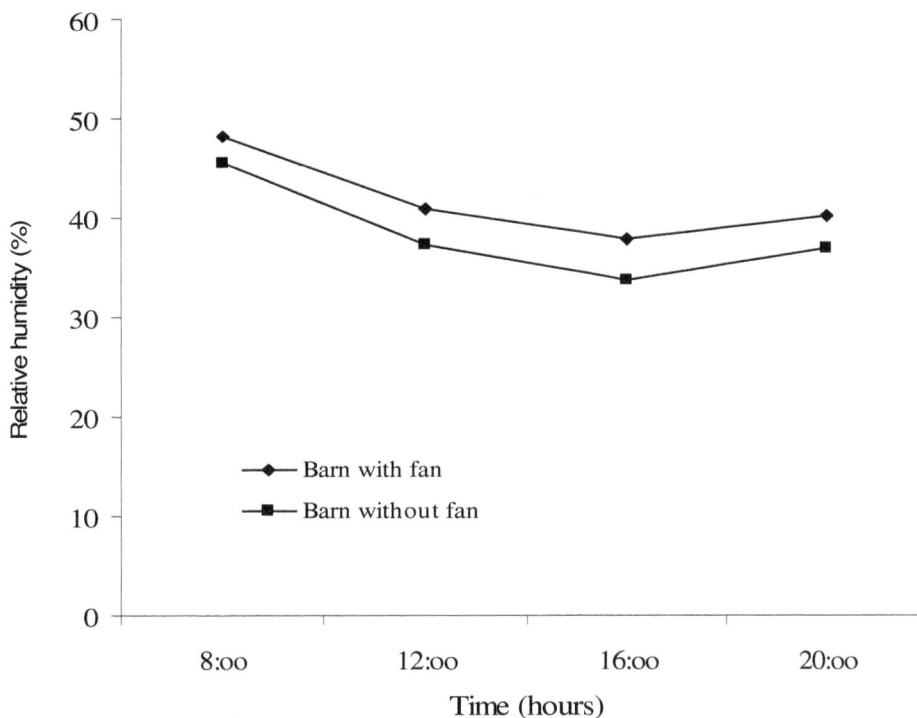

Figure 2. Average relative humidity in the two barns.

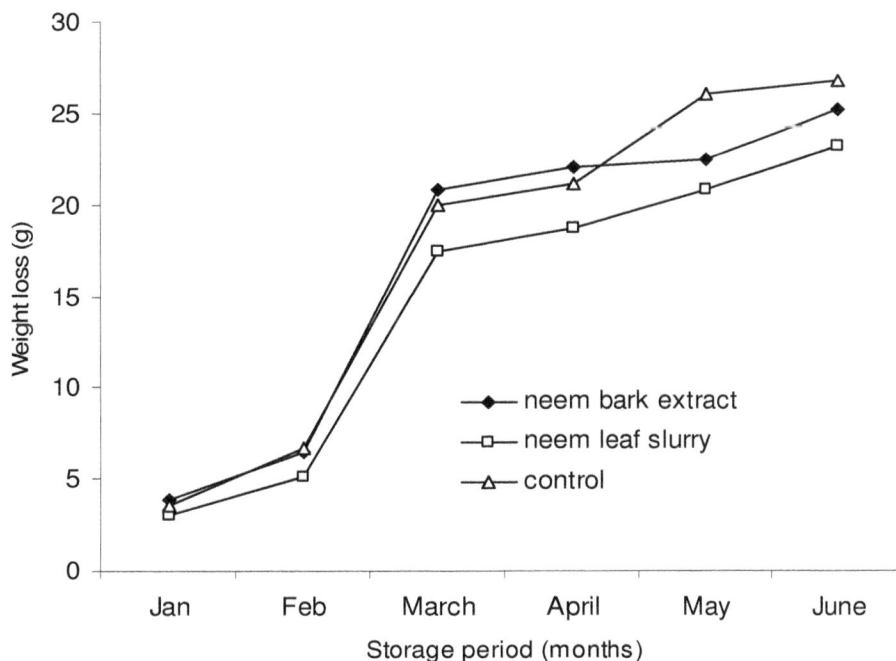

Figure 3. Monthly average percentage weight loss of yam tubers stored in barn with fan.

cases (structure and treatment) the tuber lost more than 20% of its initial weight at the end of six month storage period. After six months of storage the tubers stored in the ventilated barn showed 6.2% less weight loss compared to the tubers in barn without fan. The Analysis of Variance (ANOVA) for weight loss presented in Table 4 shows that, both the treatment and the barn type had significant effect on the weight loss and there was no interaction effect. The means for the treatment were separated by LSD to determine the factor with significant

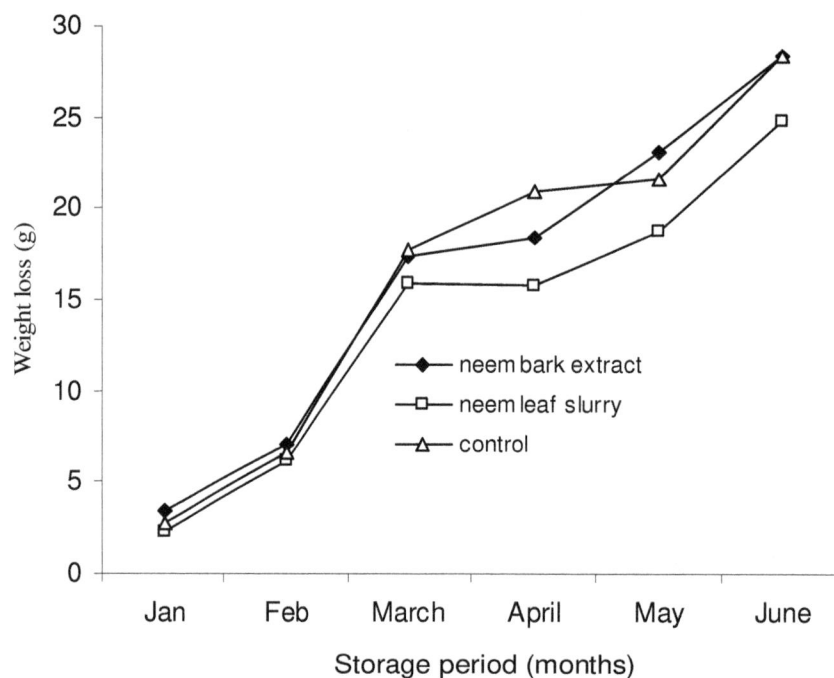

Figure 4. Monthly percentage weight loss of yam tubers stored in barn without fan.

Table 3. Average percentage weight loss of yam tubers at the end of the storage period.

	Weight loss (%)		
	Neem bark extract	Neem leaf slurry	Control
Barn with fan	26.2	21	26.8
Barn without fan	28.42	24.6	28.36

Table 4. ANOVA table for weight loss.

Sources of variation	df	ss	ms	Fcal	Ftab	Remarks
Treatment combinations	7	199.86	28.55	0.006	2.66	ns
Factor A (Neem extract)	3	27691.77	19230.59	3.738	3.21	*
Factor B (barn type)	1	43592.92	43592.92	8.473	4.49	*
Interaction AB	3	11236.75	3745.58	0.728	3.21	ns
Error	16	82321.58	5145.10			
Total	23					

ns = not significant; * = Significant.

effect on weight loss. Table 5 shows the result of the F-LSD for weight loss.

Effect of neem bark extract and neem leaf slurry treatment on sprouting of stored yam tubers

Figure 5 shows the sprouting vigor of neem bark extract and neem leaf slurry treated yam tubers and stored in barn with fan. Sprout vigor was high in the control sample compared with the neem tree extract treated tubers. The neem bark extract treated tubers generally had the lowest sprout weights followed by the neem leaf slurry treated tubers. Figure 6 shows the sprouting vigor of neem bark extract and neem leaf slurry treated yam tubers and stored in barn without fan. Similarly, as in the barn with fan, the control had the highest sprout weights; however, there was no remarkable difference between the tubers

Table 5. Result of F – LSD for weight loss.

Sources of variation	Sum of squares	Degree of freedom	Mean differences	LSD values at 0.05% variances	Remarks
Between bark extract and leaf slurry	8607.81	2	15.18	14.24	*
Between bark extract and control	9525.71	2	8.72	11.53	ns
Between leaf slurry and control	7033.42	2	7.46	6.46	*

ns = not significant.; * = significant.

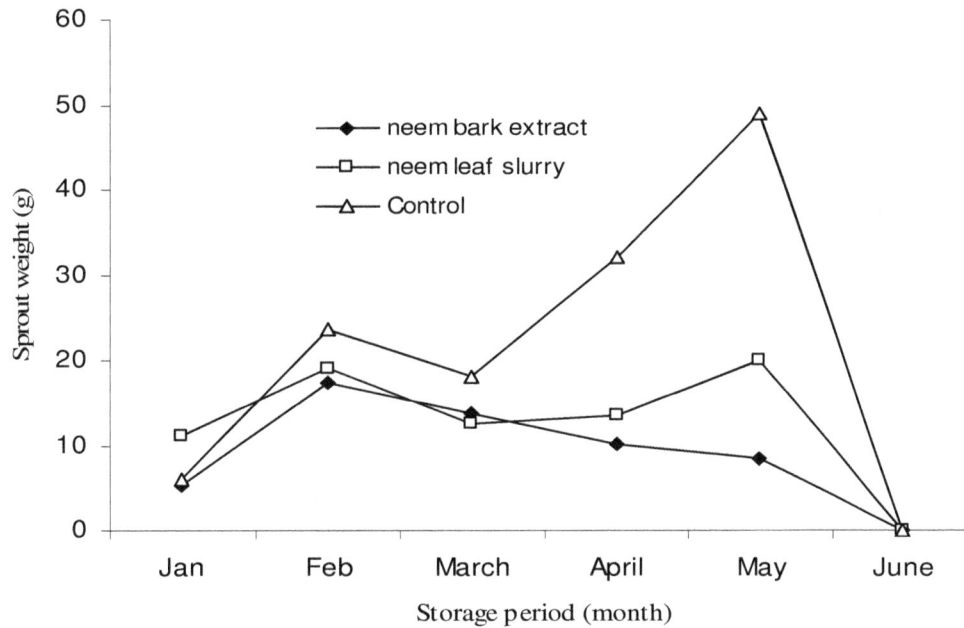

Figure 5. Average sprout weight of tubers stored in barn with fan.

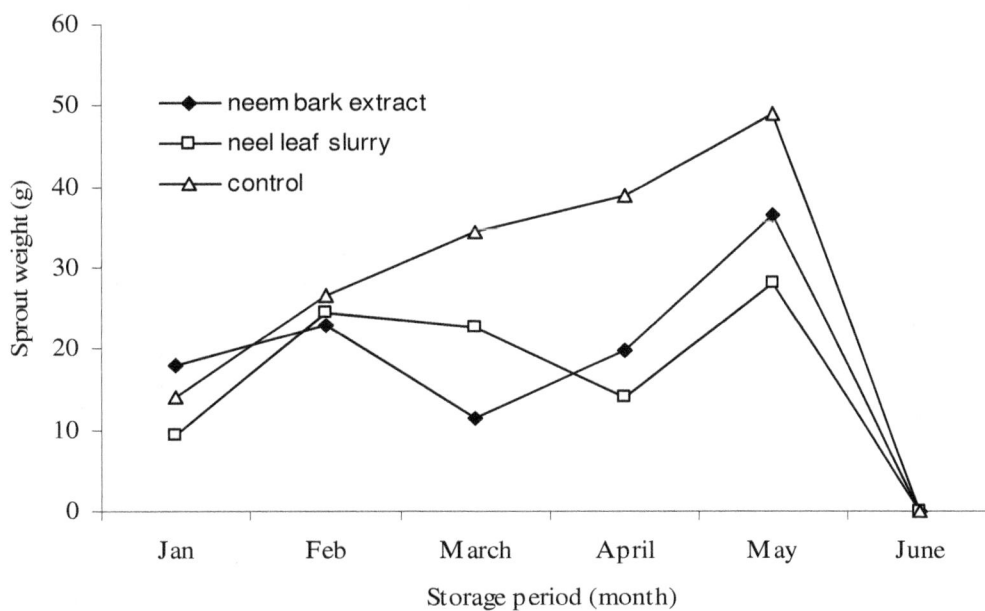

Figure 6. Average sprout weight of tubers stored in barn without fan.

Table 6. ANOVA table for sprouting.

Sources of variation	df	Ss	ms	Fcal	Ftab	Remarks
Treatment combinations	7	1100.54	157.22	0.0379	2.66	ns
Factor A (neem extract)	3	24043.53	18014.51	4.346	3.21	*
Factor B (barn type)	1	37503.64	27503.64	9.047	4.49	*
Interaction AB	3	58881.78	1960.59	0.473	3.21	ns
Error	16	66,328.41	4145.52			
Total	23					

ns = not significant; * = significant.

Table 7. Result of F – LSD for sprouting.

Sources of variation	Sum of squares	Degree of freedom	Mean differences	LSD values at 0.05% variances	Remarks
Between bark extract and leaf slurry	6090.3	2	13.76	20.44	ns
Between bark extract and control	9468.8	2	9.45	19.78	ns
Between leaf slurry and control	7403.6	2	23.20	22.92	*

ns = not significant ; * = significant.

treated with the neem leaf slurry and neem bark extract, as both treatments showed low sprout weights.

For all the treatments, the tubers stored in the barn with fan had less sprout weight compared to the barn without fan. There was an average of 7% less sprout weight in the barn with fan. The analysis of variance for the effect of neem extract on sprouting of stored yam tubers is given on Table 6. From the table both the treatment and storage conditions had an effect on sprout weight. The means for the neem extract treatment were separated by LSD, to determine the factor with significant effect on sprouting. Table 7 shows the result of the F-LSD for sprouting. From the result, it can be concluded that the leaf slurry treatment has a significant effect on sprouting of stored yam tubers.

DISCUSSION

The difference observed in temperature and relative humidity between the two barns could be attributed to the presence of fan which helped to improve airflow and dispersed accumulated heat and moisture on and around the surface of the yam tuber. This is in agreement with the findings of Mozie (1983). The sprout weight in the tubers stored in the barn with fan was lower than that of the tubers stored in the barn without fan. Also, the weight loss at the end of the storage period was less in the yam tubers stored in the barn with fan than that of tubers stored in barn without fan. This could be due to the reduction in temperature and increase in relative humidity in the barn with fan. Similar results were obtained by Mozie (1983). Gerardin et al. (1998) also reported

reduction in sprout weight and in weight loss with reduction in temperature and increase in relative humidity of the storage environment.

From the result, it can be concluded that the leaf slurry treatment has a significant effect on weight loss of stored yam tubers. This agrees with the finding of Ibrahim et al. (1987) and Schmutterer et al. (1980). Weight loss in stored yam tubers is as a result of sprouting, respiration and transpiration (Ravi et al., 1996b). Neem bark extract treated tubers had low sprouting but high weight loss; even though sprouting is one of the factors responsible for weight loss. The reason for the high weight loss when sprouting was low could be due to high rate of respiration and transpiration, as individual tubers have different respiration and transpiration pattern. Respiration and transpiration also depends on the size of the tuber. Even though sprouting may be low, other factors such as explained above may be responsible for the high weight loss of the neem bark extract treated tubers. Sprouting started at the end of January and was highest at the end of March for the neem tree extract treated tubers; the control recorded the highest sprout weight throughout the storage period. At the end of the storage period sprouting reduced and eventually stopped completely at the sixth month of storage (June), this could be due to temperature change (that is, decrease in temperature as a result of the rains) and the regular removal of sprouts. Ibrahim et al. (1987) reported that neem tree extract treatment have effect on sprouting as they were able to suppress sprouting for five months in stored yam tubers (*D. rotundata*). However, in this work none of the neem bark treatments used was able to suppress sprouting, it only reduced the sprout weight compared to non treated tubers.

Conclusions

The effect of intermittent forced air flow using a fan and pre-storage treatment using neem leaf slurry and neem bark extract on weight loss and sprouting of stored yam tuber were studied. The result showed that the temperature of the storage environment was low in barn with fan while the humidity was high compared to the barn without fan. It was also observed that the intermittent air flow in stored yam tuber reduces sprout vigor and weight loss after six months of storage. The neem tree extract treatment had significant effect on weight loss and sprouting on the stored yam tubers; with the neem leaf slurry having a greater influence than the neem bark extract. Weight loss of the yam tubers was more influenced by the treatment than sprouting.

RECOMMENDATIONS

Based on the work done, the following recommendations for further work were made:

1. Different concentration rates and application rates of neem tree extract should be further investigated to identify the optimum rate of application.
2. Farmers should be encouraged and sensitized on the effectiveness of neem extracts on sprout suppressing and weight loss reduction.
3. Effect of neem tree ash on sprouting and weight loss should be investigated, as it might be easier to use ash than slurry or paste. .

REFERENCES

Ganguli S (2002). Neem - a therapeutic for all season. Curr. Sci., June P1340, 82(11).

Gerardin O, Nindjin C, Farah Z, Escher F, Stamp P, Otokore D (1998) Effect of storage system and sprout removal on post-harvest yam (*Dioscorea* spp.) fresh weight loss. J. Agric. Sci. Cambridge, pp. 130, 329–336.

IITA (2007). Yam Research for Development. IITA Publication, 1: 1-10.

Ibrahim MH, Williams JO, Abiodun MO (1987). Assessment of parts of Neem tree for yam tuber storage. Rep. Nig. Stored Prod. Res. Inst., 4: 37–41.

Mozie O (1983). Effect of Air Flow on Weight Losses and Sprouting of White Yam Tubers Stored in the Conventional Barn. Trop. Root Tuber Crops News Lett., 13: 32-37.

Nwakiti AO, Makurdi D (1989). Traditional and Some Improved Storage Methods for Root and Tuber Crops in Nigeria. In: Deutsche Stiftung Fur Internationale Entwicklung (DSE) (ed.): Roots, Tubers Legumes, Bonn, pp. 51-67.

Ohu JO, Aviara NA, Yusuf T (2007) Effect of oil coating on the weight, moisture and dry matter losses of yam in storage. Ife J. Technol., 16(1): 1-7.

Onwueme IC (1978). The Tropical Tuber Crops: Yams, Cassava, Sweet Potato, Cocoyams. John Wiley and Sons; New York, p. 234.

Opara LU (1999). Yam storage In: Bakker –Arekema et al. (eds). CIGR Handbook of Agricultural Engineering. The American Society of Agricultural Engineers. St. Joseph. M.I., Agro Process.. IV: 182 – 214.

Orhevba BA, Osunde ZD (2006). Effect of Storage condition and CIPC treatment on sprouting and weight loss of stored yam tubers. Proceedings of the Seventh International Conference and 28th Annual General Meeting of the Nigerian Institutions of Agricultural Engineers held at Zaria, Kaduna State.

Osagie AU (1991). Sugar Composition of Yam (Dioscorea) Tubers during Post Harvest Storage. J. Food Composition.

Osagie AU (1992). The Yam Tuber in Storage. Post harvest Research Unit, University of Benin, Nigeria, pp. 107–173.

Osunde ZD, Yisa MG (2003). Effect of Storage Structures and Storage Period on Weight Loss and Sprout Growth on Stored Yams. Proc. 1st Int. Conf. NIAE, 22: 196-199.

Osunde ZD, Orhevba BA (2009). Effects of storage conditions and storage period on nutritional and other qualities of stored yam (Dioscorea spp.) tubers. AJFAND, 9(2): 678–690.

Osunde ZD (2008) Minimizing Postharvest Losses in Yam (*Dioscorea* spp.): Treatments and Techniques In: Robertson, G.L. and Lupien, J.R. (Eds), Using Food Science and Technology to Improve Nutrition and Promote National Development, © International Union of Food Science and Technology.

Ramanujam T, Nair SG (1982). Control of sprouting of edible yams (*Dioscorea* spp.) J. Root Crops, 8: 49-54.

Ravi V, Aked J, Balagopalan C (1996a). Review on tropical tuber crops: I. Storage methods and quality changes. Crit. Rev. Food Sci. Nutr., 36: 661–709.

Ravi V, Aked J, Balagopalan C (1996a). Review on tropical tuber crops: II Physiological disorders in freshly stored roots and tubers. Crit. Rev. Food Sci. Nutr., 36: 711–731.

Sahore DA, Nemlin GJ, Kamenan A (2007). Changes in nutritional properties of yam (*Dioscorea* spp), green plantain (*Musa* spp) and cassava (*Manihot esculenta*) during storage. Food Sci. Technol., 47: 81–88.

Schmutterer H, Ascher KRS, Rembold H (1980). Natural Pesticides from the Neem Tree. Proceedings of the First Neem Conference. Rottach Egern, Published by the German Agency for Technical Cooperation (G T Z) Eadhborn FRG.

3

Post harvest losses of rice from harvesting to milling in Ghana

Appiah F[1]*, Guisse R[1] and Dartey P. K. A[2]

[1]Department of Horticulture, Kwame Nkrumah University of Science and Technology, Kumasi, Ghana.
[2]Crops Research Institute of the Centre for Scientific and Industrial Research, Fumesua, Kumasi, Ghana.

Estimation of postharvest losses of rice (*Oriza sativa*) from harvesting to milling was carried out in Ejisu Juabeng District of Ghana to provide basic information important regarding the losses. Harvesting losses were higher (2.93%) in sickle-harvesting than in panicle harvesting method (1.39%). Threshing losses were also higher (6.14%) in the 'bambam' in the bag beating method (2.45%). Harvesting losses ranged between 4.07 and 12.05% at farmer's fields. Storage and drying losses were 7.02 and 1.66% respectively. SB30 milling machine was more efficient and produced 67.3% head grains compared to SB10 (50%) and the locally manufactured machine (47.3%).

Key words: Rice, postharvest losses, milling, grain quality.

INTRODUCTION

Rice (*Oriza* spp) is after wheat, the most widely cultivated cereal in the world and it is the most important food crop for almost half of the world's population (IRRI, 2009a). It is estimated that rice sustains the livelihood for 100 million people and its production has employed more than 20 million farmers in Africa (WARDA, 2005). According to Harris and Lindblad (1978) postharvest losses comprise all changes in the ability, wholesomeness or quality of food that prevents it from being consumed by people. Postharvest losses can occur during any of the stages in the postharvest operations.

Whatever the source, postharvest losses represent more than just a loss of food as it ripples through the factors (including land, water, labour, seeds, time and fertilizer). The wastes indicate that postharvest food loss translates not just into human hunger and minimizing the revenue of farmers but into tremendous environmental waste as well (Earthtrend, 2001).

The steady increase in population and a corresponding increase in demand for food have led to increased rice imports in Sub-Saharan Africa. Between 1989 and 1996 Ghana was reported to be only 15.1% self-sufficient in rice production after dropping from 48.3% between 1970 and 1974 (Oteng and SantAnna, 1999). According to

WARDA (2007) Ghana was below 25% self-sufficiency in rice production. This means that Ghana still require huge imports to augment the difference in local demand. According to Manful and Fofona (2010) as well as Aidoo (1993), quantitative postharvest losses of rice in Sub-Saharan Africa are estimated to be between 10 to 22% while qualitative losses could be as high as 50%. Reducing postharvest losses could help in reducing rice imports with its accompanied economic losses. However, there is insufficient data on postharvest losses of rice in Ghana with regards to what, where and why the losses occur in the production system. For effective reduction in losses it is important to estimate the losses and the stages at which they occur. This study therefore aimed at assessing the postharvest losses that occur in rice in Ghana from harvesting to milling with the aim of providing information for reducing postharvest losses and ultimately increasing rice supplies without increasing acreages under cultivation or imports. This study provides critical assessment of what, where and why losses occur, and what could be done to reduce such losses.

MATERIALS AND METHODS

Experimental site

The research was conducted at *Nobewam* and *Besease* in the Ejisu-Juabeng District in the Ashanti Region of Ghana between 2009 and 2010.

---
*Corresponding author. E-mail: fappiah_sp@yahoo.com.

Experimental procedure

The experiment was done in two phases: A survey and a field work.

Survey

The survey on farmers' perception and knowledge of postharvest losses of rice was conducted at Besease in the Ejisu Juabeng District of the Ashanti Region of Ghana. A semi structured questionnaire aimed at investigating some rice farmer's perception about postharvest losses of rice was administered to thirty rice farmers in "Besease" a rice farming community in the Ashanti region of Ghana. Information on farmers' perception of postharvest losses and methods of reducing such losses was collected. Other important information collected included the causes of postharvest losses, the estimation of postharvest losses.

Field experiment

Two rice varieties, Nerica 1 and Nerica 2, commonly grown by farmers were grown for assessment. For each variety, an area of 4 x 5 m was demarcated for cultivation. There were three replications per variety. Cultural practices carried out on the field included land clearing, ploughing, retovation, raising nursery for seedlings and transplanting. At maturity the crops were, harvested, threshed, dried, stored and milled to determine the postharvest losses that were involved at each stage.

Experimental design

A 2 x 2 RCBD was used comprising 2 varieties (Nerica 1 and Nerica 2) and 2 harvesting methods (panicle and sickle) for determining harvesting losses. The experimental design for milling yield was 2 x 3 CRD comprising 2 varieties and 3 milling machines.

Postharvest studies

Determination of harvesting losses

The rice plots were divided into quadrants (5 x 5 m) and skilled harvesters were allowed to harvest as per farmer practice using panicle and sickle harvesting methods. Left over rice grains on the harvested plots (both on the ground and on unharvested standing plants) were thoroughly collected, cleaned, dried, weighed and stored in a cloth bag. Percentages of harvesting losses were determined using the method described by Badawi (2003). For farmers' grown fields (5), harvesting was done using sickle. The weight of paddy rice left on the field per quadrant was determined and losses estimated using the formula:

Harvesting losses = left over paddy/ Total harvested paddy x 100

Determination of threshing losses

Two different types of threshing methods as practiced by farmers were used bag-beating (panicle) and bambam (sickle). Panicle harvested rice were put in a bag and beaten with stick to separate the grains from the stalks. Rice harvested with sickle was threshed using the "bam bam" a locally made wooden box with a tarpaulin beneath. In the bambam method, the rice stems were held and the stems together with the panicles on them and beaten against an inner side of the box. Removed grains were allowed to drop onto the tarpaulin beneath the box. After threshing, all the rice grains that fell out and was found outside the wooden box as well as the bags were collected, cleaned, dried and weighed and all the rice grains remaining on the stalks after the beating were also collected, cleaned, dried and weighed. Threshing losses were also assessed on five different farmers' fields using the bag a beating method (panicles). Threshing losses were estimated using the formula:

Threshing losses = [Weight of left over grains/ Total weight of collected grains] x 100.

Determination of Weight losses during storage

Two rice varieties (Nerica 1 and Nerica 2) were harvested, threshed, dried, weighed and stored in rice bags for 60 days in a well ventilated room at room temperature after which they were weighed at the end of the storage. At the end of the 60 day period, the pre-weighed bags of rice were reweighed. Storage losses were calculated using the formula:

By weight = [(Initial weight of paddy rice - Final weight of paddy rice)/ Initial weight of paddy rice] x 100.

Milling yield and milled rice quality

The performance of three different milling machines (One-pass type mill - SB30, SB10 and Engelberg type mill - locally made). Each machine was used to mill 25 kg of Nerica 1 and Nerica 2 paddy. Milling was done in triplicates. The resulting rice, bran and husk from each milling machine were collected and weighed. Milling yield was determined according to the procedure of IRRI (2009b) using the formula:

Milling yield = [Weight of white rice/ Weight of paddy] x 100.

The grains after milling were subjected to head grain count to determine which of the milling machines produce more breakages. Ten grams (10 g) of milled rice from each sample was taken. Head grains (unbroken grains) were separated from the broken grains and weighed. The percentages of broken and unbroken grains from each machine were determined as described by the method described by IRRI (2009):

Percentage of Head rice = (Weight of whole grains/ Weight of paddy sample) x 100
Percentage of Broken rice = (Weight of broken grains/ Weight of paddy sample) x 100

Statistical analysis

The Statistical Package for Social Sciences (SPSS) version 17 was used to analyze the responses on farmer's perception of postharvest losses. Analysis of variance (ANOVA) was performed on experiment data collected using GENSTAT Discovery Edition 3 and separation of treatment means was done using the LSD at 5% level of significance.

RESULTS AND DISCUSSION

Survey - Farmers' perception of postharvest losses of rice

Farmers' experience of postharvest losses varied. Ninety

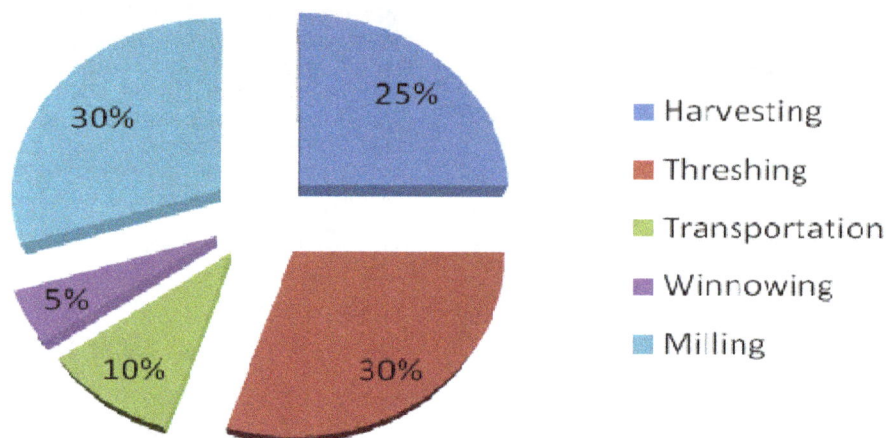

Figure 1. Stages at which most postharvest losses occur.

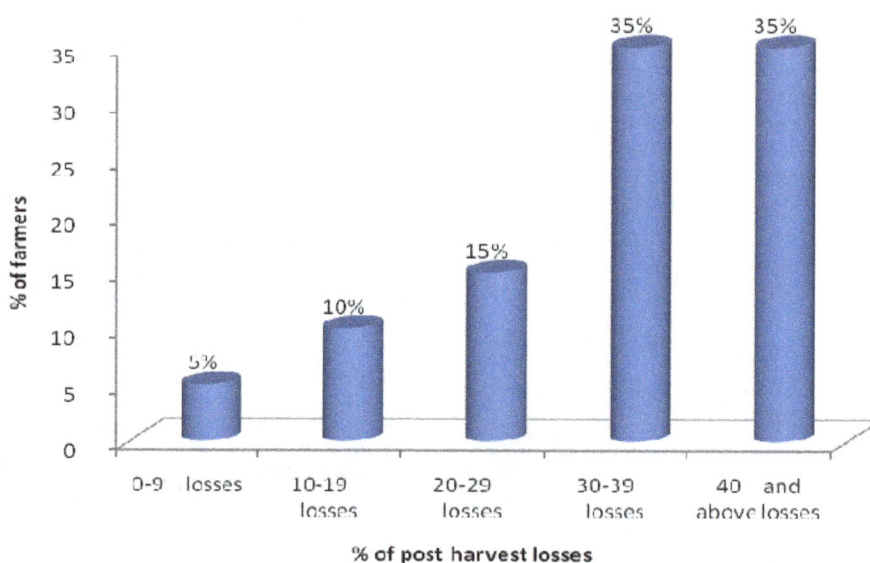

Figure 2. Perception of farmers on postharvest losses in rice from harvesting to milling.

five percent (95%) of the respondents reported that they had experienced postharvest losses of rice whilst the remaining 5% indicated they had not. According to the farmers, most postharvest losses in rice production occured from harvesting to milling and accure also at threshing and milling stages.

Thirty percent (30%) of the respondents indicated that the highest losses occur during threshing while another 30% reported that the highest losses were at milling. The results also showed that 25% of the farmers experienced the highest postharvest losses at the harvesting stage, 10% of farmers at the transportation stage and the remaining 5% at the winnowing stage as shown in Figure 1. The farmers also reported that the causes of losses were flooding of rice fields during harvesting when there were heavy rains, insufficient availability of postharvest machinery, other reported causes included rice shattering at harvesting, rice paddy getting mouldy during drying as well as rice grain breakage during milling.

As to how much is lost during the entire production chain, there were varying responses Thirty five percent (35%) of the respondents reported that they incur a total postharvest losses of 40% and above, 35% indicated that losses ranged between 30 and 39%, while 15% reported 20 to 29% losses (Figure 1). The remaining 15% of the farmers lost between 0 and 19% of their produce. These loss figures were unacceptable to majority (90%) of the respondents. However, the remaining 10% considered the such losses as normal (Figure 2). From the responses, it is obvious that the perceived losses were unacceptably high. The implication is that the rice farmers lose huge amounts since 70% of the rice farmers

Table 1. Harvesting losses in of Nerica 1 and Nerica 2 rice varieties using two kind of harvesting methods: Panicle and sickle.

Treatments variety	Total weight of harvested rice (g)	Harvesting losses (g)	Harvest weight loss (%)
Nerica 1	6688	132	2.19
Nerica 2	6926	148	2.13
Panicle	6430	83	1.39
Sickle	7184	196	2.93
Lsd	1692.4	59.7	1.338
Nerica 1 x Panicle	6450	66	1.13
Nerica 1 x Sickle	6925	197	3.25
Nerica 2 x Panicle	6409	100	1.64
Nerica 2 x Sickle	7443	195	2.62
Lsd	2393.4	84.4	1.89
CV (%)	21.8	11.4	32.3

Table 2. Threshing losses of Nerica 1 and Nerica 2 rice varieties under two different threshing methods: Bambam and bag beating method.

Treatments	Total weight of harvested rice	Threshing losses (g)	Threshing losses (%)
Varieties			
Nerica 1	6688	294	4.65
Nerica 2	6926	288	3.94
Threshing methods			
Bag beating (Panicle)	6430	148	2.45
Bambam (Sickle)	7184	434	6.14
Lsd	1692.4	80.4	1.47
Variety x Threshing method			
Nerica 1 Panicle	6450	239	3.98
Nerica 1 Sickle	6925	349	5.33
Nerica 2 Panicle	6409	57	0.92
Nerica 2 Sickle	7443	519	6.96
Lsd	2393.4	113.6	2.1
CV (%)	21.8	11.2	12.8

reported losses of at least 30% and above. It is therefore important for stakeholders in the local rice industry to discuss this great losses to stimulate the growth of the industry and alleviate poverty among the farmers.

Field studies - Harvesting losses

The results of harvesting losses of rice have been presented in Table 1. There was no significant difference between Nerica 1 and Nerica 2 with respect to harvesting losses. However, losses due to the methods of harvesting (panicle and sickle) were significantly different (P < 0.05) among the varieties. The use panicle harvesting method resulted in 1.39% grain loss. On the other hand, there was 2.94% loss when sickle harvesting

method was used. This indicates that panicle harvesting should be the method of choice based on the figures for harvesting losses observed in this study.

The interaction between variety and method of harvesting impacted significantly on harvesting losses. Panicle harvesting of Nerica 1 showed the least harvesting loss of 1.13% while sickle harvesting of Nerica 1 resulted in the highest loss of 3.25%.

Generally, harvesting loss values of (1.13 to 3.25%) reported in this study falls far below than the 12.05% (Table 2) harvesting losses that were found on some farmers' fields during this study but falls within range (1 to 3%) reported for South East Asia (World Resources Institute, 1998). Even though panicle harvesting resulted in minimum postharvest losses when compared to sickle harvesting, harvesting by sickle method is twice faster

Table 3. Harvesting and threshing losses at farmer's fields.

Parameter	Harvesting losses (g)	Threshing losses (g)	Total weight of harvested rice (g)	Harvesting losses (%)	Threshing losses (%)	Total losses (%)
Farmer 1	382	35	4837	7.91	0.73	8.65
Farmer 2	135	50	1164	12.05	4.07	16.14
Farmer 3	198	211	7773	2.60	3.00	5.60
Farmer 4	299	144	3723	8.20	3.73	11.93
Farmer 5	177	36	7124	3.03	0.53	3.57
Lsd	171.6	72.5	3421.5	5.44	1.95	4.57
CV (%)	39.6	43.4	38.2	44.2	44.5	27.4

and less laborious than panicle harvesting (Pingali and Hossain, 1998). Consequently, harvesting using panicle method on a large field might not be practical. The cost of the extra man hours, time spent as well as other extra resources spent on harvesting does not make the gain in panicle harvesting economical. Farmers therefore might still be better off harvesting with sickle in the absence of improved mechanized harvesting.

Threshing losses

Sickle harvested rice from both Nerica 1 and Nerica 2 were threshed using the locally made wooden box commonly known in Ghana as "bambam" method. Since during panicle harvesting only the panicles are cut, the harvested rice does not come together with the stalks as in the case of sickle harvesting to enable threshing using the wooden box (bambam) method. Threshing for the panicle harvested rice was therefore done using the bag-beating method instead (Table 2).

Even though the Nerica 2 variety had lower threshing losses (3.94%) compared to the Nerica 1 variety (4.65%), the difference was not significant. Threshing losses were higher (6.14%) in the sickle-harvested rice that used the "bambam" than in the panicle-harvested rice (2.45%) that used the bag beating method. This outcome is attributable to the different methods of harvesting (sickle or panicle) which dictated the threshing methods (bambam or bag-beating) that had to be used. The losses were lower in the bag beating method because very little grains dropped from the bag during threshing. However, in the "bambam" method, some grains scattered and were lost during threshing using the threshing box. Threshing in an enclosed room where escaping grains could be trapped on tarpaulin, may help reduce threshing losseswhich especially in the "bambam" method since scattered grains can be collected.

The interaction between variety and threshing method resulted in significant differences in the threshing losses. When the bag-beating method was used for threshing, Nerica 2 had significantly lower losses (0.92%) than

Nerica 1 (3.98%). On the other hand when the bambam method was used there were no significant difference between threshing losses of Nerica 1 and Nerica 2. Generally the bambam method resulted in higher threshing losses (between 5.33 and 6.96%) than the bag beating method (between 0.92 and 3.98), regardless of the variety. These values are lower than the 4 to 6% threshing losses reported for South-East Asian countries (IRRI, 1997). These lower values (0.92 to 3.98%) contradict the perception of the rice farmers (30%) that the highest loss occur at threshing (Figure 1).

Harvesting and threshing losses at five different farmers' fields

Five different rice farmers cultivating different rice varieties, mostly the Nericas and Sikamo rice varieties were also assessed for harvesting and threshing losses.

The results of the survey showed that harvesting losses ranged between 3.03 and 12.05% while threshing losses varied from 0.53 to 4.07% (Table 3). Total losses due to only harvesting and threshing losses at farmer's fields ranged between 5.60 and 16.14%. The differences in postharvest losses among the farmers were due to different level of skill of harvesting and threshing as well as poor weed control. During the study it was observed that the farmers' fields were engulfed in a lot of weeds. Poor weed control is known to interfere with effectiveness of harvesting (Al-Khatib, 1995).

Weight losses during storage

There were weight losses in both varieties (Nerica 1 and Nerica 2) ranging between 6.19 and 9.35% (Table 4). The reduction in the weight could also be due to moisture losses from the grains during storage as well as pest and insect infestations. The storage losses observed were higher than the 2 to 6% reported for South East Asia (IRRI, 1997). Proper drying and pest control are important to minimize storage losses.

Table 4. Weight losses during storage losses of Nerica 1 and Nerica 2.

Variety	Initial weight (g)	Final weight (g)	(%) Loss
Nerica 1 (Panicle harvest)	6450	6053	6.07
Nerica 1 (Sickle harvest)	6925	6277	9.26
Nerica 2 (Panicle harvest)	6409	6019	5.97
Nerica 2 (Sickle harvest)	7443	6933	6.79
Lsd	3415.5	3088.0	3.29
CV (%)	26.6	25.9	24.9

Table 5. Total postharvest losses studied.

Activity	Percentage losses
Harvesting losses	3.03 to 12.05
Threshing losses	0.53 to 4.07
Drying losses	1.57 to1.76
Total	4.60 to 17.88

Table 6. Milling efficiency of different milling machines used by the farmers.

Machine	Milling yield (%)	Bran weight (%)	Husk weight (%)
SB30	67.30	14.53	18.13
SB10	66.0	17.87	16.13
LLM	63.33	36.67	0
Lsd	4.37	3.4	1.64
CV (%)	3.3	8.7	7.2

Total postharvest losses of rice at harvesting, threshing and drying, operations

The total losses of rice in this study from harvesting to milling have been presented in Table 4 to 6. There were variations at each stage. Harvesting losses ranged between 3.03 and 12.05%. Threshing losses varied between 0.53 and 4.07% while drying losses ranged narrowly between 1.57 and 1.76%. Losses during storage varied between 5.97 and 9.26% (Table 5).

The total loss estimate up to 18% observed in the field under experiment is similar to the report indicating 15% of the farmers (Figure 2) who mention total losses of ranged between 0 and 19%. The observed overall losses of up to 18% between harvesting and drying indicated that much is lost since that means a revenue loss of 18% of lost revenue, labor, man hours, food (rice), land as well as the other factors of production employed during production. All stakeholders should discuss these high losses. Capacity building inputs and machinery are crucial in redressing these losses. Implementation of appropriate safeguards should be encouraged (Balasubramanian et al., 2007) to reduce milling and threshing losses.

According to Saunders et al. (1978), a reduction of 2% of postharvest losses in developing countries would provide at least 4 million metric tonnes equivalent to the annual caloric requirement of 10 million people.

Milling efficiency of used machines

The results of the milling analysis have been presented in Table 6. The results showed that SB30 had marginally higher milling yield (67.32%) than SB10 (66%).

However, the differences between the milling yields of either the SB30 machine and the locally manufactured milling machine or that of the SB10 machine and the locally manufactured milling machine were significant. The locally manufactured milling machine had the lowest milling yield (63.33%). This implies that the locally manufactured milling machine was less efficient as it also resulted in higher percentage (52.7%) of broken grains (Table 6). This implies that the local machine produced less white rice per unit weight of paddy. This resulted in less recoverable rice and therefore less revenue. SB30 is

Table 7. Effect of milling machine type on grain breakages.

Machine	Weight of broken grains (g)	Weight of unbroken grains (g)	Unbroken grains (%)	Broken grains (%)
(SB30)	3.60	6.73	67.3	32.7
(SB10)	5.00	5.00	50.0	50.0
(LMM)	5.27	4.73	47.3	52.7
Lsd	1.37	1.30	13.00	13.01
CV (%)	14.8	11.8	11.8	14.8

therefore superior to both SB10 and the Local machine in terms of milling yield. In spite of the higher milling yield of SB30, it is still lower than the 67.5% (2010 data) considered to the lowest on record in the USA (Robinson, 2010). According to the author, this could lead to a shift to selling more broken grains, consequently mean lower prices. Norman and Otoo (2003) recommended that improved dehulling and whitening machines should be developed and tested for use in rice processing.

Grain quality of rice from the three milling machines

There were significant differences (Table 7) between the milling machines to produce unbroken grains (head grains). The results showed that SB30 produced the highest (67.3%) percentage of unbroken grain, followed by SB10 (50%). The locally manufactured machine (LMM) produced the least (47.1%) percentage of unbroken rice grains after milling. The differences between SB30 and SB10 as well as between SB30 and LMM were significant. The performance of SB10 and the LMM were not significantly different from each other.

In the Bangladesh inspection standard for completely milled rice, the upper limit of broken grains was 12% for big broken and 4% for small broken aromatic rice (Afsar et al., 2001). On the other hand, in the United States of America, the upper limit of 25% is taken for the fourth grade rice (Schmidt, 2010).

Clearly, the levels of broken grains obtained in this study with SB30, SB10 and LMM were much higher than the internationally acceptable limits, although according to Sakurai et al. (2006) milling machines in Ghana are efficient. This indicates that rice milled using the SB30 machine falls below the international standard although it produced higher percentage of unbroken grains among the machines assessed. The results suggest that rice milled using the available milling machines in Ghana might not qualify even for the lowest grades in the international market (15% Japan, 25% USA). Accordingly the milling machines used in Ghana do not produce milled rice that falls in the acceptable grading standard of America and Japan. In this respect, Ghana milled rice might not compete very well with Japanese or American rice if they are all at the same market where unbroken grains is mostly demanded. Consequently, Ghanaian rice

farmers might not get competitive prices for their rice on the international market.

Conclusion

The study has highlighted the fact that rice farmers are aware of the postharvest losses involved in rice cultivation in Ghana. Harvesting losses were higher when the sickle method of harvesting was used, compared to the panicle harvesting method. Threshing losses were also higher in sickle harvesting where threshing was done by the "bambam" method. SB30 milling machine performed significantly better than SB10 and the local manufactured machines in terms of both milling yield and head grain quality. From the results obtained during this study, postharvest losses of rice during harvesting, threshing and drying ranged between 4.6 and 17.88%, with the exclusion of losses resulted from storage, transportation, winnowing and handling losses were not included.

ACKNOWLEDGEMENT

The authors would like to appreciate the support of Strengthening Capacity for Agricultural Research and Development in Africa (SCARDA) Programme for providing funding for this study.

REFERENCES

Afsar AKMN, Baqw M, Rawman M, Rouf MA (2001). Grades, Standards And Inspection Procedures Of Fuce In Bangladesh. Bangladesh Food Management and Research Support Project. Ministry of Food, Government of the People's Republic of Bangladesh. International Food Policy Research Institute. FMRsP Working Pap., (20). Retrieved December 28, 2010 from http://pdf.usaid.gov/pdf_docs/PNACN994.pdf.
Aidoo KE (1993). Postharvest storage and preservation of tropical crops. Int. Biodeterioration Biodegrad., 32(1-3): 161-173.
Al-Khatib K (1995). Weed control in wheat. Washington State University. Extension Bull., p. 1803. Retrieved December 28, 2010 from http://cru.cahe.wsu.edu/CEPublications/eb1803/eb1803.html.
Badawi AT (2003). A proposal on the assessment of rice post-harvest losses. CIHEAM - Options Mediteraneennes. Conference Paper Document. Workshop on rice technology Alexandria (Egypt). Hote Conrad Resort and Casino, pp. 10–13. March. Retrieved December 28, 2010 from http://ressources.ciheam.org/om/pdf/c58/03400069.pdf

Balasubramanian V, Sie M, Hijmans RJ, Otsuka K (2007). Increasing Rice Production in sub-Saharan Africa: Challenges and Opportunities. Adv. Agron., 94: 55-133.

Harris KL, Lindblad CJ (1978). Postharvest Grain Loss Assessment. Methods. Minnesota, Am. Assoc. Cereal Chem., St. Paul, Minnesota, p. 193.

IRRI (1997). Disappearing food. How Big are the postharvest losses. Retrieved May 26, 2010, from http://www.earthtrends.wri.org/features/view_feature.php?theme.

IRRI (2009a). Rice Policy- World Rice Statistics (WRS). Retrieved May 28, 2010 from http://www.irri.org/science/ricestat.

IRRI (2009b). Procedures for Measuring quality of milled rice. Retrived December 20, 2010 from http://www.knowledgebank.irri.org/rkb/index.php/procedures-for-measuring-quality-of-milled-rice.

Manful J, Fofana M (2010). Postharvest practices and the quality of rice in West Africa. CORAF/WECARD. 2nd Science Week. 24-29 May, 2010, Cotounou, Benin.

Norman JC, Otoo E (2003). Rice development strategies for food security in Africa. Proceeding of the 20th Session of the International Rice commission. Bangkok, Thailand. 23-26 July, 2002.

Oteng JW, SantAnna R (1999). Rice production in Africa: Current situation and issues. International Rice Commission Newsletter.

Pingali P, Hossain M, (1998). Impact of rice research. Proceedings of the International Conference on the Impact of Rice Research, 3-5 June, 1996, Bangkok, Thailand. Thailand Development Research Institute, Bangkok, Thailand, and International Rice Research Institute, P.O. Box 933, Manila, Philippines, p. 428.

Robinson R (2010). Low milling yield could impact U.S. rice markets. Retrieved December 29, 2010 from http://westernfarmpress.com/rice/low-milling-yield-could-impact-us-rice-markets.

Sukurai T, Furuya J, Futakuchi K (2006). Rice Miller Cluster in Ghana and Its Effects on Efficiency and Quality Improvement. Contributed paper prepared for presentation at the International Association of Agricultural Economists Conference, Gold Coast, Australia, August 12-18, 2006. Retrieved December 28, 2010 from http://ageconsearch.umn.edu/bitstream/25683/1/cp060481.pdf.

Saunders RM, Mossman AP, Wasserman T, Beagle BC (1980). Rice Postharvest losses in Developing countries. Science and Education Administration. Agricultural Reviews and Manuals. Wester Series (ARM-W-12 April 1980). Retrieved December 23, 2010 from http://www.archive.org/stream/ricepostharvestl022689mbp/ricepostharvestl022689mbp_djvu.txt.

Schmidt H (2010). USA Rice Federation. African Buyer Conference. Lagos, Nigeria. 28 April, 2010. Retrieved December 28, 2010 from http://www.africanbuyerconference.com/USARice_Hartwig.pdf.

WARDA (2005). Rice, a strategic crop for Food Security and Poverty Alleviation. Retrieved May 26, 2010 from www2.slu.se/cigar/CGIAR_WARDAppt

WARDA (2007). Overview of Recent Developments in sub-Saharan Africa Rice Sector. Africa Rice Trends, Cotounou, Benin, p. 10.

World Resources Institute (WRI) (1998). Disappearing Food: How big are the postharvest losses. EarthTrends: Environmental Information. Washington DC: World Resources Institute. Retrieved May 10, 2010 from http://earthtrends.wri.org/pdf_library/feature/agr_fea_disappear.pdf.

4

Effect of perforated blue polyethylene bunch covers on selected postharvest quality parameters of tissue-cultured bananas (*Musa* spp.) cv. Williams in Central Kenya

Muchui M. N.[1], Mathooko F. M.[2], Njoroge C. K.[3] Kahangi E. M.[3] Onyango C. A.[3]
and Kimani E. M.[1]

[1]Kenya Agricultural Research Institute-Thika, P.O. Box 220-01000, Thika, Kenya.
[2]South Eastern University College (A constituent College of the University of Nairobi), P.O.Box 170-90200, Kitui, Kenya.
[3]Jomo Kenyatta University of Agriculture and Technology, P. O. Box 62000-00200, Nairobi, Kenya.

Banana farming in Kenya has recently moved from subsistence to commercial farming. There is therefore the need to produce high quality fruits that are visually acceptable, have good postharvest quality attributes and therefore sell well in both local and international markets. Technologies such as pre-harvest bunch covers have been shown to improve postharvest quality of banana fruits. However, earlier reports on the effect of bunch covers on postharvest quality of banana fruits in the tropics have been contradictory. This study therefore aimed at understanding the effect of perforated polyethylene bunch covers on the postharvest quality characteristics of tissue-cultured bananas using banana (*Musa* spp.) cv. Williams as the test variety. The trial was carried out in Maragwa region in central province of Kenya, in a complete randomized design and was replicated three times. Perforated dull and shiny blue polyethylene covers were placed when the hands had started to turn upwards. Fruits were harvested at full three quarter maturity. Parameters measured were; bunch weight, finger grade and length, starch, total soluble solids (TSS), sugars, total titratable acidity (TTA), pulp/peel ratio, colour, chlorophyll content, firmness, moisture content, weight loss, green life and shelflife. Banana bunches were also evaluated for cleanliness and bruise marks at harvest. Data were subjected to analysis of variance (ANOVA) using the general linear model (GLM) procedure of SAS statistical programme. The means were compared using Student Newman Keuls' test (SNK) and least significant difference (LSD) at significance level of 5%. Results showed that bunch covers did not influence finger grade, length and bunch weight at harvest. Colour, chlorophyll content, firmness, starch content, TSS, TTA, moisture content, weight loss, individual and total sugars and pulp/peel ratio at harvest and during ripening were not influenced by bunch covers. Bunch covers did not influence greenlife and shelflife significantly. Fruits grown under cover were more visually appealing, cleaner and had minimal bruises compared to the unbagged fruits. However, bunch covers had some detrimental effects on postharvest quality characteristics of banana fruits of the covered bunches compared to the fruits from the control, where few fingers of top hands of some bunches suffered sun burn. The study has shown that perforated dull and shiny blue bunch covers may be used in commercial banana orchards in Kenya to produce high quality fruits especially in the cooler areas.

Key words: Musa *spp*, polyethylene bunch covers, postharvest quality.

INTRODUCTION

In Kenya, banana is a major food and cash crop grown in almost all provinces (MOA, 2006). However, production has mainly been by subsistence farmers who rarely produce high quality fruits due to constraints such as diseases and pests coupled with traditional agronomic practices (Qaim, 1999; Acharya and Mackey, 2008). With the introduction of the tissue culture technology which availed virus disease-free plantlets, commercial banana production has greatly increased with a resultant increase in its volume and value. In 2006, 1,058,018 metric tonnes (MT) valued at KES 9,298,122,000 was produced in the country (MOA, 2006) compared to 1,024,360 tonnes (T) produced in 1996 to 1997 period (Qaim, 1999). Recently, farmers have been uprooting coffee due to problems in marketing in central and eastern provinces of Kenya. In Rift Valley of Kenya, farmers have also been uprooting citrus due to the greening disease to replace with banana orchards (Acharya and Mackey, 2008). This has translated into large volumes of the crop being produced at a commercial level. Establishing a tissue culture banana orchard is expensive compared to establishing an orchard using conventional suckers. This is attributed mainly to the high cost of the tissue culture planting material which is about seven times the average cost that the growers incur in acquiring conventional suckers (Qaim, 1999). There is therefore the need to produce high quality fruits of banana that are visually acceptable, have good postharvest quality attributes and therefore sell well in both local and international markets.

External appearance, internal quality and market quality of bananas are influenced by several factors, including pre-harvest production practices. The external appearance includes key attributes such as colour, shape, size and freedom from defects. The internal attributes such as taste, texture, sweetness, aroma, acidity, flavour, shelflife and presumed nutritional values of the fruit are important in ensuring repeat buys for sustained repeat purchase (Hewett, 2006; Shewfelt, 2009). The physical appearance of the peel is especially important in the highly competitive export markets and in some local niche upmarkets like the supermarkets. Buyers in these prime markets require consistent supplies of uniform coloured fruit with blemish-free peels. Banana bunch covers allow for production of high quality banana fruits that are not bruised, and hence have acceptable visual appearance. Consumers use visual quality to purchase fresh produce (Shewfelt, 1999; Shewfelt, 2009). Market returns for bananas in international markets are generally greatest for large fruit that are blemish-free (Johns, 1996).

The supply of blemish-free fruit is difficult due to various types of mechanical injury and insect damage imparted on the delicate peel surface during growth and development, with wind and insects being the principal agents of this damage (Anon, 2003). Pre-harvest insect feeding has been shown to be a main cause of peel damage to banana fruits (Shanmugasundaram and

Manavalan, 2002). However, bagging of bananas with bags impregnated with insecticides has been shown to protect fruits from insect attack (Amarante et al., 2002). Wind blows dust and debris which hits the delicate outer skin causing cellular damage and subsequent fruit scarring. Considerable physical injury and damage to the fruit peels can also be caused by the blowing of adjacent leaves and rubbing of leaf petioles onto the developing bunch (Anon, 2003). This chaffing from leaves during growth has also been reported to be eliminated by bunch covers (Weerasinghe and Ruwapathirana, 2002). Bunch covers of various colours and conditions (perforated and non-perforated) have been extensively used in both tropical and subtropical banana growing countries with the aim of improving yield and quality (Stover and Simmonds, 1987; Robinson, 1996). Improved quality includes appealing skin colour, reduced sunburn, reduced fruit splitting, increased finger length and bunch weight among others (Robinson, 1996; Amarante et al., 2002). Bunch covers have also been used to protect bunches from low temperatures, especially in temperate countries (Gowen, 1995; Robinson, 1996; Harhash and Al-Obeed, 2010). Indeed bagging has been shown to reduce winter stress under supra-optimal condition which resulted in early fruit maturation (Jia et al., 2005). This is due to enhanced physiological and metabolic activities provided by the microclimate created by bagging (Johns and Scott, 1989a).

However, the effect of fruit bagging, especially in the tropics, on size, maturity, skin colour among other postharvest parameters has been contradictory, which may reflect differences in the type of bag used, fruit age at bagging, fruit and cultivar response, prevailing climatic conditions and conditions of holding fruit after harvest (Johns and Scott, 1989a; Amarante et al., 2002; Weerasinghe and Ruwapathirana, 2002; Narayana et al., 2004). Technologies such as bunch covering that enhance production and help realize the benefits of tissue culture technology would go a long way in boosting commercial banana farming in Kenya where bunch covering has not been extensively practiced. Recently, a few farmers have attempted this practice in collaboration with importers of the bunch covers (K. Njiba, commercial banana farmer, personal communication). However, the effect of the covers on the postharvest quality of tissue-cultured bananas in Kenya has not been studied. The objective of this study therefore was to investigate the effect of bunch covering on postharvest quality of tissue-cultured banana fruits using cv. Williams as the test variety.

RESEARCH AREA AND MATERIALS

The trial was carried out in an already existing banana orchard in Maragwa District, in central province of Kenya. Nine bunches of banana cultivar (cv.) Williams were randomly selected and tagged. The fruits were grown using the recommended banana growing

*Corresponding author. E-mail: margaretmuchui@yahoo.com. Tel: +254 722 615590.

procedures (Anon, 2002). Perforated dull blue and shiny blue bunch covers were applied to the bunches when the flower bracts had hardened and the hands had started to curl upwards. The bunch covers had perforations measuring 8 mm spaced at 10.5 x 9 cm and a thickness of 5μm and were left to hang for about 150 mm below the distal hands and were securely attached to the bunch stalk above the proximal hand using a rubber band. Some of the tagged bunches were not covered and they served as a control.

Experimental layout and design

The treatments were applied randomly and were replicated three times. The banana fruits were allowed to grow to full ¾ maturity stage and were harvested, dehanded, placed in plastic crates and then transported to the postharvest laboratory of Jomo Kenyatta University of Agriculture and Technology.

Data collection and analysis

Parameters measured at harvest were bunch weights, finger grade and finger length. The fruits were also assessed for general visual appearance, dirt, bruises (blemishes), spider webs and bird droppings. The fruits were then washed with tap water and dipped in 100 ppm sodium hypochlorite (Jik, Reckitt Benckiser-East Africa Limited, Kenya) in order to control postharvest rots such as anthracnose and crown rot, and then air dried. The fruits were then ripened in a ripening chamber at 18˚C and 95% RH using ripe purple passion fruit as the ethylene source. Five fingers per replicate were placed on the bench for green life at ambient conditions of temperature (24 ± 1˚C) and humidity (60± 5%). Parameters measured during ripening were: starch, total soluble solids (TSS), sugars, total titratable acidity (TTA), pulp/peel ratio, colour, chlorophyll content, firmness, moisture content and weight loss. The fruits were also evaluated for green life and shelflife. Data were subjected to analysis of variance (ANOVA) using the General Linear Model (GLM) procedure of SAS statistical programme (SAS, 2001). All the means were compared using Student Newman Keuls' test (SNK) except for green life, shelflife and dirt which were compared using Least Significant Difference (LSD). All means were compared at 5% significance level.

Determination of fruit weight, length and diameter (grade)

All three bunches were dehanded and the hands weighed to give the bunch weights. Three fruits from the second hand in the three bunches per plot were weighed with an electronic balance (Type 1240, Shimadzu, Japan) to give the finger weights. Finger length was measured using a tape measure while finger grade was measured with a caliper (CD-20C, Mitutoyo, Japan) as diameter of the middle finger of the outer whorl of the second hand (Stover and Simmonds, 1987).

Evaluation of visual appearance

The fruits were checked for incidences of dirt, which included, dust, bird droppings and spider webs and mechanical injuries (blemishes). They were also checked for general visual appearance. Percentage surface area covered was rated based on the Merz 0 to 6 scale (Merz, 2000), adopted for surface area covered by dirt instead of lesions where, 1=0 to 2%, 2=2 to 5%, 3=5 to 10%, 4=10 to 25%, 5=25 to 50% and 6=>50% of the surface area covered by the blemishes, dust and spider webs.

Determination of pulp:peel ratio

Pulp:peel ratio was calculated after measuring the pulp and the peel weights with an electronic balance (Type 1240, Shimadzu, Japan) for both green and ripe fruits. Three fingers from the equatorial region hands per bunch were peeled and the pulp and peel weighed separately. The ratio was calculated as weight of pulp (g) per weight of peel (g).

Determination of starch content

Starch staining was done by cutting the banana fruits across at the equatorial region of the fruit, applying iodine/potassium iodide (I/KI) (2g/10g) solution and waiting for at least one minute for starch patterns to develop and rating using the Cornell Starch Chart (Watkins, 2006) for comparison. This chart has a scale of 3 to 8 with 3 = all starch and 8 = no starch.

Ripening

Two to three hands per bunch from the equatorial region were ripened in a ripening chamber at 18˚C and 95% RH using ripe passion fruit as ethylene source until ripeness stage 6 when the fruit was fully yellow (CSIRO, 1972; Marin et al., 1996; Paull, 1996; Jiang et al., 1999).

Determination of firmness

This was determined at harvest and during ripening. Both subjective and objective methods were used. Hand firmness assessment was carried out using the scale 1 = hard, 2 = firm, 3 = slightly soft, 4 = moderately soft, 5 = soft and 6 = very soft (Joyce et al., 1993; Jiang et al., 1999). Objective fruit firmness measurement was determined along three equatorial regions of the fruit: at the base, middle and apical sections of the fruit using a rheometer (Model CR-1000, Sun Scientific Co. Ltd, Japan) with an 8 mm probe for fruit at harvest and during ripening. The average of these three measurements was considered as one replicate firmness was expressed as Newton (N) (Joyce et al., 1993; Jiang et al., 1999).

Determination of moisture content

The pulp and the peel were analysed for moisture content from green stage through to fully ripe stage using oven drying method (AOAC, 1996).

Determination of weight loss

Three fingers per treatment were placed on the bench at ambient conditions of temperature (24 ± 1˚C) and humidity (60 ± 5%). Weight loss was determined by weighing the fingers every day from green to yellow stage. The initial weight (W1) of the fruit at day 0 and the weight of the same fruit (W2) at each sampling day were noted. Weight loss in percentage was then calculated as 100X (W1-W2)/W1.

Colour assessment

Subjective colour was determined at harvest and during ripening. Visual colour assessment was carried out using the scale of 1 to 8 where 1 = green, 2 = light green, 3 = half yellow half green, 4 = 3/4 yellow with green, 5 = yellow with green tip, 6 = fully yellow, 7 =

Table 1. Effect of bunch covers on finger grade (mm), finger length (cm) and bunch weight (kg) of tissue culture banana cv. Williams.

Treatment	Grade (mm)	Finger length (cm)	Bunch weight (kg)
Control	30.94 [a]	19.07 [a]	8.62 [a]
Dull blue	33.33 [a]	20.03 [a]	10.44 [a]
Shiny blue	33.44 [a]	20.37 [a]	9.16 [a]
LSD	ns	ns	ns

Values in the column followed by the same letter are not significantly different at p=0.05. Values are means of 3 replicates.

yellow with spots and 8 = yellow with coalesced black spots (CSIRO, 1972; Marin et al., 1996; Paull, 1996; Jiang et al., 1999). Colour of both the pulp and peel at ripeness stage 1 to 6 were measured using a Minolta colour difference meter (Model CR-200, Osaka, Japan) calibrated with a white and black standard tile. Measurements were made on three spots along the equatorial region and the average of these considered as a replicate. The L*, a* and b* coordinates were recorded and, a* and b* values converted to hue angle (H°), where H°= (arc tan(b/a), for first quadrant +a and +b, 180+arc tan (b/a) for second quadrant (-a, +b) and third quadrant (-a,-b) and hue=360+arc tan (b/a) for fourth quadrant (Mclellan et al., 1995).

Determination of chlorophyll content

This was determined using the method of Arnon (1949) with a uv-visible spectrophotometer (Model UV mini 1240, Shimadzu Corp. Kyoto, Japan). Total chlorophyll in the crude extract was calculated using MacKinney's coefficients after measuring absorbances at 645 and 663 nm and calculated as follows:

Total chlorophyll content (µg/g) = $20.2A_{645} + 8.02A_{663}$

Determination of total soluble solids content

Total soluble solids content was measured at harvest and during ripening until colour stage 6. Three fingers from the equatorial region hands were used for TSS determination. Total soluble solids content was determined using Atago hand refractometer (Type 500, Atago, Tokyo, Japan). A scoop of banana pulp from the apical, middle and basal part of the fruit was placed on a muslin cloth separately, and a drop of it squeezed out onto the refractometer. Readings were taken in °Brix.

Determination of sucrose, fructose and glucose contents

Sugars were analysed using the AOAC method (1996). Sugars were measured at harvest and during ripening to the fully ripe stage. Ten grams of the fruit was refluxed in ethanol for one hour. The sample was then concentrated with rotary evaporator and diluted with 75% acetonitrile. The individual sugars were analyzed using a high performance liquid chromatograph (HPLC) (Model LC-10AS, Shimadzu Corp., Kyoto, Japan) using a refractive index (RI) detector. Conditions included oven temperature, 35°C, recorder speed: 3, attenuation: 2, range: 4 and flow rate: 0.5 ml/min.

Determination of total titratable acidity

Total titratable acidity (TTA) was measured at harvest and during ripening using the AOAC method (1996). Total titratable acidity was

determined by titration with 0.1N NaOH in the presence of phenolphthalein indicator and expressed as percent malic acid.

Determination of green life

Fifteen fingers from the equatorial region hands were placed on a bench at ambient conditions of temperature (24 ± 1°C) and humidity (60 ± 5%). Five fingers served as a replicate. Green life was determined as the number of days taken by half of the fruits of one hand to progress from green stage to turning to a yellow tinge as described by Peacock and Blake, (1970) and Dadzie and Orchard, (1997).

Determination of shelflife

Fifteen fingers from the equatorial region hands were placed on a bench at ambient conditions of temperature (24±1°C) and humidity (60± 5%). Five fingers served as a replicate. Shelflife was then determined as the number of days taken by the fruit to progress from ripeness stage 6 to 8 (CSIRO, 1972; Marin et al., 1996; Paull, 1996; Jiang et al., 1999).

RESULTS AND DISCUSSION

Effect of bunch covers on grade, finger length and bunch weight

Bunch bagging had no significant (p>0.05) effect on grade, finger length and bunch weight (Table 1). This confirms earlier findings (ShihChao et al., 2004) that polyethylene bunch covers do not influence grade, finger length and bunch weights. Similar observations had earlier been reported (Vilela et al., 2001). This contradicts earlier findings where, for banana cv. robusta grown under high density production system, finger diameter (grade) and weight were significantly increased by polyethylene bunch covers (Reddy, 1989). Also, in South Africa, a 16.5% increase in 'Williams' bunch mass was recorded due to a 10% increase in finger length. This may have been due to increased temperatures (0.5°C) under blue covers that favoured growth (Robinson, 1996). Banana bunches sealed with polyethylene bags had increased fruit size at harvest (Amarante et al., 2002; Weerasinghe and Ruwapathirana, 2002). However, bagging some fruits such as lychee and mangoes had no effect on fruit weights (Amarante et al., 2002). Research

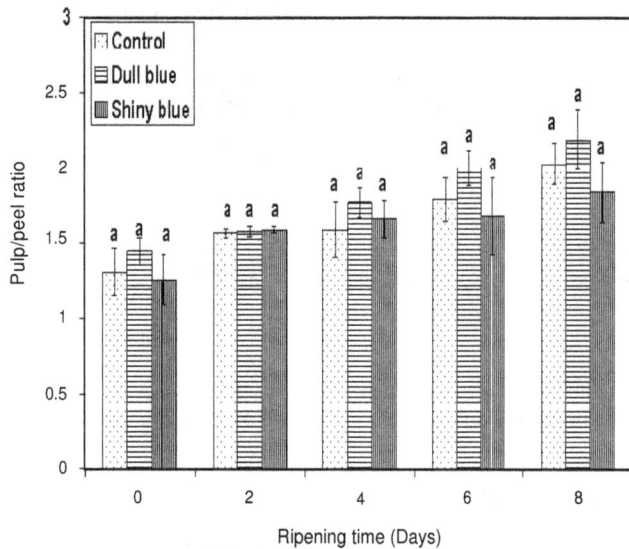

Figure 1. Pulp/peel ratio during ripening of cv. Williams fruits. Vertical bars represent SE of the mean of 3 replicates. Means denoted by the same letters in the same day have no significant difference at p=0.05.

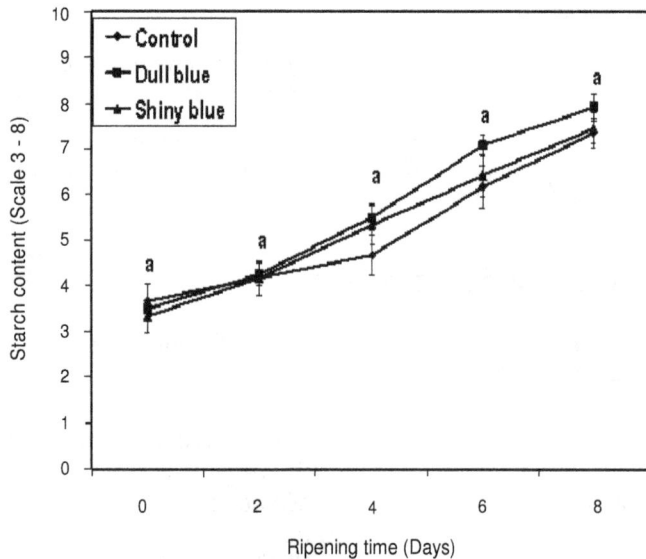

Figure 3. Effect of bunch covers on total soluble solids content (º Brix) of cv. Williams banana fruits during ripening. Vertical bars represent SE of the mean of 3 replicates. Same letters at different periods indicate no significant difference at p=0.05.

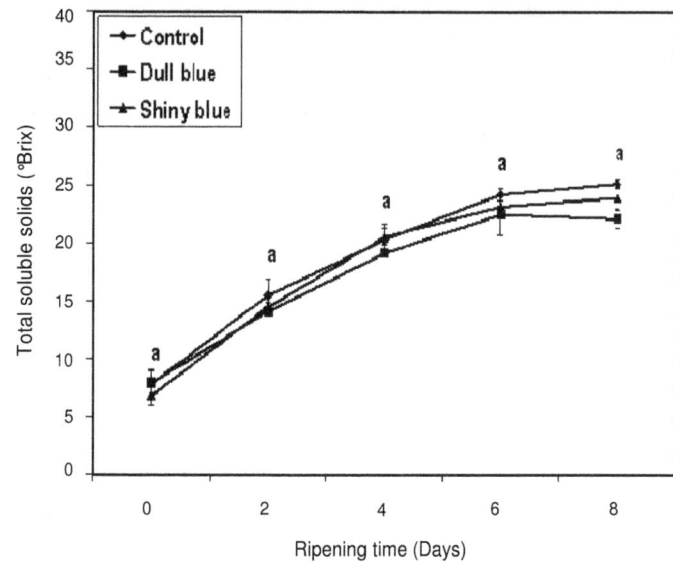

Figure 2. Effect of bunch covers on starch content of cv. Williams banana fruits using the Cornell Starch Chart scale 3 to 8 where 3= all starch and 8= no starch. Vertical bars represent SE of the mean of 3 replicates. Same letters at different periods indicate no significant difference at p=0.05.

Effect of pre-harvest bunch covers on pulp/peel ratios

Bunch covering had no significant (p>0.05) effect on the pulp/peel ratios of fruits of cv. Williams at harvest and during ripening (Figure 1). In bananas, the pulp portion continues to grow even in the later stages of maturation (Turner, 1997; Nakasone and Paull, 1998).

Effect of pre-harvest bunch covers on starch and total soluble solids

Both starch and total soluble solids (TSS) at harvest and during ripening were not influenced significantly (p>0.05) by bunch covers (Figures 2 and 3). Starch reduced as ripening progressed while TSS increased as expected in ripening banana (Stover and Simmonds, 1987). Unripe bananas have large amount of starch, with a content of 20 to 25% found in the pulp of the fruit (Nascimento et al., 2006). During the climacteric stage, the accumulated polysaccharide is rapidly degraded and most of it is converted into soluble sugars which form a large proportion of TSS in the banana (Marriot, 1980; Seymour et al., 1993).

Bagging, however, did not influence the starch formation during banana growth and starch degradation during ripening considerably in this study. Starch degradation in control fruits grown covered and uncovered proceeded normally in this study. However, in apples, bagging reduced starch content and fruit soluble solids at harvest (Proctor and Lougheed, 1976; Mattheis

reports on bagging of fruits have given contradictory information on the effect of bagging on both physical and compositional quality of fruits (Amarante et al., 2002) which may reflect differences in cultivar, bagging material and climatic conditions. In the current study, the covers were perforated and may not have increased the temperatures considerably to influence the parameters.

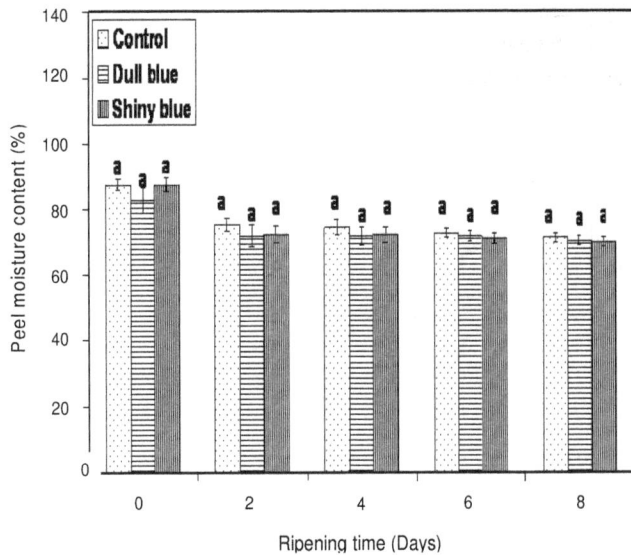

Figure 4. Effect of bunch covers on peel moisture content (%) of cv. Williams banana fruits during ripening. Vertical bars represent SE of the mean of 3 replicates. Means denoted by the same letters in the same day have no significant difference at p=0.05.

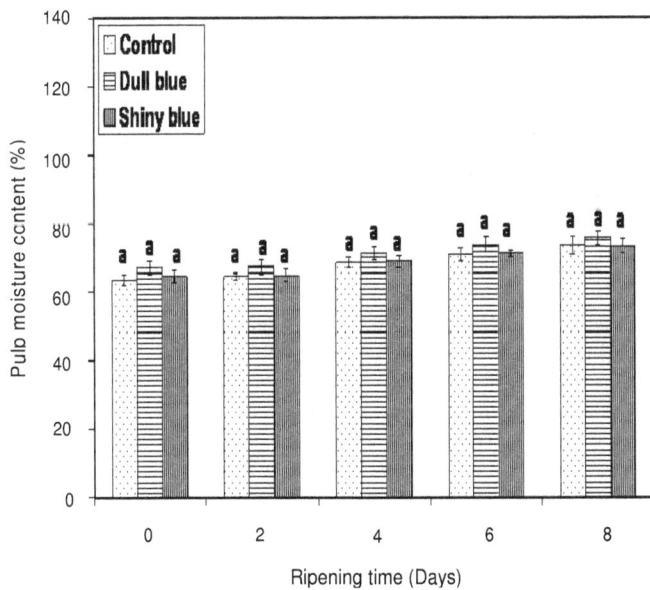

Figure 6. Effect of bunch covers on changes in percentage weight loss of cv. Williams banana fruits during ripening. Vertical bars represent SE of the mean of 3 replicates. Same letters at different periods indicate no significant difference at p=0.05.

Effect of pre-harvest bunch covers on fruit moisture content and weight loss

Fruits from the bagged and non bagged treatments had similar moisture contents for peel (Figure 4) and pulp (Figure 5) at harvest and during ripening. Changes in weight loss of fruits during ripening were not significantly (p>0.05) influenced by bunch covers (Figure 6). Moisture content of the peel reduced gradually during ripening while that of the pulp increased with ripening. Percentage fruit weight loss increased with days of storage in all the treatments. During normal ripening, the banana peel loses water to both the pulp and the atmosphere (Stover and Simmonds, 1987; Burdon et al., 1994).

Fruit weight loss is attributed to physiological weight loss due to respiration, transpiration and other biological changes taking place in the fruit during ripening (Rathore et al., 2007). Fruit surfaces are covered by cuticle covers which restrict water loss through transpiration, also. Fruits from the bagged and control bunches may have had similar cuticle structures (Amarante et al., 2002). Since the bunch covers in the current study had perforations, it is possible that the control and fruits grown under cover had similar humidity environment during growth and after harvesting. Similar observations were recorded in pears between fruits grown under perforated covers and control ones where both the moisture content and weight loss were not significantly (p>0.05) affected by pre-harvest bagging as they had similar skin permeability due to similar wax content of the cuticle (Amarante et al., 2002).

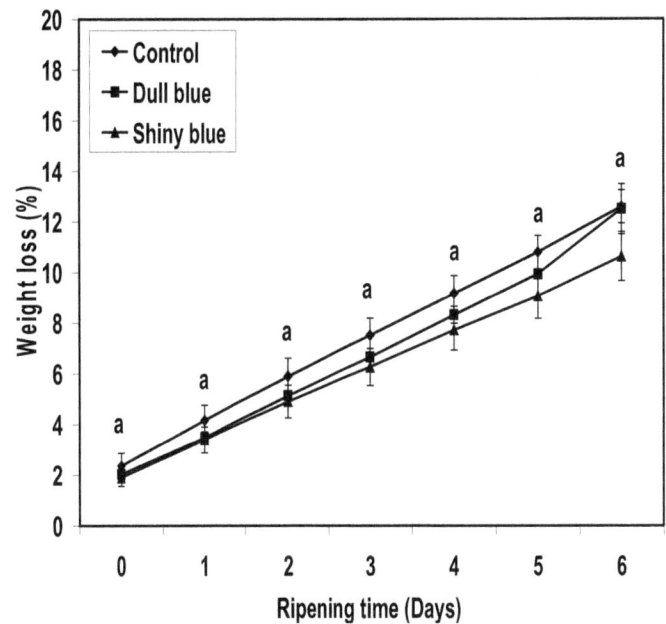

Figure 5. Effect of bunch covers on pulp moisture content (%) of cv. Williams banana fruits during ripening. Vertical bars represent SE of the mean of 3 replicates. Means denoted by the same letters in the same day have no significant difference at p=0.05.

and Fellman, 1999). In other reports, panicle bagging of lychee was found to have no effect on total soluble solids (Tyas et al., 1998). Elsewhere, fruit ripening for mangoes was enhanced by preharvest bagging although there was no effect on TSS and sensory quality at the postharvest stage for the bagged and unbagged fruits (Hoffman et al., 1997).

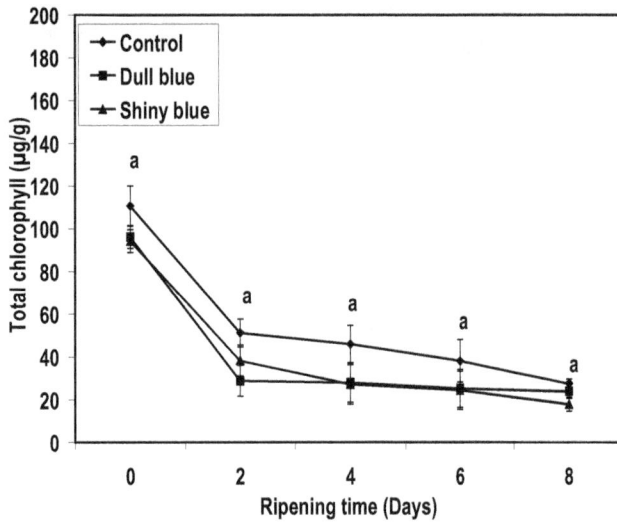

Figure 7. Effect of bunch covers on changes in peel chlorophyll degradation (µg/g) of cv. Williams banana fruits during ripening. Vertical bars represent SE of the mean of 3 replicates. Same letters at different periods indicate no significant difference at p=0.05.

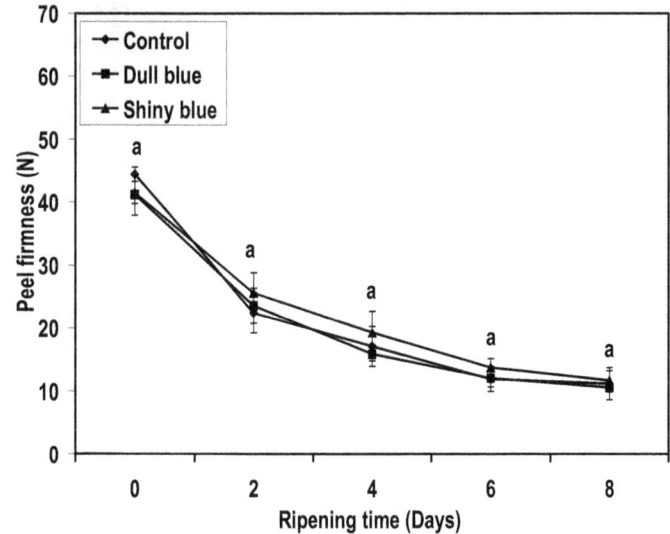

Figure 9. Effect of bunch covers on changes in objective peel firmness (N) of cv. Williams banana fruits during ripening. Vertical bars represent SE of the mean of 3 replicates. Same letters at different periods indicate no significant difference at p=0.05.

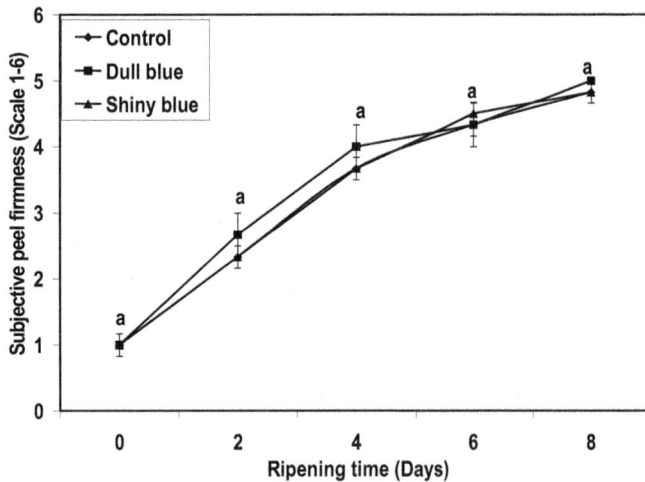

Figure 8. Effect of bunch covers on subjective firmness of cv. Williams banana fruits during ripening using the scale 1 to 6 where 1 = hard, 2 = firm, 3 = slightly soft, 4 = moderately soft, 5 = soft and 6 = very soft (Joyce et al., 1993). Vertical bars represent SE of the mean of 3 replicates. When absent, the SE fall within the dimensions of the symbol. Same letters at different periods indicate no significant difference at p=0.05.

Effect of pre-harvest bunch covers on total chlorophyll content

Effect of bunch covers on total chlorophyll content was not significant at harvest and during ripening (Figure 7). Chlorophyll content generally decreased on ripening as the fruits turned yellow. This is as a result of chlorophyll degradation and/or unmasking of the yellow carotenoids

or synthesis of new pigments (Gray et al., 2004). Bunch bagging had no effect on the chlorophyll degradation. The pigment has been shown to be converted to colourless non-fluorescent chlorophyll catabolites in a pathway that is probably active in all higher plants (Gray et al., 2004). Variable results in pigment development in fruits due to bagging have been reported. Bananas grown under non-perforated blue transparent polyethylene, non-transparent blue polythene, non-transparent black polythene and without covers had green, pale green, glossy white and dark green peel which probably affected the chlorophyll content of the peel (Shanmugasundaram and Manavalan, 2002).

Anthocyanin accumulation and red colour development of the skin was reduced by bagging (Hoffman et al., 1997; Joyce et al., 1997; Fan and Mattheis, 1998) while other reports indicate increased red colour development in apples (Wang et al., 2000) and pears (Amarante et al., 2002). This may reflect differences in the type of bagging material and whether perforated or not perforated. In this study, the bags were translucent blue and were perforated and hence allowed light penetration which may explain why bagging did not affect the chlorophyll content.

Effect of pre-harvest bunch covers on peel and pulp firmness

Peel and pulp firmness measured objectively and subjectively were not significantly different (p>0.05) in all the treatments at harvest and during ripening (Figures 8, 9 and 10). Firmness decreased rapidly during ripening, and

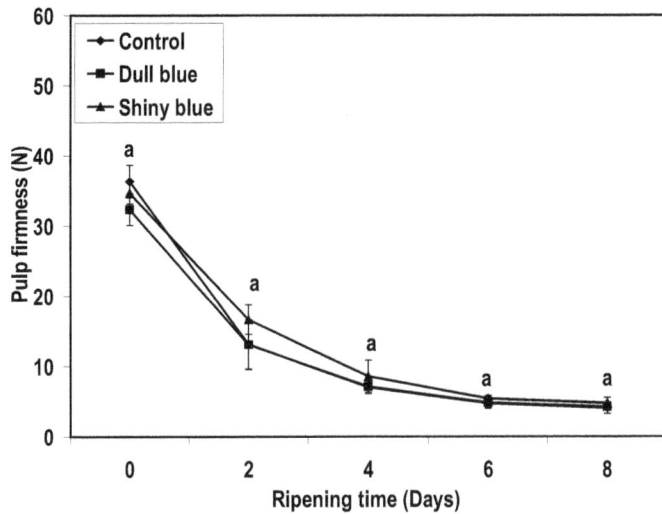

Figure 10. Effect of bunch covers on changes in objective pulp firmness (N) of cv. Williams banana fruits during ripening. Vertical bars represent SE of the mean of 3 replicates. When absent, the SE fall within the dimensions of the symbol. Same letters at different periods indicate no significant difference at p=0.05.

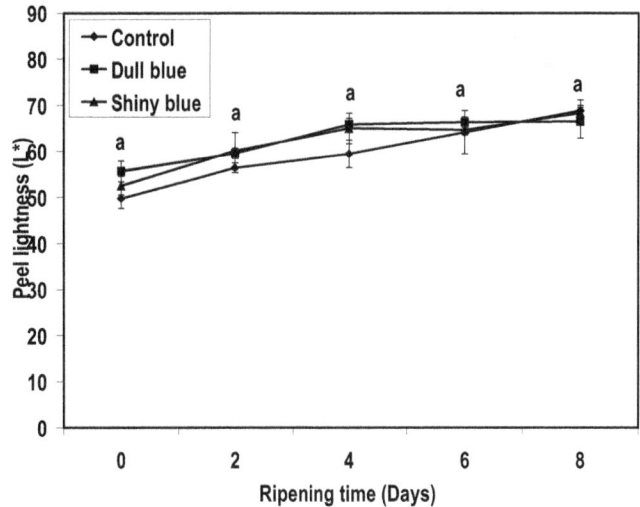

Figure 12. Effect of bunch covers on changes in peel lightness (L*) of cv. Williams banana fruits during ripening. Vertical bars represent SE of the mean of 3 replicates. When absent, the SE fall within the dimensions of the symbol. Same letters at different periods indicate no significant difference at p=0.05.

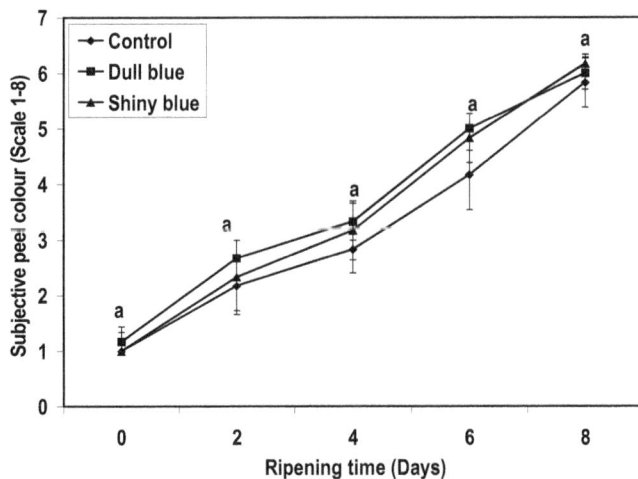

Figure 11. Effect of bunch covers on subjective peel colour of cv. Williams fruits during ripening using the scale of 1 to 8 where 1=green, 2=light green, 3=half yellow half green, 4=3/4 yellow with green, 5=yellow with green tip, 6=fully yellow, 7=yellow with spots and 8=yellow with coalesced black spots (CSIRO, 1972; Turner, 1997). Vertical bars represent SE of the mean of 3 replicates. When absent, the SE fall within the dimensions of the symbol. Same letters at different periods indicate no significant difference at p=0.05.

gradually after ripening of the fruits. Bagging did not change the peel and pulp properties considerably in this study. However, bagging of fruit reduced fruit firmness in the postharvest stage for bananas (Berill, 1956) while it had no effect on firmness at harvest although it enhanced loss of firmness during cold storage for pears (Amarante et al., 2002). The variable results reported on the effect of

bagging on fruit firmness at harvest and postharvest stage may reflect differences in the cultivar, type of bag, duration of cover, storage conditions and methods of testing for fruit firmness. In mangoes, opaque white plastic bags hastened softening of the skin while white waterproof paper bags did not have this effect (Joyce et al., 1997). When non-destructive methods of assessing peel firmness are performed over the fruit skin, they mainly reflect the changes in skin properties. Differences in softening may reflect differences in skin composition and structure between treatments affecting loss of cell wall integrity (Amarante et al., 2002). In this study, the bags were perforated and translucent and probably did not change the skin properties compared to the control.

Effect of pre-harvest bunch covers on colour

Subjective colour at harvest and during ripening was not influenced significantly (p>0.05) by bunch covers for both banana cultivars (Figure 11). Likewise, objective colour (L* and hue angle values) of both peel and pulp were not affected by bagging in the current experiment (Figures 12, 13, 14 and 15). The peel changed from green to yellow as the chlorophyll was degraded to unmask the yellow carotenoids (Gray et al., 2004) hence influencing the lightness of the peel positively on ripening. Therefore, L* value increased for the peel but decreased for the pulp on ripening as the peel degreened and the pulp turned from whitish to cream. Hue angle decreased for the peel also due to the change of the peel colour from green to yellow. Several reports have documented that bagging fruit increased skin lightness (Fan and Mattheis, 1998)

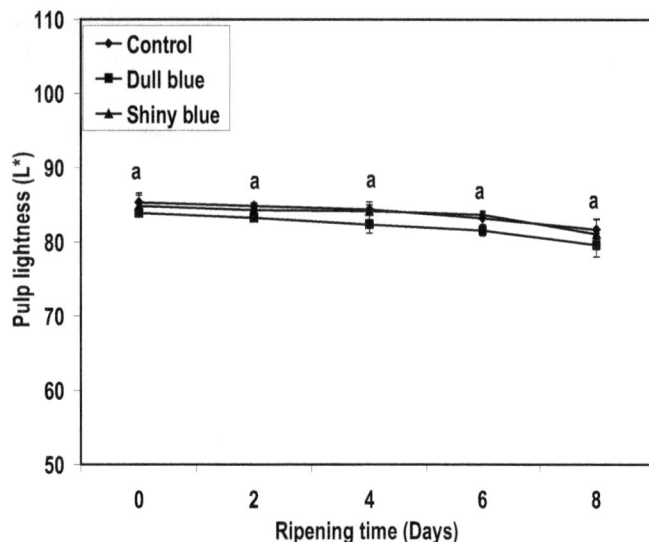

Figure 13. Effect of bunch covers on changes in pulp lightness (L*) of cv. Williams banana fruits during ripening. Vertical bars represent SE of the mean of 3 replicates. When absent, the SE fall within the dimensions of the symbol. Same letters at different periods indicate no significant difference at p=0.05.

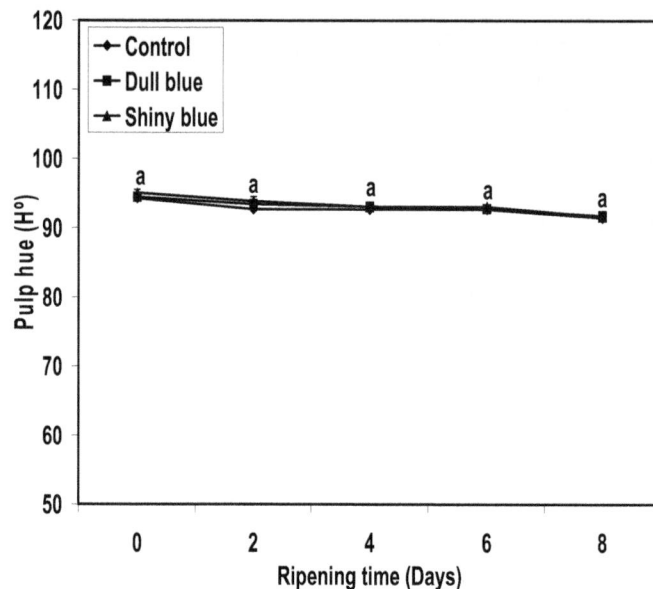

Figure 14. Effect of bunch covers on changes in peel hue (H^{o}) of cv. Williams banana fruits during ripening. Vertical bars represent SE of the mean of 3 replicates. When absent, the SE fall within the dimensions of the symbol. Same letters at different periods indicate no significant difference at p=0.05.

which shows that bagging has different effects on different fruit cultivars.

The difference in effects on colour may also be dependent on type and duration of bagging. Other workers showed that preharvest bagging of pears with micro-perforated polypropylene bags resulted in fruits with a more attractive light green colour and did not reduce blush on the exposed side of the skin (Amarante

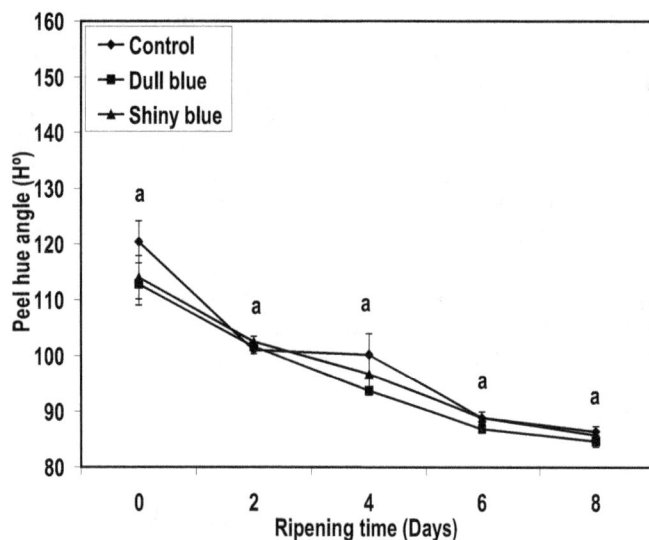

Figure 15. Effect of bunch covers on changes in pulp hue (H^{o}) of cv. Williams banana fruits during ripening. Vertical bars represent SE of the mean of 3 replicates. When absent, the SE fall within the dimensions of the symbol. Same letters at different periods indicate no significant difference at p=0.05.

et al., 2002). Unbagged lychee fruits had had lower intensity of colour (lower C*) than those bagged for 80 days but not different from those bagged for 20 and 42 days. The covers applied to the banana bunches in the current study were translucent and perforated and therefore did not cause substantial modification of bag internal atmosphere to reduce chlorophyll accumulation and hence colour. Pear fruits bagged with micro-perforated transparent plastic bags had similar anthocynin content hence similar skin colour with control fruits probably due to the fact that the bags did not cause significant changes in internal atmosphere to reduce anthocyanin accumulation (Amarante et al., 2002). When bagging affects fruit colour components significantly, then the visual colour is also affected probably due to the influence of the bag on radiation and temperature and consequently pigment production (Tyas et al., 1998).

Effect of pre-harvest bunch covers on sugar content

Both individual and total sugar contents were not significantly (p>0.05) influenced by covers (Figure 16 and 17). Noro et al. (1989) reported results where only fructose was affected by bagging in apples with bagged fruits having higher content while other main sugars were not affected. Watson et al. (2002) have reported that pre-harvest shading of strawberry fruits caused a significant reduction in sucrose and glucose/fructose contents compared to fruits from unshaded treatments. In the later experiment, shade netting was used which blocked some

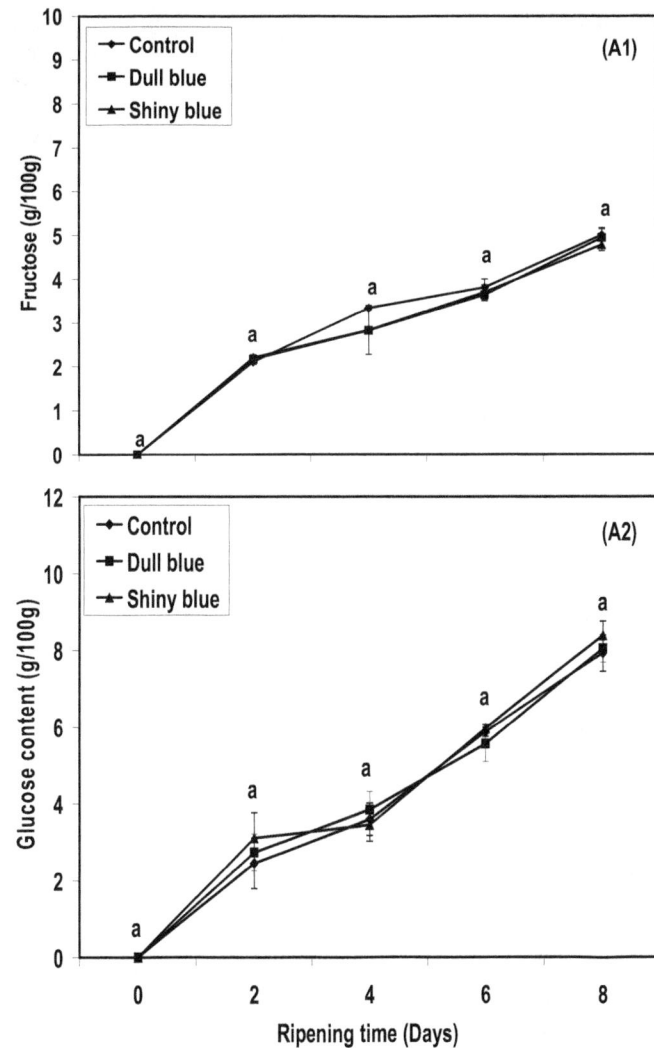

Figure 16. Effect of bunch covers on changes in fructose (A1) and glucose (A2) content of cv. Williams banana fruits during ripening. Vertical bars represent SE of the mean of 3 replicates. When absent, the SE fall within the dimensions of the symbol. Same letters at different periods indicate no significant difference at p=0.05.

Figure 17. Effect of bunch covers on changes in sucrose (B3) and total sugars (B4) content of cv. Williams banana fruits during ripening. Vertical bars represent SE of the mean of 3 replicates. When absent, the SE fall within the dimensions of the symbol. Same letters at different periods indicate no significant difference at p=0.05.

percentage of light from reaching the crop and hence may have affected such processes as photosynthesis and ultimately sugar synthesis. In the current study, the covers probably allowed enough light and hence did not interfere with starch/sugar synthesis. Blue polyethylene covers have been shown to allow blue-green and ultraviolet lights and also infrared rays (ShihChao et al., 2004). Light exposure of 'Sunscrest'/GF 677peaches resulted in increased reducing sugars content (Watson et al., 2002). Covering grapes with cellulose bags was shown to reduce sugar content in the fruits compared to the uncovered control (Signes et al., 2007). The inconsistent result in effect of bagging on sugar content may be due to different cover materials, fruit cultivars and holding environment after harvest.

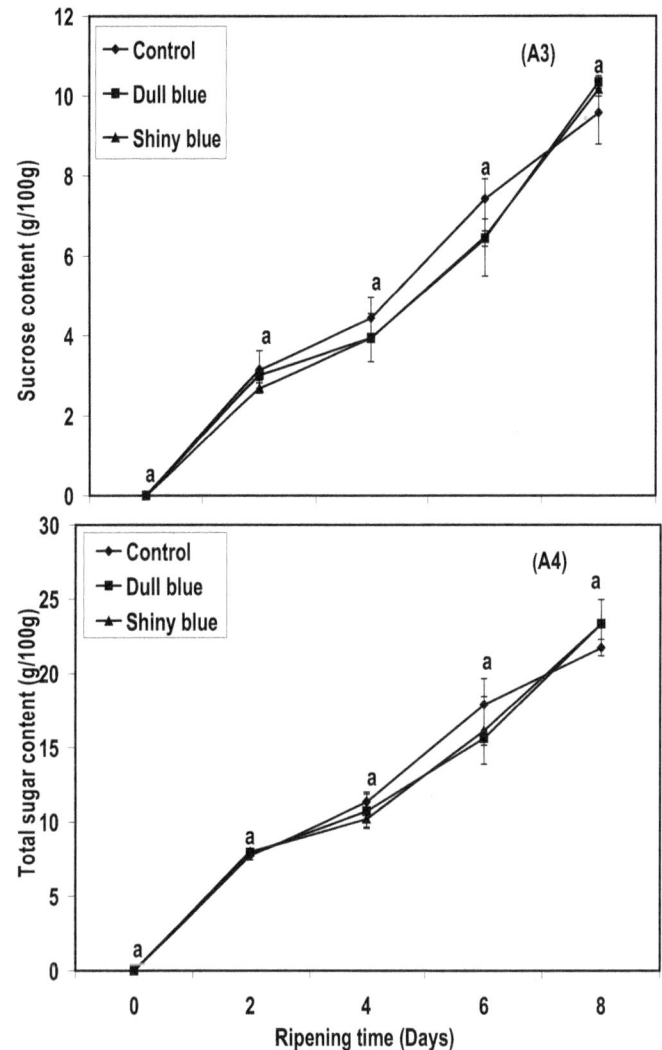

Effect of pre-harvest bunch covers on green life and shelflife

Bunch covers did not influence green life and shelflife significantly (p>0.05) (Table 2). Research reports on bagging of fruits have given contradictory information on the effect of bagging on both physical and compositional quality of fruits. Narayama et al., (2004) found bagging of bananas coupled with postharvest hot water treatment and storing with ethylene absorbent to be beneficial in extending shelflife. Elsewhere, banana grown under bunch covers had delayed ripening (Scott et al., 1971; Johns and Scott, 1989a) which may have possibly influenced green life. Fruit bagging has also been shown to adversely affect fruit quality. Sealed plastic covers

Table 2. Storage of tissue-cultured banana cultivar Williams as influenced by bunch covers.

Treatment	Green life (Days)	Shelf life (Days)
Control	14.67 [a]	5.33 [a]
Dull blue	10.33 [a]	3.67 [a]
Shiny blue	11.67 [a]	4.33 [a]
LSD	ns	ns

Values in the column followed by the same letter are not significantly different at p=0.05. Values are means of 3 replicates.

Table 3. Effect of bunch covers on bunch area covered by blemishes, dust, spider webs and bird droppings of tissue-cultured banana cv. Williams using Merz 0-6 scale (Merz, 2000), adopted for surface area covered by dirt instead of lesions where, 1=0 to 2%, 2=2 to 5%, 3=5 to 10%, 4=10 to 25%, 5=25 to 50% and 6= >50% of the affected surface area.

Treatment	Area covered by blemishes (scale 0 to 6)
Control	6 [a]
Dull blue	2 [b]
Shiny blue	2 [b]
LSD	1.15

Values in the column followed by the same letter are not significantly different at p=0.05. Values are means of 3 replicates.

delayed bunch maturity of bananas (Scott et al., 1971). Fruit ripening for mangoes was enhanced by bagging (Hoffman et al., 1997) which may have affected the green life. Banana bunches sealed with polyethylene covers during fruit growth delayed ripening (John and Scott, 1989b) probably due to delayed fruit development as a result of modification of atmosphere inside the sealed covers (John and Scott, 1989a). The covers used in the current study were perforated and translucent and hence did modify the atmosphere considerably to affect the green life and shelflife significantly.

Effect of pre-harvest bunch covers on visual appeal

Bagged banana fruits in the current experiment had minimal bruises (2 to 5%) and were significantly cleaner from dust, spider webs and bird droppings at harvest compared to the unbagged fruits (>50%) (Table 3) based on the Merz 0 to 6 scale (Merz, 2000). The covered fruits were therefore more visually appealing, cleaner compared to the unbagged fruits (Plate 1). This agrees with Weerasinghe and Ruwapathirana (2002) who found out that banana fruits grown under covers had no blemishes at all and were attractive to consumers at a glance while unbagged fruits had black spots and blemishes caused by thrips and freckle fungi attacks.

Similarly, postharvest fungal attack on lychee fruit was

Plate 1. Visual appearance of banana cultivar Williams fruits grown unbagged (A) and bagged (B) at harvest.

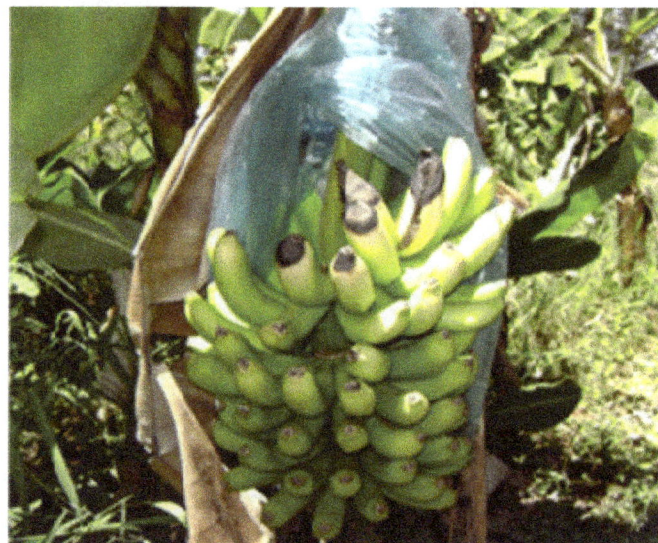

Plate 2. Sunburn of fruits of top hands caused by shiny blue covers of banana cultivar Williams during growth.

also reduced by bagging (Kooariyakul and Sardsud, 1997). However, a few of the covered fruits suffered sunburn which adversely affected fruit quality (Plate 2) especially during the hot season. This affected the bunches which were not well covered by leaves during growth. Top hands were mainly affected especially those of bunches covered with dull blue covers probably due to more heat absorbed inside the cover compared to the shiny blue covers which may have reflected some heat away.

Elsewhere, bagging of bananas resulted in sun scorching of the fruits irrespective of the colour of the bunch covers (Weerasinghe and Ruwapathirana, 2002). However, this can be overcome by maintaining enough leaves on the plant to shade the plant and also by using reflective blue covers (Anon, 2003). Pulling leaves over the covered bunches during growth may also reduce/

prevent sunburn. Inserting a newspaper on the inside of the bunch covers to cover top hands to prevent them from sun scorch has been shown to work (Linbing et al., 2004). The blue polyethylene covers have been shown to absorb more blue-green and ultraviolet lights which may cause sunburn to banana fruits (ShihChao et al., 2004)

Conclusion

The study has shown that perforated dull and shiny blue bunch covers may be used in commercial banana orchards in Kenya to produce high quality fruits. The physical and biochemical properties of the banana fruits were not adversely affected by the bunch covers. Also, the fruits grown covered were more visually appealing as they were clean and had minimal bruises compared to those grown uncovered which implies reduced water usage during postharvest preparation of the fruits. However, the bunch covers caused sun scald of a few top hands during the hot months. Bunch covers may therefore be useful mainly in the cooler months of the year and also in cooler climates where sunburn may not be a major concern. However, the use of bunch covers should be coupled with proper postharvest handling procedures to ensure that the clean, visually appealing fruits are not bruised during the postharvest period. Such fruits could also be targeted for the export market where they may fetch better prices as the consumer clientele appreciates the visually appealing fruits and are willing to pay more for such fruits. A cost benefit analysis also needs to be done to find out whether banana bagging is profitable in Kenya. This work should also be carried out in other agro-ecological zones in the country such as in Upper midland zones 1 and 2 (UM1 and UM2), especially the cooler banana growing areas in the Meru region in Eastern Province. Other banana varieties may also be tested as they may exhibit differences in the way they are affected by the sun.

ACKNOWLEDGEMENTS

The authors acknowledge the financial support received from Kenya Agricultural Productivity Project (KAPP) under KARI through a PhD study grant to Margaret Muchui. The authors are grateful to the family of Eng. Gitoho for allowing them to use their farm as the experimental site.

REFERENCES

Acharya SS, Mackey MGA, (2008). Africa Harvest Biotech Foundation International (AHBFI). Socio-economic impact assessment of the tissue culture banana industry in Kenya. Nairobi, Kenya.
Amarante C, Banks NH, Max S (2002). Effect of preharvest bagging on fruit quality and postharvest physiology of pears (Pyrus communis). New Zealand J. Crop and Hortic. Sci., 30: 99-107.
Anon (2002). Kenya Agric. Res. Inst., (KARI). Banana growing guide
Anon (2003). Bunch covers for improving plantain and banana peel quality. National Agricultural Research Institute. Technical Bulletin No. 4. www.agrinetguyana.org.gy/moa.mfcl
AOAC (Association of Official Analytical Chemists). (1996). Official methods of analysis. Washington, D.C., 37: 1-53.
Arnon D (1949). Copper enzymes in isolated chlroplasts: Polyphenoloxidase in Beta vulgaris. Plant Physiol., 24: 1-15.
Berill FW (1956). Bunch covers for bananas. Queensland Agric. J., 82: 435-439.
Burdon JN, Dori S, Lomaniec E, Marinansky R, Pesis E (1994). The post-harvest ripening of water stressed banana fruits. J. Hortic. Sci., 69 (5): 799-804
CSIRO (Commonwealth Scientific and Industrial Research Organisation). (1972). Australia Division of Research, 1972. Banana ripening guide, Circular 8. Melbourne NSW.
Dadzie BK, Orchard JE (1997). Routine postharvest screening of banana/plantain hybrids: Criteria and methods. INIBAP Technical guidelines 2. International Plant Genetic Resources Institute, Rome, Italy.
Fan X, Mattheis JP (1998). Bagging 'Fuji" apples during fruit development affects colour development and storage quality. Hort Sci., 33: 1235-1238.
Gowen S R (1995). Pests. In: Gowen, S. (ed) Bananas and Plantains. Chapman and Hall: London, pp. 382-402
Gray J, Wardzala E, Yang M, Reinbothe, S, Haller S, Paul F (2004). A small family of LLSI-related non-heme oxygenases in plants with an origin amongst oxygenic photosynthesizers. Plant Mol. Biol., 54: 39-54.
Hewett EW (2006). An overview of preharvest factors influencing postharvest quality of horticultural products. Intern. J. Postharv. Technol. Innov., 1(1): 4-15.
Hoffman PJ, Smith LG, Joyce DC, Johnson GI, Meiburg GF (1997). Bagging of mango (Mangifera indica cv. Keitt) fruit influences fruit quality and mineral composition. Postharv. Biol. Techn., 12: 83-91.
Jia H, Araki, Okamato G (2005). Influence of fruit bagging on aroma volatiles and skin coloration of 'Hakuho' peach (Prunus paesicha Batsch). Postharv. Biol. Technol., 35: 61-68.
Jiang Y, Joyce DC, Macnish AJ (1999). Extension of the shelf-life of banana fruit by 1-methylcyclopropene in combination with polyethylene bags. Postharv. Biol. Technol., 16: 187-193.
Johns GG, Scott K J (1989a). Delaying harvesting of banana with 'sealed' covers on bunches. 1. Modified atmosphere and microclimate inside sealed covers. Austr. J. Exp. Agric., 29: 719-726.
Johns GG, Scott KJ (1989b). Delaying harvesting of banana with 'sealed' covers on bunches . 2. Effect on fruit yield and quality. Austr. J. Exp. Agric., 29: 727-733.
Johns GG (1996). Effects of bunch trimming and double bunch covering on yield of bananas during winter in New South Wales. Austr. J. Exp. Agric., 36: 229-235.
Joyce DC, Beasly DR, Shorter DJ (1997). Effect of preharvest bagging on fruit calcium levels, and storage and ripening characteristics of 'Sensation' mangoes. Austr. J. Expt. Agric., 37: 383-389.
Joyce DC, Hockings PD, Mazucco RA, Shoeter AJ, Brereton IM (1993). Heat treatment injury of mango fruit revealed by non-destructive magnetic resonance imaging. Postharv. Biol. Technol., 3(4): 305-311.
Kooariyakul S, Sardsud V (1997). Bagging of Lychee fruit to reduce postharvest diseases. In: Coates, L.M. Hofman, P.J. and Johnson, G.I. (ed). Disease control and storage life extension in fruit. Australian Centre for International Agricultural Research (ACIAR) Proceedings, 81: 92-100.
Linbing X, Hu Y, Bingzhi H, Yuerong W (2004). Production and R&D of banana in China In: Proceedings of the 21st BASNET steering committee meeting, Jakarta (IDN), pp. 49-59.
Marin DH, Blankenship SM, Sutton TB, Swallow WH (1996). Physiological and chemical changes during ripening of Costa Rican banana harvested in different seasons. J. Ameri. Soc. Hortic. Sci., 121: 1157-1161.
Marriot J (1980). Bananas-Physiology and biochemistry of storage and ripening for optimum quality. CRC Critical reviews in Food Sci. Nutri., 13: 41-88.

Mattheis JP, Fellman JK (1999). Preharvest factors affecting flavor of fresh fruit and vegetables. Postharv. Biol. Technol., 15: 227-232.

Mclellan M R, Lind LR, Kime RW (1995). Hue angle determination and statistical analysis for multiquadrant Hunter L, a, b data. J. Food Quality, 18(3): 235-240.

Merz U (2000). Powdery scab. Research in Switzerland. In the proceedings of the first European powdery scab workshop, ed. U. Merz and A.K. Lees, Aberdeen, Scotland, pp. 67-72

MOA (Ministry of Agriculture). (2006). Horticultural annual Report. Ministry of Agriculture, Horticulture Division. Nairobi, Kenya, pp. 118-136.

Nakasone HY, Paull RE (1998). Tropical fruits. CAB International New York.

Narayana CK, Krishnan P, Satiamoorthy S (2004). Effect of bunch covering and postharvest treatments on quality of banana during storage at low temperature. In international congress on Musa harnessing research to improve livelihoods.

Nascimento JRO, Junior AV, Bassinello PZ, Cordenunsi BR, Mainardi JA, Purgatto E, Lajolo FM (2006). Postharv. Biol. Technol., 40 (1): 41-47.

Noro S, Kudo N, Kitsuva T (1989). Differences in sugar and organic acid contents between bagged and unbagged fruits of yellow apple cultivars, and the effect on development on anthocyanin. J. Japanese Soc. For Horti. Sci., 58: 17-24.

Paull RE (1996). Ethylene, storage and ripening temperatures affect Dwarf Brazilian banana finger drop. Postharv. Biol. Technol., 8: 65-74.

Peacock BC, Blake JR (1970). Some effects of non-damaging temperatures on the life and respiratory behaviour of bananas. Queensland J. Agric. and Anim. Sci., 27: 147-168

Proctor JTA, Lougheed EC (1976). The effect of covering apples during development. Hort Sci., 11: 2.

Qaim M (1999). Assessing the impact of Banana Biotechnology in Kenya. ISAAA Briefs no. 10. Ithaca, New York, USA.

Rathore HA, Masud T, Sammi S, Soomro AH (2007). Effcet of storage on physico-chemical composition and sensory properties of mango (*Mangifera indica* L.) variety Dosehari. Pakistan J. Nutr., 6: 143-148.

Reddy SA (1989). Effect of bunch covers on bunch maturity and fruit size in high density robusta banana orchards. J. Res. (APAU), 17: 81-82.

Robinson JC (1996). Bananas and plantains. Institute for Tropical and Subtropical Crops, University Press, Cambridge, pp.172-174.

SAS (2001). SAS Institute Inc., Cary NC, USA. Release 8.2.

Scott KJ, Wills RBH, Rippon LE (1971). The use of sealed polyethylene bunch covers during growth as a retardant to the ripening of bananas. Tropical Agric., 48: 163-165.

Seymour GB, Taylor JE, Tucker GA (1993). Biochemistry of fruit ripening Seymour G.B., (ed). Chapman and Hall, London, pp. 83-106.

Shanmugasundaran KA, Manavalan RSA (2002). Postharvest characteristics of 'Rasthali' bananas grown under different polyethylene covers. Infomusa, 11(2):43-45.

Shewfelt RL (1999). What is quality?. Postharvest Biol. Technol., 15: 197-200.

Shewfelt RL (2009). Measuring quality and maturity, In: Postharvest Handling – A systems approach. W.J. Florkowski, R.L. Shewfelt, B. Brueckner and S.E. Prussia (eds) Academic press, Inc.London, pp. 461-481.

ShihChao C, DinFon K, ChunMei C, MeiJen C (2004). Comparison in the bunch development and post-harvest quality of banana as affected by kraft paper cover and polyethylene cover. J. Chinese Soc. Hortic. Sci., 50(3): 245-252.

Signes AJ, Burlo F, Martinez-Sanchez, F, Carbonell-Barrachina AA (2007). Effcets of preharvest bagging on quality of black table grapes. World J. Agric. Sci., 3(1): 32-38.

Stover RH, Simmonds NW (1987). Bananas 3rd ed. Longman Scientific and Technical, Essex.

Turner DW (1997). In: Postharvest Physiology and Storage of Tropical and Subtropical Fruits. CAB International, Mitra S.K. (ed.), New York, NY, pp. 47-77.

Tyas JA, Hoffman PJ, Underhill SJR, Bell KL (1998). Fruit canopy position and panicle bagging affects yield and quality of 'Tai So' lychee. Sci. Horti., 72: 203-213.

Vilela RMG, Ferreira SR, Joao MLP (2001). Influence of polyethylene banana bunch cover for irrigated banana tree in the north of Minas Gerais State. Rev. Bras. Frutic., 23(3): 559-562.

Wang H, Arakwa O, Motomura Y (2000). Influence of maturity and bagging on the relationship between anthocyanin accumulation and phenylalanine ammonia-lyase (PAL) activity in 'Jonathan' apples. Postharv. Biol. Technol., 19: 123-128.

Watkins CB (2006). 1-methylcyclopropene (1-MCP) based technologies for storage and shelf-life extension. Int. J. Postharv. Technol. Innov., 1: 62-68.

Watson R, Wright CJ, McBurney T, Taylor AJ, Linforth RST (2002). Influence of harvest date and light integral on development of strawberry flavour compounds. J. Expt Bot., 53 (377): 2121-2129.

Weerasinghe SS, Ruwanpathirana KH (2002). Influence of bagging material on bunch development of banana (*Musa* spp.) under high density planting system. Annals of the Sri Lankan Dept of Agric., 4: 47-53.

5

Evaluation of price linkages within the supply chain of rice markets in Cross River State, Nigeria

Ohen S. B.* and S. O. Abang

Department of Agricultural Economics and Extension, University of Calabar, Calabar, Cross River State, Nigeria.

This paper evaluates price linkages within the supply chain of rice markets in Cross River State using weekly prices in three urban markets located in major rice producing areas of the State. The Johansen cointegration test indicated one cointegrating vector both at the 1 and 5% levels of significance. The results of the study indicate that the supply chain (farmgate-assembler-wholesaler-retailer) in Cross River is integrated. Though the price changes may vary in the short run between the different levels (farmgate-assembler-wholesaler-retailer), they were expected to move together as a system in the long run. The study recommends that facilitative policies that will enhance the provision of infrastructures such as good roads, market structures and efficient market information network systems should be formulated and implemented. Also, the government should provide price regulatory services to enhance market integration and reduce market exploitation by intermediaries especially in the short run.

Key words: Cointegration, supply chain, price linkages, rice markets.

INTRODUCTION

Market integration of agricultural products has retained importance in developing countries due to its potential application to policy making (Heman and Fateh, 2005). The extent of integration gives the government a direction on how to formulate policies of providing infrastructure and regulatory services to avoid market exploitation. Price behaviour along supply chains is an important indicator of overall market performance. Markets that are not integrated may convey inaccurate price information distorting the marketing decisions of rice producers and contributing to inefficient product movements. More so, rice has been cultivated consumed and marketed by women and men worldwide for more than 10,000 years (Kenmore, 2003) longer than any other crop. The total area under rice cultivation is globally estimated to be 150,000,000 ha with annual production averaging 500,000,000 metric tons (Tsuboi, 2005). This represents 29% of the total output of grain crops worldwide, (Xu and Guofang, 2003). FAO (2001) asserts that in Nigeria, the demand for rice has been increasing at a much faster rate than in any other African country, since the mid

1970s. The average Nigerian consumes 24.8 kg of rice per year which represents 9% of total caloric intake.

This increase in consumption according to Akande (2004), Ogundele and Okoruwa (2006) and Daramola (2005) is largely due to urbanization, population growth, increased income levels, and the fact that rice is easy to prepare when compared to other traditional cereals, thereby reducing the chore of food preparation and fitting more easily in the urban lifestyles of the rich and poor alike. Rice is produced in all the agro ecological zones of Nigeria. Production is primarily by small-scale producers, with average farm sizes of 1 to 2 ha. According to Daramola (2005), there are three major rice production systems in Nigeria namely upland rain-fed, lowland rain-fed and irrigated. Rice cultivation is widespread within the country extending from the northern to southern zones with most rice grown in the eastern (Enugu, Cross River and Ebonyi States) and middle belt (Benue, Kaduna, Niger and Taraba States) of the country. In Nigeria, the enterprise provides employment for more than 80% of their inhabitants in various activities along the production/ distribution chains from cultivation to consumption. Also, marketing is a cardinal determinant of the frequency and intensity of product distribution. Ihene (1996) opines that rice marketing covers the performance of all business

*Corresponding author. E-mail: suzyben01@yahoo.com.

activities in the flow of paddy and milled rice, from the point of mutual production until they are in the hands of the ultimate consumers. This must be at the right time, in the right place and as convenient as possible, at a profit margin that will keep the marketer in operation.

Rice marketing involves various intermediaries between the producers and the consumers who facilitate exchange among trading partners to move rice to consumers. These intermediaries function in environment constrained by low investments in marketing and market infrastructure, shortage of food supply and the limited progression toward more visible market arrangements. However, there is need for price information to flow accurately within the supply chain. Markets that are not integrated may convey inaccurate price information, distorting the marketing decisions of rice producers and contributing to inefficient product movements. Therefore, it is important to analyse price integration within the different levels within the marketing system. The objective of this paper is therefore to empirically evaluate the price linkages within the supply chain of rice markets in Cross River State, using weekly prices in three urban markets located in major rice producing areas of the State. The specific objectives are to:

1. Assess the level of price stationarity;
2. Test for cointegration of the price series.

METHODOLOGY

Study area

Cross River State is located within the tropical rainforest belt of Nigeria. It lies between latitude 4° 28' and 6° 55' north of the equator and longitude 7° 50' and 9° 28' east of the Greenwich meridian. It shares common boundaries with the Republic of Cameroon in the east, Benue State in the North, Ebonyi and Abia States in the West, Akwa Ibom State in the Southwest and the Atlantic Ocean in the South. It has a total landmass of about 23,000 km^2. At least five distinct ecological zones are represented in the State ranging from mangrove and swamp forest towards the coast, tropical rain forests further inland, and savannah woodlands in the Northern parts of the State. The highlands of Obudu Plateau offer montane type vegetation. The favourable climate of tropical, humid, dry and wet seasons gives rise to rich agricultural lands, thus encouraging both perennial and annual crop cultivation. The population of the State in 2001 stood at 2,526,542 giving a population density of 110 persons / km^2. The gender distribution of the population is 1,263,915 (50.03%) males and 1,262627 (49.97%) females. Cross River State has two distinct wet and dry seasons occurring in April to November, and December to March respectively. The varied ecological zones of the State makes it rich in a variety of crops such as rice, rubber, cocoa, cashew, yam, cocoyam, plantain, banana, groundnut and assorted vegetables.

Sampling procedure

The study adopted a multistage sampling method. In the first stage, Cross River State was purposively selected. Secondly, three markets comprising one urban and two rural markets were chosen.

This was based on the fact that the rural markets are located within communities were intensive rice cultivation is done. Thus, a total of three markets were purposively selected and used for the study. The markets selected were Watt market, Ofudua rice market and Yala rice market. The sampling frame which comprised all the registered members of the different rice associations in the selected markets were obtained from the presidents of the different rice associations. These lists were further upgraded to remove names of participants who had left the trade either by death or relocation. Using the simple random technique, assemblers, wholesalers and retailers were selected. At least 50% of the traders' population at each level was selected. The sizes were guided by the homogeneity in the population and the time and resources available to the researcher (Ndiyo, 2005). The assemblers were selected at the processing markets within the rice markets. It was ensured that assemblers chosen, performed only assembling functions. Data on farmers' price, rice assemblers' price, wholesalers' price and retailers' price were collected from 100 respondents comprising 15 assemblers, 40 wholesalers and 45 retailers for a period of 44 weeks starting from March, 2007 to January, 2008 using questionnaires and price data forms.

Analytical technique

The use of ordinary least square (OLS) to estimate price integration has the shortcoming of assuming that data series are stationary though most agricultural time series data tend to be non stationary and also the inability of the technique to give the short and long run adjustments, thus using OLS with non stationary data may result to spurious regression (Granger and Newbold, 1974). To avoid these problems, co-integration analysis is used to check for the relationship among prices in different levels. When a long-run linear relation exists among different price series, these series are said to be co-integrated. In addition, to make a clear distinction between short-run and long-run integration, the study uses an error correction model (ECM). This allows the researcher to derive the speed of price transmission from one level to another. In this study, we apply the Johansen procedure to test for long run and short run price integration. The approach adopted is to estimate a vector auto-regressive model (VAR) in which market prices of rice at a level are explained by its own lagged prices and lagged prices on other market levels. The price series used for this test were time series collected for a period of 44 weeks.

A prerequisite for undertaking cointegration tests is to verify that the series is nonstationary and to ascertain the variables' integration order. The most commonly used test for determining whether a series is nonstationary is the augmented Dickey-Fuller (ADF) unit root test. In this test, a null hypothesis is imposed that the data are non-stationary (that is contain a unit root) against the alternative hypothesis of being a stationary variable. Differencing a non stationary variable generally results in a stationary variable. If a series is differenced d times before it becomes stationary, thus containing d unit roots, it is said to be integrated of order d and is denoted as being $I(d)$. Variables that are stationary in their levels, that is $I(0)$ should be discarded from cointegration analysis. In most cases it is not strictly necessary for all the variables in question to have the same order of integration (Harris, 1995). Another important implication of cointegration and the error correction representation is that cointegration between two variables implies the existence of causality (in the Granger sense) between them in at least one direction (Granger, 1988). Cointegration itself cannot be used to make inferences about the direction of causation between the variables, and thus causality tests are necessary. Granger (1969) proposed an empirical definition of causality based only on its forecasting content: if x_t causes y_t then y_{t+1} is better forecast if the information in x_t is used, since there will be a smaller

Table 1. Augmented Dickey Fuller (ADF) test on price series.

Price	No. of observation	Unit root on price levels		Unit of root on first difference	
		ADF with constant	ADF with constant and trend	ADF with constant	ADF with constant and trend
LNASPI	44	-1.6559	-4.2811	-5.4097*	-7.2629*
LNFGPI	44	-1.3284	-2.4777	-7.0410*	-6.9946*
LNRLPI	44	-0.7185	-1.7047	-7.0409*	-7.5677*
LNWSPI	44	-1.1916	-2.9765	-75239*	-7.5677*

Estimated from field survey (2007/2008); * indicates significance at 1%; LNASP1 = natural log for average assemblers' selling price of rice sold in three markets selected from Cross River State; LNWSP1 = natural log for average wholsalers' selling price of rice sold in three markets selected from Cross River State; LNRLP1 = natural log for average retail selling price of rice sold in three markets selected from Cross River State; LNFGP1 = natural log for average buying price purchased by assemblers' in three markets selected from Cross River State. This price was used to represent the farm gate price; ADF = Augmented Dickey Fuller.

variance of forecast error. More so, if two markets are integrated, the price in one market, p_1, would commonly be found to Granger-cause the price in the other market, p_2 and/or vice versa. Therefore, Granger causality provides additional evidence as to whether, and in which direction, price integration and transmission is occurring between two price series or market levels. In line with this, this study will therefore carry out a Granger causality test to make inferences about the direction of causation between the price series under study.

Mathematically, the cointegration test is specified as follows:

$$X_t = A_1 X_{t-1} + A_2 X_{t-2} + A_{p-1} X_{t-(p-1)} + A_p X_{t-p} + \varepsilon_t$$

(1)

Where: t = 1,2....n refers to the weeks of prices considered; p = an a prori unknown integer, whose value is determined in the estimation process; X_t is an nx1 vector of variables (X_{1t} ,X_{2t}..........X_{nt}) (prices at n market levels); A_t is an (nxn) matrix of coefficients; ε_t is an (nx1) vector of error terms; n is the number of prices included in the analysis.

With $\Delta X_t = X_t - X_{t-1}$ Equation (1) can be put in a more suitable form as:

$$\Delta X_t = \sum \Pi_i \Delta X_{t-i} + \Pi X_{t-1} + \varepsilon_t$$

(2)

Where Π and Π_i are defined as:

$$\Pi = \sum_{i-1}^{p} A_i - 1 \text{ and } \Pi_i = \sum_{j=i+1}^{p} A_j$$

The Johansen approach defines two matrices α and β, both of dimension nxr, where r is the rank of Π such that $\Pi \beta$. The matrix β is the matrix of cointegrating relations and the matrix α is the matrix of weights with which each cointegrating vector enters the n equations of the vector error correction model (VECM). α can be viewed as the matrix of the speed of adjustment parameters. The Johansen procedure allows for a wide range of hypothesis testing

on the coefficients α and β, using likelihood ratio test (Johansen, 1990).

RESULTS AND DISCUSSION

Testing for stationarity in the price series

A unit roots analysis of each of the time series of the chosen variables were undertaken to ascertain the order of integration or test for the stationarity of the prices. This is to ensure that the variables are not integrated of order greater than one. Different test such as the Phillip-Perron and augumented Dickey Fuller (ADF) could be used. This study used the ADF unit root test. The results are presented in Table 1. Using the ADF test, the results presented in Table 1 indicate that the price series were stationary at first difference 1(1). This result implied that inclusion of first differences as variables in the model, instead of normal price series, will eliminate the stochastic trend to which the nominal series are exposed.

Testing for cointegration in the price series

The Johansen cointegration test indicated one cointegrating vector both at the 1 and 5% levels of significance (Table 2). This result implied that the supply chain in Cross River is integrated. Though the price changes may vary in the short run between the different levels, they were expected to move together as a system in the long run. This result gave the opportunity to estimate the movement of prices in the long and short run, using a vector error correction mechanism (Table 3). The long and short run equations were selected using the Schwarz criterion and Akaike information criterion (AIC). The results revealed that the prices (wholsaler, assembler, retailer and farmgate) were significantly integrated in the long run (Table 3). The short run dynamics revealed that 55% of the deviations from the long run equilibrium corrected per week .This is explained

Table 2. Johansen cointegration test for prices in three markets.

Null hypothesis	Alternative	Test value	95% critical value	99% critical value
Trace test				
r = 0	r > 0	72.20*	47.21	54.46
r < 1	r > 1	18.33	29.68	35.65
r < 2	r > 2	6.97	15.41	20.04
r < 3	r > 3	2.84	3.76	6.65
Max test				
r = 0	r > 0	53.86*	27.07	32.24
r < 1	r > 1	11.36	20.97	25.52
r < 2	r > 2	4.13	14.07	18.63
r < 3	r > 3	2.84	3.76	6.65

Estimated from field survey (2007/2008); * indicates significance at 5%; r = rank or the number of cointegrating equations.

Table 3. Estimates for the short and long run price integration in three selected markets

Variable	Dependent variable LNFGP1					
	Short run estimates			**Long run estimates**		
	Coefficient	Standard error	t-statistic	Coefficient	Standard error	t-statistic
ECM(-1)	-0.5509	0.092	-5.9880*			
D(LNFGP1(-1))	-0.177	0.0898	-1.9716**			
D(LNASP1(-1))	0.2096	0.117	1.791			
D(LNWSP1(-1))	-0.20644	0.101	-2.6164*			
D(LNRLP1(-1))	-0.3372	0.1044	-3.2298*			
LNASP1				1.7535	0.2222	-7.8926*
LNWSP1				0.3394	0.4113	0.8251
LNRLP1				0.5409	0.2756	1.9629**
R^2	0.6482					
ADJUSTED R^2	0.6182					
F-Statistic	22,597					
Log likelihood	292.1225					
Akaike AIC	-2.7972					
Schwarz SC	-2.6165					

*Indicates significance at 1% and ** at 5%; LNASP1 = natural log for average assemblers' selling price of rice sold in three markets selected from Cross River State; LNWSP1 = natural log for average wholesalers' selling price of rice sold in three markets selected from Cross River State; LNRLP1 = natural log for average retail selling price of rice sold in three markets selected from Cross River State; LNFGP1 = natural log for average buying price purchased by assemblers' in three markets selected from Cross River State, this price is used to represent the farm gate price; D(LNASP1(-1)) = first difference of the natural log for average assemblers' selling price of rice sold in three markets in Cross River State; D(LNWSP1(-1)) = first difference of the natural log for average wholesalers' selling price of rice sold in three markets in Cross River State; D(LNRLP1(-1)) = first difference of the natural log for average retail selling price of rice sold in three markets in Cross River State; D(LNFGP1(-1)) = First difference of the natural log for average buying price purchased by assemblers' three markets in Cross River State, this price is used to represent the farm gate price; R^2 = coefficient of determination.

by the coefficient of the error correction mechanism (ecm(-1)). The coefficient of the error correction term (ecm(-1)) is also significant at 1% and carries the expected negative sign. The significance of the error correction mechanism (ECM) supports cointegration in the price series and suggests the existence of long-run steady state of equilibrium between the prices. The significant coefficients (t-statistics) indicated that in the long-run the different identified levels (farmgate-assembler-wholesaler-retailer) within the marketing system of rice in Cross River were highly co-integrated with the farm gate level. The implication is that since rice is a major staple crop produced within the study area, the price formation process in these areas highly depends on the farmgate prices.

CONCLUSIONS AND RECOMMENDATIONS

The study concludes that the supply chain of rice in Cross River State, Nigeria is significantly integrated in the long run. The short run dynamics revealed that 55% of the deviations from the long run equilibrium are corrected per week. This implies that facilitative policies that will enhance the provision of infrastructures such as good roads, market structures and efficient market information network systems are necessary. Also, the government should provide price regulatory services to enhance market integration and reduce market exploitation by intermediaries especially in the short run.

REFERENCES

Akande T (2004). An Overview of the Nigeria Rice Economy. A paper presented at a workshop organized by The Nig. Inst. Soc. Eco. Res., (NISER) Ibadan, pp. 1-15.

Daramola B (2005). Government policies and competitiveness of Nigeria rice economy. Paper presented at the workshop on rice policy and food security in Sub-Saharan Africa Organized by WARDA, Cotonou, Republic of Benin.

Food and Agriculture Organization of the United Nations (2001). FAOSTAT Online Statistical Service. Available online at: http://faostat.fao.org.

Granger CW, Newbold P (1974). Spurious regressions in econometrics. J. Econ., 2(5): 1-120.

Granger CWJ (1988). Some recent developments in the concept of causality. J. Econ., 39: 99-211.

Granger CWJ (1969). Investigating causal relationships by econometric models and cross spectral methods. Econometrica, 37: 24-438.

Harris RID (1995). Using cointegration analysis in econometric modeling. Hemel, Hempstead, Hertfordshire, Prentice Hall-Harvester Wheatsheaf.

Heman DL, Fateh MM (2005). Spatial price linkages in Regional Onion markets of Pakistan. J. Agric. Soc. Sci., 1(4): 318-321.

Ihene DA (1996). The Marketing of Staple Food Crops in Enugu State, Nigeria: A case study of Rice, Maize and Beans, M.Sc. Thesis University of Nigeria, Nsukka.

Kenmore P (2003). Sustainable rice production, food security and enhanced livelihoods in "Rice Science: Innovations and Impact for Livelihood, pp. 27-34. Edited by Mew T.W,

Ndiyo NA (2005). Fundamentals of Research in Behavioural Sciences and Humanities. Wusen Publishers, pp. 182-189.

Ogundele OO, Victor OO (2006). Technical efficiency differentials in rice production technologies in Nigeria. AERC Research Paper 1554. Afr. Econo. Res. Consortium, Nairobi.

Tsuboi, T (2005). Paper presented at the WARDA – NERICA rice Workshop, Ivory Coast, 8th October, 2005.

Xu K, Guofang S (2003). Promoting Chinese rice production through innovative science and technology, edited by Mew T.W. et al. Proceedings of the International Rice Research Conference, 16-19 September 2002, Beijing, China, pp. 11-18

6

Hot water and chitosan treatment for the control of postharvest decay in sweet cherry (*Prunus avium* L.) cv. Napoleon (Napolyon)

Mir Javad Chailoo* and Mohammad Reza Asghari

Department of Horticulture, Faculty of Agriculture, Urmia University, Iran.

The effectiveness of chitosan and hot water treatments, alone or in combination, to control storage decay of sweet cherries, (*Prunus avium* L.) was investigated. In single treatments, chitosan was applied by postharvest dipping or preharvest spraying at 0.5 and 1.0% concentrations; hot water treatments at were applied for 5 and 10 min. Sweet cherries were kept in storage at and 98% relative humidity. Rot incidence was evaluated after 30 and 60 days storage life. Chitosan and hot water treatments applied alone significantly reduced decay. A combined treatment with 0.5% chitosan and hot water at for 5 and 10 min after 30 days of storage life was the best in controlling decay. In addition to reducing postharvest decay, combined treatment with 0.5% chitosan and hot water at for 5 min after 60 days of storage life showed good result in compartion to other treatments. The results indicate that the combination of hot water and chitosan treatments is a valid strategy to improve the existing ones already used in controlling postharvest decay of sweet cherries.

Key words: Hot water, chitosan, sweet cherry, decay, storage.

INTRODUCTION

Postharvest decay may result in serious economic losses to sweet cherries, a commodity of economic importance in many production areas worldwide. The use of synthetic fungicides to control postharvest diseases of sweet cherries is not allowed by European legislation, and there is a clear need for alternative natural materials for postharvest disease control that reduce fungal decay and carry lower risks for consumers. Biological control with yeast antagonists (Dündar and Göçer, 2001), hot water and hot air treatments (Wild, 1990; Schirra and D'Hallewin, 1997; Özdemir and Dündar, 2001), modified atmosphere packaging (Özdemir and Kahraman, 2004), sodium bicarbonate (Smilanick et al., 2005), and chitosan treatment (Chien et al., 2007) are natural alternatives to synthetic chemical postharvest treatments for disease control in sweet cherry. Short-duration (as brief as 20 s) hot water treatment (HWT) is one physical method that can effectively reduce postharvest decay on fresh fruits

and vegetables (Ben-Yehoshua et al., 2000; Lanza et al., 2000). For example, Lanza et al. (2000) reported that hot water dip at 52°C for 180s was as effective as non-heated imazalil in controlling postharvest decay of lemon. In addition, brushing grapefruit for 20s with 56, 59 or 62°C water reduced decay by 20, 5 or 1%, respectively, compared to the control (Porat et al., 2000). A wide range of fruit ripening processes are affected by heat, such as color (Cheng et al., 1988; Tian et al., 1996), ethylene synthesis (Ketsa et al., 1999), respiration (Inaba and Chachin, 1988), fruit softening and cell wall metabolism (Lurie and Nussinovich, 1996), volatile production (McDonald et al., 1999). Postharvest heat treatment also can reduce chilling injury in many kinds of fruits during subsequent low temperature storage as well as reduce pathogen level and disease development (McDonald et al., 1999; Lurie, 1997). Heat treatments may affect postharvest quality in several ways. It has a direct effect on fungal growth, it may induce antifungal substances and the wax layer may melt into wounds and stomata (Schirra et al., 2000). Ferguson et al. (2000) present a survey of studies on the effect of heat treatments on

*Corresponding author. E-mail: J-chailoo@yahoo.com

postharvest quality in fruits.

Edible coatings such as chitosan has antifungal (Wojdyla et al., 2001) and eliciting properties (Hirano, 1999). The polymer has generally been applied in postharvest treatments (Cheah et al., 1997; Romanazzi et al., 2001b), and there are a few examples of preharvest application (Romanazzi et al., 1999c; Reddy et al., 2000). Coating citrus fruit with chitosan was effective in controlling fruit decay caused by *Penicillium digitatum* Sacc. and *Penicillium expansum* Link (Chien et al., 2007) and rots including gray mould and blue mould caused by *Botrytis cinerea* and *P. expansum* in sweet cherry fruit were reduced by preharvest spraying or postharvest dipping of chitosan (Romanazzi et al., 2003). In addition chitosan, have shown antimicrobial functions against the growth of certain microorganisms (Park et al., 2005; Zhang and Quantick, 1998). Edible coatings can possibly control the internal gas atmosphere of the fruit, minimizing fruit respiration rate (Park, 1999) and may serve as a barrier to water vapor, reducing moisture loss and delaying fruit dehydration (Baldwin et al., 1995). Chitosan-treated strawberries have shown a range of changes that are related to a slowed ripening, such as increased titratable acidity (TA) (El Ghaouth et al., 1991; Zhang and Quantick, 1998; Reddy et al., 2000; Han et al., 2004; Chaiprasart et al., 2006; Hernandez-Munoz et al., 2006; Vargas et al., 2006; Mazaro et al., 2008), with delayed changesin pH (Han et al., 2004; Hernandez-Munoz et al., 2006; Vargas et al., 2006), antocyanin content (El Ghaouth et al., 1991; Zhang and Quantick, 1998; Reddy et al., 2000; Vargas et al., 2006), soluble solids content (SSC) (Chaiprasart et al., 2006; Vargas et al., 2006; Ribeiro et al., 2007), and with reduced ethylene production (Mazaro et al., 2008). Changes in enzyme activities in coated strawberries has been shown to involve chitinase, chitosanase and -1,3-glucanase (El Ghaouth et al., 1992a; Zhang and Quantick, 1998), and phenylalanine ammonia-lyase (PAL), which increased three-fold in treated berries (Romanazzi et al., 2000). Moreover, a decreased respiration rate (El Ghaouth et al., 1991; Devlieghere et al., 2004; Vargas et al., 2006) and hydrogen peroxide production (Romanazzi et al., 2007b) have been shown in fruit treated with the biopolymer.

MATERIALS AND METHODS

Sweet cherries (*Prunus avium* L., cv "Napoleon") were picked in the khoy orchards. Immediately after harvest the fruits were brought to the laboratory. Safe and unwounded fruits were selected for preliminary tests. Fruit were surface-sterilized with 2% sodium hypochlorite for 2 min at room temperature rinsed with tap water in order to remove the heavy dirt, pesticides and fungal spores that are covering the fresh harvested produce and allowed to dry at room temperature and then the fruits were dipped in hot water of wide range of temperatures from various exposure times from 5 and 10 min, and sweet cherries were dipped in 0.5 and 1.0% concentrations chitosan and a total treated combinations were examined. After such hot water dipping and chitosan coating, both

Table 1. Influence of hot water dipping temperatures on decay reduction percent of sweet cherry measured.

Treatment condition	Decay reduction (%)
Control	2.5[a]
50 °C for 5 min after 30 days	1.2[b]
50 °C for 10 min after 30 days	1.0[b]
50 °C for 5 min after 60 days	2.5[a]
50 °C for 10 min after 60 days	2.7[a]

treated and non-treated (control) fruits were stored for months in the cold storage at 98% relative humidity. The following fruits with pitting, decay or abnormal softening during storage were defined as heat or chitosan damaged fruits.

For each fruit, texture was determined using a TA.XT.Plus Texture Analyzer (Stable Microsystems, Godalming, UK) interfaced to a personal computer. Interfaced to a personal computer, content of soluble solids by an Atago PR-101 refractometer (Japan) at 22 °C and titratable acidity was determined by titrating diluted juice samples to pH 8.2 using 0.1 N NaOH. A panel of five trained judges gave scores for overall quality on a scale from disease severity according to the following empirical scale: 0 = healthy berry; 1 = one very small lesion (beginning of infection); 2 = one lesion 10 mm^2 in size; 3 = several lesions or 25% of the berry infected; 4 = 26 to 50% of the berry surface infected, sporulation present; and 5 = more than 50% of the berry surface infected, sporulation present (Romanazzi et al., 2006).

Statistical analysis

A completely randomized factorial design with ten replications was used. Each treatment was applied to 10 replicates of 25 fruits per each. An analysis of variance (ANOVA) was used to analyze difference between means and the Duncan test was applied for mean separation at $P \leq 0.05$. All analyses were done with SPSS and MSTAT-C statistical software.

RESULTS

Effect of hot treatment on decay sweet cherry fruits

Table 1 shows the decay reduction percent of sweet cherry fruits treated at hot water.

The results are better at 5 °C for 5 and 10 min after 30 days than control fruits. At 5 °C water treatment for 5 and 10 min after 60 days of storage had the same level of decay as control.

Effect of chitosan treatment on decay sweet cherry fruits

Table 2 shows the decay reduction percent of sweet cherry fruits treated at chitosan. Chitosan 0.5 and 1% after 30 days storage had better results than control fruits and chitosan 1% after 60 days had worst result than the other treatments.

Table 2. Influence of chitosan on decay reduction percent of sweet cherry measured.

Treatment condition	Decay reduction (%)
Control after 30days	1.3[d]
Chitosan 1% after 30 days	1.4[d]
Control after 60 days	2.5[a]
Chitosan 0.5% after 60 days	2.0[b]
Chitosan 1% after 60 days	2.5[a]

Table 3. Influence combination of hot water and chitosan on decay reduction percent of sweet cherry measured.

Treatment condition	Decay reduction (%)
Control after 30days	3.0[a]
Chitosan 0.5% after 30 days and 5°C for 5 min	1.0[d]
50°C for 10 min and chitosan 0.5% after 30 days	1.0[d]
50°C for 5 min and chitosan 1% after 30 days	1.5[c]
50°C for 10 min and chitosan 1% after 30 days	1.7[c]
Control after 60 days	3.0[a]
50°C for 5 min and chitosan 0.5% after 60 days	3.0[a]
50°C for 10 min and chitosan 0.5% after 60 days	2.8[a]
50°C for 5 min and chitosan 1% after 60 days	2.3[b]
50°C for 10 min and chitosan 1% after 60 days	2.8[a]

Effect combination of hot water and chitosan treatmentes on decay reduction of sweet cherries

Table 3 shows the decay reduction percent of sweet cherry fruits treated at combination of hot water and chitosan. After 30 weeks of storage at 50°C for 5 and 10 min of the fruit dipped in water, 0.5% of chitosan trearment had least decay of sweet cherry fruits. The Table 3 show, combination of hot water and chitosan treatment had better results than alone treatments.

Conclusion

Chitosan and hot water treatments were effective in reducing decay of sweet cherries; the interaction between chitosan and hot water treatments was significant in reducing decay, and total rots in 30 and 60 days after storage. Several examples of combinations of alternative means for controlling postharvest decay have been reported in the literature. The inhibitory effect of chitosan on decay derives from the combination of its antifungal and eliciting properties. Indeed, chitosan inhibits the *in vitro* growth of many fungi, including some species causing decay on fruits and vegetables (Allan and Hadwiger, 1979); in tests performed with *B. cinerea*,

Monilinia laxa, and *Alternaria alternata* on sweet cherries, a reduction of radial growth was observed (Romanazzi et al., 2001b). Chitosan and hot water were also able to induce resistance in the host by increasing chitinase and b-1,3-glucanase in oranges, strawberries and raspberries (Fajardo et al., 1998; Zhang and Quantick, 1998), and phenylalanine ammonia-lyase (PAL) activity in table grapes (Romanazzi et al., 2000, 2002). Moreover, it increased the level of 6- methoxymellein, the principal phytoalexin of carrots (Reddy et al., 1999). Physical treatments elicit responses in harvested commodities, increasing enzymatic activity (e.g. PAL, and peroxidase) related to host resistance against pathogens (Wilson et al., 1994); resistance responses in sweet cherries, strawberries, and table grapes have also been suggested for short hot water treatments(Romanazzi et al., 2001a). In addition, subatmospheric pressures remove ethylene from the tissues (Burg and Burg, 1965), thus delaying senescence of commodities and, indirectly, reducing their susceptibility to pathogens (Lougheed et al., 1978). A direct effect of short hot water treatments on the main decay-causing fungi of sweet cherries can be excluded, however (Romanazzi et al., 2001a).

DISCUSSION

The purpose of the present study was to evaluate the feasibility of combined application of a chemical edible coatings chitosan treatment and physical hot water treatment to control postharvest rots of sweet cherries. These results are in agreement with previous investigations; as reported in the introduction of the present. Postharvest chitosan and heat treatments of sweet cherries did not influence important quality factors like colour and content of soluble solids and titratable acidity. The heated fruits were less firm than the controls. The chitosan and heat treatments had no negative effect on sensoric quality. The heated fruits scored slightly better in the panel tests then unheated fruits. However, the difference was not statistically significant. Postharvest chitosan and heat treatments of sweet cherries reduced the fruits susceptability to decay, chitosan had the stronger effect. However, the fruits treated with hot water chitosan solution had the least decay.

ACKNOWLEDGEMENTS

The authors acknowledge the Urmia University (Urmia, Iran) for providing equipment for this study. The authors also thank Dr. L. Naseri and Dr. R. Jalili Marandi for valuable technical assistance.

REFERENCES

Allan CR, Hadwiger LA (1979). The fungicidal effect of chitosan on fungi of varying cell wall composition. Exp. Mycol., 3: 285-287.

Hot water and chitosan treatment for the control of postharvest decay in sweet cherry...

43

Baldwin EA, Nisperos-Carriedo MO, Baker RA (1995). Edible coating for lightly processed fruits and vegetables. Hortscience, 30: 35-38.

Ben-Yehoshua S, Peretz J, Rodov V, Nafussi B (2000). Postharvest application of hot water treatment in citrus fruits: The road from laboratory to the packing-house. Acta Hort., 518: 19-28.

Burg SP, Burg EA (1965). Ethylene action and the ripening of fruit. Science, 148: 1190-1196.

Fajardo JE, McCollum TG, McDonald RE, Mayer RT (1998). Differential induction of proteins in orange flavedo by biologically based elicitors and challenged by Penicillium digitatum Sacc. Biol. Control, 13: 143-151.

Chaiprasart P, Hansawasdi C, Pipattanawong N (2006). The effect of chitosan coating and calcium chloride treatment on postharvest qualities of strawberry fruit (Fragaria x ananassa). Acta Hortic., 708: 337-342.

Cheah LH, Page BBC, Sheperd R (1997). Chitosan coating for inhibition of sclerotina carrots. N. Z. J. Crop Hort. Sci., 25: 89-92.

Cheng TS, Floros JD, Shewfelft RL, Chang CJ (1988). The effect of high temperature-stress on ripening of tomatoes (Lycopersicon esculentum). J. Plant Phisiol., 132, 459-464.

Chien PJ, Sheu F, Lin HR (2007). Coating citrus (Murcott tangor) fruit with low molecular weight chitosan increases postharvest quality and shelf life. Food Chem., 100: 1160-1164.

Chien PJ, Yang FH, Sheu F (2007). Effects of edible chitosan coating on quality and shelf life.

Devlieghere F, Vermeulen A, Debevere J (2004). Chitosan: antimicrobial activity, interactions with food components and applicability as a coating on fruit and vegetables. Food Microbiol., 21, 703-714.

Dündar Ö, Göçer S (2001). Control of storage rots of Washington Navel oranges and Minneola by a combination of yeast antagonist and Thiabendazole. Acta Hort., 553: 399-402.

El Ghaouth A, Arul J, Ponnampalam R, Boulet M (1991). Chitosan coating effect on storability and quality of fresh strawberries. J. Food Science 56, 1618-1620 of sliced mango fruit. J. Food Eng., 78: 225-229.

El Ghaouth A, Arul J, Grenier J, Asselin A (1992a). Antifungal activity of chitosan on two postharvest pathogens of strawberry fruits. Phytopatology, 82: 398-402.

Ferguson B, Ben-Yehoshua S, Mitcham E, McDonald R, Lurie S (2000). Postharvest heat treatments: Introduction and workshop summary. Postharvest Biol. Technol., 21(1): 1-6.

Han C, Zhao Y, Leonard SW, Traber MG (2004). Edible coating to improve storability and enhance nutritional value of fresh and frozen strawberry (Fragaria x ananassa) and raspberries (Rubus ideaus). Postharvest Biol. Technol., 33: 67-78.

Hernandez-Munoz P, Almenar E, Ocio MJ, Gavara R (2006). Effect of calcium dips and chitosan coatings on postharvest life of strawberries (Fragaria x ananassa). Postharvest Biol. Technol., 39: 247-253.

Hirano S (1999). Chitin and chitosan as novel biotechnological materials. Polym. Int., 48: 732_/734.

Inaba M, Chachin K (1988). Influence of and recovery from high-temperature stresss on harvested mature green tomatoes. HortScience, 23(1): 190-192.

Ketsa S, Childtragool S, Klein JD, Lurie S (1999). Ethylene synthesis in mango fruit following heat treatment. Postharvest Biol. Technol., 15: 65-72.

Lanza GE, di Martino Aleppo, Strano MC, Reforgiato Recupero G (2000). Evaluation of hot water treatments to control postharvest green mold in organic lemon fruit, pp. 1167-1168. In: Proc. Intl. Soc. Citricult XI Congr., Orlando, 3-7 Lougheed EC, Murr DP, Berard L (1978). Low pressure storage for horticultural crops. HortScience, 13: 21-27.

Lurie S, Nussinovitch A (1996). Compression characteristics, firmness, and texture perception of heat treated and unheated apples. Int. J. Food. Sci. Technol., 31: 1-5.

Mazaro SM, Deschamps C, May De Mio LL, Biasi LA, De Gouvea A, Kaehler Sautter C (2008). Post harvest behavior of strawberry fruits after pre harvest treatment with chitosan and acibenzolar-s-methyl. Revista Brasileira de Fruticultura, 30: 185-190.

McDonald RE, McCollum TG, Baldwin EA (1999). Temperature of hot water treatments influences tomato fruit quality following low-temperature storage. Postharvest Biol. Technol. 16: 147–155.

Özdemir AE, Dündar Ö (2001). Effect of different postharvest application on storage of Valencia oranges. 4th International Conference on Postharvest Science, Jerusalem, Israel, Acta. Hort., 553: 561-564.

Park SI, Stan SD, Daeschel MA, Zhao Y (2005). Antifungal coatings on fresh strawberries (Fragaria x ananassa) to control mold growth during cold storage. J. Food Sci., 70: 202-207.

Park HJ (1999). Development of advanced edible coatings for fruits. Trends Food Sci. Tech., 10: 254-260.

Porat R, Daus A, Weiss B, Cohen L, Fallik E, Droby S (2000). Reduction of postharvest decay in organic citrus fruit by a short hot water brushing treatment. Postharvest Biol. Technol., 18: 151-157.

Ribeiro C, Vicente AA, Teixeira JA, Miranda C (2007). Optimization of edible coating composition to retard strawberry fruit senescence. Postharvest Biol. Technol., 44: 63-70.

Reddy MVB, Belkacemi K, Corcuff R, Castaigne F, Arul J (2000). Effect of pre-harvest chitosan sprays on post-harvest infection by Botrytis cinerea and quality of strawberry fruit. Postharvest Biol. Technol., 20: 39-51.

Reddy BMV, Corcuff R, Kasaai MR, Castaigne F, Arul J (1999). Induction of resistance against gray mold rot in carrot roots by chitosan. Phytopathology, 98: S6.

Romanazzi G, Santini M, Murolo S, Landi L (2007b). Antimicrobial and eliciting activity of chitosan in the control of grey mold and Rhizopus rot of strawberries in storage. Proceedings of COST 924 "Novel approaches for the control of postharvest diseases and disorders", Bologna, pp. 403-408.

Romanazzi G, Nigro F, Ippolito A (2003). Short hypobaric treatments potentiate the effect of chitosan in reduction storage decay of sweet cherries. Postharvest Biol. Technol., 29: 73-80.

Romanazzi G, Nigro F, Ippolito A, Salerno M (2001a). Effect of short hypobaric treatments on postharvest rots of sweet cherries, strawberries and table grapes. Postharvest Biol. Technol., 22: 1-6.

Romanazzi G, Nigro F, Ippolito A (2000). Effetto di trattamenti pre e postraccolta con chitosano sui marciumi della fragola in conservazione. Rivista di Frutticoltura, 62(5): 71-75 .

Schirra M, D"Hallewin G, Ben-Yehoshua S, Fallik E (2000). Host-pathogen interactions modulated by heat treatment. Postharvest Biol. Technol., 21: 71-85.

Schirra M, D'Hallewin G (1997). Storage performance of Fortune mandarins following hot water dips. Postharvest Biol. Technol., 10: 229-238.

Tian MS, Woolf AB, Bowen JH, Ferguson IB (1996). Changes in color and chlorophyll fluorescence of broccoli florets following hot water treatment. J. Am. Soc. Sci., 121(2): 310-313.

Vargas M, Albors A, Chiralt A, Gonzalez-Martinez C (2006) Quality of coldstored strawberries as affected by chitosan-oleic acid edible coatings. Postharvest Biol. Technol., 41: 164-171.

Wild BL (1990). Postharvest treatments which reduce chilling injury. Australian Citrus News, 5: 10-13.

Wojdyla AT, Orlikowski LB, Struszczyk H (2001). Chitosan for the control of leaf pathogens. In: Muzzarelli, R.A.A. (Ed.), Chitin Enzymology 2001, Atec, Grottammare, Italy, ISBN 88-86889-06-2, pp. 191-196.

Zhang D, Quantick PC (1998). Antifungal effects of chitosan coating on fresh strawberries and raspberries during storage J. Hort. Sci. Biotechnol., 73: 763-767.

7

Physical properties of noodles enriched with whey protein concentrate (WPC) and skim milk powder (SMP)

D. Baskaran[1]*, K. Muthupandian[1], K. S. Gnanalakshmi[1], T. R. Pugazenthi[1], S. Jothylingam[1] and K. Ayyadurai[2]

[1]Department. of Dairy Science, Madras Veterinary College, Chennai- 600 007, India.
[2]Department of Veterinary Biochemistry, MVC, Chennai- 600 007, India.

The physical properties of noodles enriched with skim milk powder, whey protein concentrate and a combination of skim milk powder and whey protein concentrate at 5, 7.5 and 10% levels were studied. Volume increase, weight increase and swelling ratio of the enriched noodles were reduced at increasing levels of substitution. Total solids loss in gruel showed increasing trend as the levels of substitution increased. It was also found that, the loss of total solids was higher in noodles supplemented with SMP compared to WPC and a combination of SMP and WPC.

Key words: Noodles, skim milk powder, fortification.

INTRODUCTION

Incorporation of whey protein in pasta and macaroni like products had been used to improve the nutritional value without adversely altering their flavour and textural qualities (Morr, 1982). The Chinese National standards for starch noodles showed 10% solid loss during cooking was acceptable (CSB, 1963). The cooking loss and the specific gravity of uncooked macaroni were found to increase with increasing level of fortification with ultra filtered whey protein concentrate (Fayed et al., 1993). Conforti and Lupano (2004) stated that, the addition of whey protein concentrate decreased the firmness and consistency of dough. Geoffrey (2007) stated that, the whey proteins possess nutraceutical properties in the emerging nutritional status of the population. In this investigation, preparation of enriched noodles, physical parameters such as cooking loss, volume increase, weight decrease and breaking force were discussed.

MATERIALS AND METHODS

Whey protein concentrate (M/s. Procon) and skim milk powder (M/s Amul) with the following percentage of chemical composition were used for fortification. The preparations of the enriched noodles were discussed elsewhere (Muthu, 2005). The flow diagram depicting the technique for noodle manufacture adapted in this study is shown in

Figure 1.

Physical properties were studied using Indian Standards - IS 1485 (1993) for cooking loss, volume increase and weight increase. The breaking force was calculated following the method of Peri et al. (1983). The physical parameters were analyzed statistically as per the procedures adopted by Snedecor and Cochran (1994).

RESULTS AND DISCUSSION

The volume, weight increase, swelling ratio and total solids loss for the fortified noodles with SMP, WPC and combination of SMP and WPC for 5, 7.5 and 10% levels are depicted in Table 1.

Volume increase

A highly significant difference (P<0.01) was noticed in noodles enriched with WPC and combination of SMP and WPC within various levels of substitution and between different groups of milk protein supplements. The volume increase of noodles enriched with 5% SMP decreased from 37.40 to 31.53 at 10% SMP. These results were in accordance with Fayed et al. (1993) and Ilkay et al. (2004) who made similar observation for the combination of SMP and WPC. The volume increase of the protein enriched noodles was higher than that of the control after cooking. The highest volume was obtained with 5% addition of SMP compared to 7.5 at 10% levels of

*Corresponding author. E-mail: dharmarbaskaran@yahoo.co.in.

Raw materials used: refined wheat flour, salt

Whey protein concentrate and Skim milk powder (1:1) @ 7.5% level

↓

Addition of Salt in water Salt @ 2 g/100 g flour

↓

Dough making with the required level of water

↓

Roller compression: Dough passed through the noodle extruder

↓

Sheeting

↓

1.   I Stage (4 mm thickness)

2.   II Stage (2 mm thickness)

↓

Cutting: dough sheet passed through noodle cutting machine

↓

Drying in two stages

↓

1. Sun drying for 12 h.

2. Drying in oven at 55°C for 5 h.

↓

Packaging

1. Ordinary packaging

2. Modified atmospheric packaging

3. Vacuum packaging

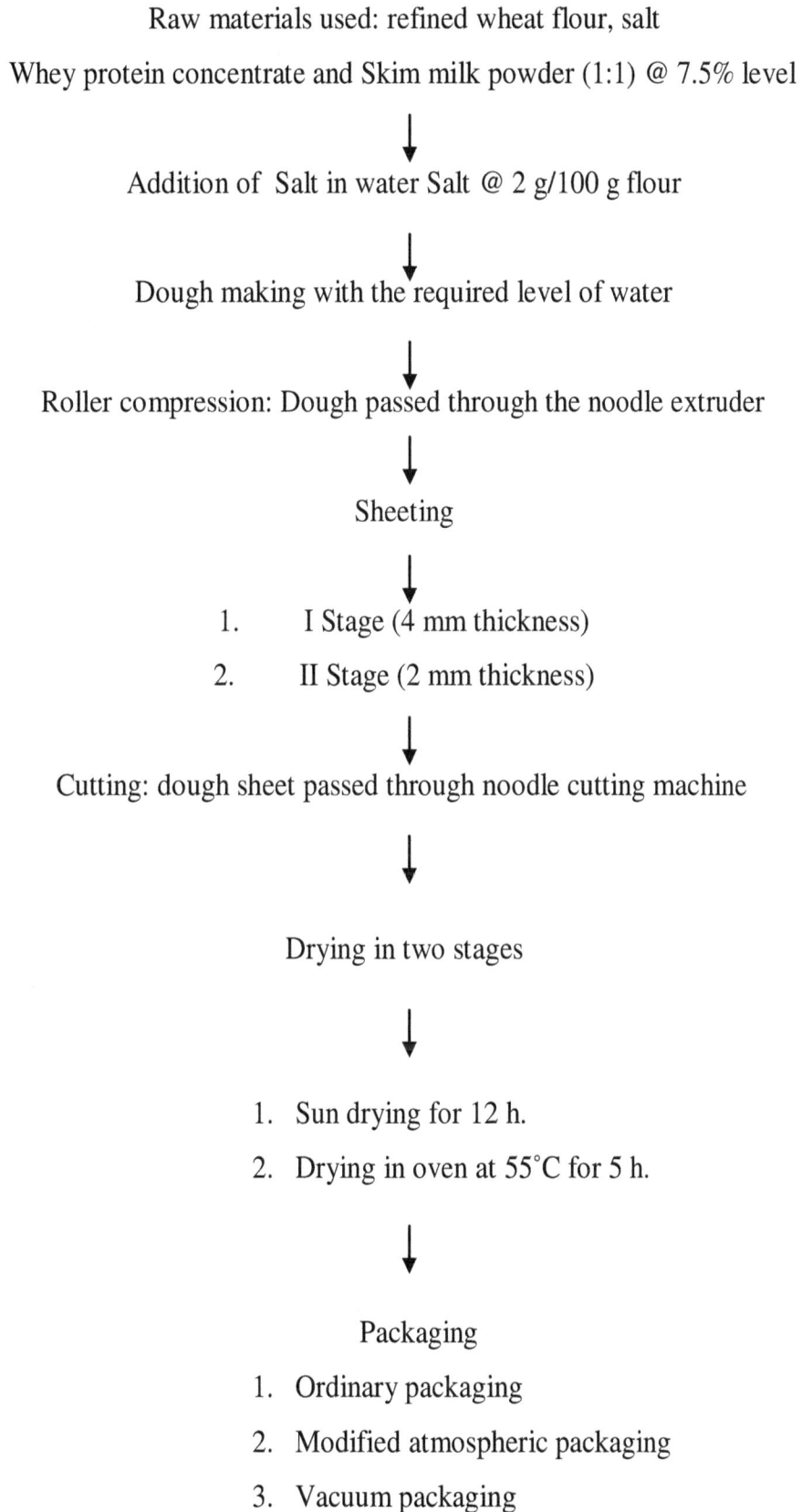

Figure 1. Flow diagram depicting the technique for noodle manufacture adapted in the present study.

Table 1. Physical qualities of noodles enriched with SMP and WPC and combination of SMP and WPC*.

Physical parameter	Control	Skim milk powder(SMP)			Whey protein concentrate(WPC)			Skim milk powder and whey protein concentrate			F-Value
		5%	7.5%	10%	5%	7.5%	10%	5%	7.5%	10%	
Volume increase (%)	32.49[ab] ± 0.31	37.40[d] ±0.19	32.07[ab] ±0.16	31.54[a] ±0.19	46.73[g] ±0.19	39.45[f] ±0.42	38.75[e] ±0.36	37.61[de] ±0.41	35.64[c] ±0.35	32.9[b] ±0.23	222.96*** CD = 1.18
Weight increase (%)	122.80[a] ± 0.52	166.26[g] ±1.03	160.35[f] ±0.46	141.89[c] ±0.54	216.18[i] ±0.45	156.85[ef] ±1.37	148.11[d] ±0.39	190.2[h] ±1.88	156.2[h] ±1.88	128.61[b] ±0.51	848.13** CD = 3.61
Swelling ratio	3.23[a] ± 0.00	3.66[g] ±0.01	3.60[e] ±0.01	3.42[c] ±0.01	4.16[h] ± 0.00	3.57[c] ± 0.01	3.48[d] ± 0.01	3.90[f] ± 0.02	3.58[b] ± 0.02	3.29[b] ± 0.01	550.87** CD = 0.40
Total solids loss (%)	8.07[a] ± 0.24	10.53[c] ± 0.28	11.93[d] ± 0.28	13.21[e] ±0.16	8.14[a] ± 0.15	8.79[a] ± 0.10	9.86[b] ± 0.18	8.80[a] ± 0.21	9.48[bc] ± 0.23	12.09[d] ± 0.10	77.35** CD =0.04

*Average of six trials. (Mean± S.E); Mean Values bearing different superscripts in a row differ significantly (P<0.01). ** Highly significant (P<0.01).

Table 2. Chemical composition and physical characteristic of Milk based supplements.

Supplement powder	Moisture	Protein	Fat	Lactose	Minerals	Solubility index (ml)
Whey protein concentrate	4.41	70.00	6.70	15.00	6.70	0.50
Skim milk powder	4.00	35.30	1.25	51.00	8.00	1.20

substitution. These results are in conformity with Morad et al. (1980).

Swelling ratio

A highly significant difference (P<0.01) was noticed in noodles enriched with SMP and WPC and the combination of SMP and WPC within various levels of substitution and between different groups of milk protein supplements. The swelling ratio of the protein enriched noodles decreased with increasing levels of supplementation of SMP and WPC. It is related to the weight increase due to water uptake and hence, as the increase in weight decreased and the swelling ratio also decreased. As the substitution level of milk protein increased with the reduction in the gluten content, the swelling ratio of the final product showed a decreasing trend. Volume increase, weight increase and swelling ratio of the noodles enriched with milk protein sources reduced at increasing levels of substitution (Table 2).

Total solid loss in gruel

Statistical analysis revealed a highly significant difference in the total solids loss between the control and the noodles enriched with SMP, as well as the combination of SMP and WPC. The total solid loss in gruel increased as the level of substitution increased. The increase in loss due to enrichment may be related to gluten dilution and protein solubility fraction of wheat germ. The results were in conformity with Olfat et al. (1993) and Fayed et al. (1993). The total solids loss was higher in the noodles substituted with SMP when compared to WPC. This may be attributed to the compact structure of WPC and the porous nature of the SMP enriched noodles. Total solids loss in gruel increases as the level of substitution increases. It was also found that, the loss of total solids was higher in noodles supplemented with skim milk powder compared to whey protein concentrate and a combination of SMP and WPC.

ACKNOWLEDGEMENTS

The authors are grateful to the Secretary, Tamilnadu State Council for Science and Technology, Chennai for having financially assisted the Scheme and the Dean, Madras Veterinary College, Chennai for having granted permission to conduct this project.

REFERENCES

Conforti PA, Lupano CE (2004). Functional properties of biscuits with whey protein concentrate and Honey. Int. J. Food Sci. Technol., 39: 745-753.

CSB (1963). Chinese National Standards No.1485. Central Standards Bureau, Ministry of Economics, ROC.

Fayed AE, Shatanovi GA, Kerolles SY (1993). Ultra filtered whey concentrate as protein fortifier in macaroni manufacture from wheat flour. Ann. Agric. Sci., 1: 259-269.

Geoffrey WK (2007). Emerging Health properties of whey proteins and their clinical implications. J. Am. Coll. Nutr., 26(6): 713-723.

Ilkay P, Ibanoglu S, Orier MD (2004). Effect of storage on the selected properties of macaroni enriched with wheat germ. J. Food Eng., 64: 249-256.

Indian Standards (1993). IS 1485: Macaroni, Spaghetti, vermicelli and egg noodles specification. ISI, New Delhi.

Morad MM, Mayoli SB, Afigi, SA (1980). Macaroni supplemented with lupin and defatted soybean flours. J. Food Sci., 45: 404-405.

Morr CV (1982). Functionality of whey protein products. Review: Newzealand J. Dairy Sci. Technol., 17: 185-194.

Olfat YM, Yaseen AAE, Aziza IA (1993). Enrichment of macaroni with cellulose derivative protein complex from whey and corn steep liquor. Nahrung, 37(6): 544-552.

Peri C, Barbieri R, Casiraghi EM (1983). Physical, chemical and nutritional quality of extruded corn germ flour and sweet potato. Int. J. Food Sci. Technol., 35: 235-242.

Snedecor GW, Cochran WG (1994). Statistical methods. Oxford and IBH publishing Co. Kolkata.

8

Genetic variability studies between released varieties of cassava and central Kerala cassava collections using SSR markers

Sree S. Lekha[1], Jaime A. Teixeira da Silva[2] and Santha V. Pillai[1]*

[1]Central Tuber Crops Research Institute, Sreekariyam, Trivandrum 695017, Kerala, India.
[2]Faculty of Agriculture and Graduate School of Agriculture, Kagawa University, Miki-cho, Ikenobe 2393, Kagawa-ken, 761-0795, Japan.

Twelve released varieties of cassava and 24 central Kerala collections were assessed at the genomic DNA level with 36 SSR primers for genetic diversity study. The minimum number of SSR primers that could readily be used for identification of the 36 cassava genotypes was also determined. For the genetic diversity study, the similarity coefficients generated between released varieties and central Kerala varieties ranged from 40 to 95% and two separate DNA cluster groups were formed at 0.60 coefficients using "numerical taxonomy" and "multivariate analysis system software package". The similarity index for released varieties ranged from 60 to 93% and in the case of central Kerala varieties it ranged from 70 to 98%. The mean fixation index (F) for released varieties was 0.0688 and that for central Kerala collections was 0.1337, indicating an overall conformance to Hardy-Weinberg equilibrium. Principal component analysis helped in identifying primers which contributed much to the variation present in the population and reduce the cost and time of research for genetic diversity and genotype identification studies for cassava genetic improvement programs.

Key words: Cassava, genetic diversity, genotypes, microsatellites, principal component analysis, similarity index, simple sequence repeats primers.

INTRODUCTION

The genus *Manihot* originates from Latin America where 98 species are found (Rogers and Appan, 1973). *Manihot esculenta* Crantz (cassava) was initially introduced to Africa 400 years ago, where its cultivation for food spread throughout tropical and subtropical regions. The second *Manihot* species present in Africa, *M. glaziovii* Mueller Von Argau, was introduced 200 years ago as a source of rubber, although its distribution was less extensive (Jones, 1959). Cassava, which is generally propagated vegetatively, is one of the major sources of food in Africa (Cock, 1982). The roots, which are an excellent source of carbohydrates, have a very low protein content. In addition, the roots have a high content of cyanogenic glucosides (de Bruijn, 1971) which often necessitates extensive processing before cassava is edible. Cassava has the advantage of being well adapted to a wide range of environmental stresses. It grows very well in less fertile soil in contrast to many other crops that are highly vulnerable to environmental stresses during critical stages of plant development (Ugorji, 1998). Current economy advancement has also turned cassava into a cash crop, since several items are processed from it, which find various end uses. One of the best methods to increase cassava production to serve as the main food security and cash crop in Africa and developing countries is by the development of better varieties that are resistant to diseases, pests, and drought (Ugorji, 1998).

Genetic improvement of cassava is to a certain extent limited by a poor knowledge of genetic diversity within the

*Corresponding author. E-mail: santhavp@gmail.com.

Table 1a. Morphological characters of CTCRI released varieties of cassava.

Characteristics	RV1	RV2	RV3	RV4	RV5	RV6	RV7	RV8	RV9	RV10	RV11	RV12
Plant type	E.B	E.B	E.B	E.B	E.B	E.B	E.B	E.B	E.B	EB	MH	T
Stem colour	D.G	L.G	G	D.S	D.G	R.B	G	RB	GB	BW	LG	GG
Leaf colour	L.S	L.B	Gr	L.B	S	L.S	L.P	S	LP	LS	LS	LG
Leaf type	B	B	B	B	B	D.G	B	B	B	B	M	M
Petiole colour	D.G	L.G	G	D.G	P	P	L.P	LP	LG	P	P	DG
Flowering	F	F	F	F	F	SF	S F	F	F	SF	F	F
Tuber shape	C	Fu	C	Fu	C	C	C	Co	Co	Co	Co	C
Tuber skin colour	L.B	G.B	Cr	Br	L.B	Br	L.B	B	B	LB	B	SW
Tuber rind colour	Cr	Cr	L.P	Cr	Cr	Cr	Cr	P	Cr	Cr	LY	W
Tuber flesh colour	W	W	W	L.Y	W	W	W	W	LY	Cr	LY	W
Tuber neck	A	A	A	A	L.N	S.N	A	A	A	A	A	A

EB-Erect branching, DG-dark green, LS- light sepia, B- brown, F- flowering, Co-conical, C- cylindrical, LB- light brown, Cr-cream, W-white, A-absent, lg-light green, Fu- fusiform, GB- greyish brown, Gr- grey, G-green, LP-light pink, DS-dark sepia, S-sepia, P-pink, LY-light yellow, LN-long neck, SN-small neck, SF- shy flowering, RB-reddish brown, M-medium, MH-medium height, T-tall, GG-grayish green and SW- silvery white.

species. Isoenzymes have been used as a method to estimate genetic diversity within cassava, but low polymorphism was detected and the technique was not reproducible (Hussain et al., 1987; Ramírez et al., 1987; Lefèvre and Charrier, 1993). Studies have been conducted earlier to assess the variability based on biometrical characters as well as RAPD (randomly amplified polymorphic DNA) markers (Pillai, 2002; Pillai et al., 2004). Studies were conducted earlier to study the variability of cassava in Kerala using simple sequence repeats (SSR) markers (Sree Lekha and Pillai., 2008, 2010). DNA-based molecular markers such as RAPDs, nuclear RFLPs (restriction fragment length polymorphism) and microsatellites (= SSR markers) were used to develop the cassava molecular genetic map (Fregene et al., 1997). There is a wide range of molecular techniques available to assess genetic variability of a species. Due to their co-dominant inheritance, robustness and amenability to high throughput, SSRs or microsatellites have become a tool of choice for investigating important crop germplasm (Hokanson et al., 1998). SSR markers have been confirmed to be the most informative and appropriate for cassava (Mba et al., 2000). Perera et al. (2001) also supported SSR markers as the most informative for plants.

Valuable attributes of all SSR markers are co-dominance (many alleles are found among closely related individuals), technical simplicity, sensitivity, analytical simplicity (data are unambiguously scored, and highly reproducible) and are high abundance (markers are uniformly dispersed throughout genome as frequently as every 10 kb and therefore are ideal tools for many genetic applications. Microsatellites are short stretches of tandemly repeated, 1 to 5 nucleotide sequences, such as (G-A) n. They are ubiquitously present in eukaryotic genomes and are highly polymorphic (Tautz 1989).

Conservation of microsatellite flanking sequences allows the design of primers for PCR amplification. In cassava, SSR markers have been used to search for duplicates in the CIAT (International Centre for Tropical Agriculture, Cali, Colombia) core collection (Chavarriaga-Aguirre et al., 1999) and to analyze variation in natural populations of putative progenitors of cassava (Olsen and Schaal, 2001). At present more than 500 SSR markers are available in cassava which will provide genetic tags for various phenotypes in cassava.

The objective of the present study was to: 1) quantify the genetic variability and diversity available in the land races of central Kerala and released varieties and 2) to assess the minimum number of SSR primers that could readily be used for the identification of 36 cassava genotypes in order to reduce the time and cost of research studies.

MATERIALS AND METHODS

Plant material

Twelve varieties of cassava which were released from our institute CTCRI to the farmers and twenty four cassava cultivars that were collected from central part of kerala were selected for this study. The varieties were planted at the CTCRI farm and were evaluated for plant type; stem colour, leaf colour, leaf type, petiole colour, flowering, tuber shape, skin colour, rind colour and flesh colour (Table 1a and b).

DNA extraction

DNA was extracted according to Dellaporta et al. (1983). Plants 3 to 4 weeks old were selected and approximately 2 g of fresh and young leaf tissue was used for DNA extraction. After crushing the fresh leaf tissue in a porcelain pestle using liquid nitrogen, 5 ml of extraction buffer was added then incubated for 30 min at 60°C.

Table 1b. Morphological characters of Central Kerala varieties of cassava.

Characteristics	CK1	CK2	CK3	CK4	CK5	CK6	CK7	CK8	CK9	CK10	CK11	CK12
Plant type	EB	EB	EB	EB	EB	EB	EB	EB	EB	EB	EB	EB
Stem colour	W	W	LP	W	D	W	W	LB	LB	W	LB	W
Leaf colour	LP	LP	LP	LP	LP	P	LP	P	G	P	G	G
Leaf type	B	B	M	B	M	B	B	B	B	B	B	M
Petiole colour	R	R	G	R	R	P	G	G	P	LP	P	P
Flowering	F	F	F	F	F	F	F	F	F	F	SF	SF
Tuber shape	C	C	C	C	C	C	C	Co	C	C	C	C
Tuber skin colour	LB	LB	B	L	B	B	LB	LB	LB	LB	LB	LB
Tuber rind colour	C	C	C	C	C	C	C	P	LP	C	P	C
Tuber flesh colour	W	W	W	W	C	W	W	W	C	C	W	W
Tuber neck	A	A	A	A	A	A	A	A	A	A	A	A

Table 1b. Contd.

Characteristics	CK13	CK14	CK15	CK16	CK17	CK18	CK19	CK20	CK21	CK22	CK23	CK24
Plant type	EB	EB	EB	EB	EB	EB	EB	EB	EB	EB	EB	EB
Stem colour	W	W	W	W	W	DB	W	W	LB	D	LB	W
Leaf colour	G	LP	P	G	G	G	LP	G	LP	P	G	G
Leaf type	B	B	B	B	B	B	B	B	M	B	B	B
Petiole colour	P	P	P	P	P	P	P	P	G	P	P	P
Flowering	F	SF	SF	F	F	F	F	F	F	F	F	SF
Tuber shape	Co	Fu	Fu	C	C	C	C	C	C	Fu	C	C
Tuber skin colour	LB	LB	LB	LB	LB	LB	B	LB	LB	LB	LB	B
Tuber rind colour	LP	C	C	C	C	C	C	LP	C	P	C	C
Tuber flesh colour	W	W	C	W	W	W	C	W	W	W	W	W
Tuber neck	A	A	A	A	A	A	L	MN	LN	MN	LN	LN

EB-Erect branching, W-white, G-green, B-broad, P-pink, F-flowering, Co-conical, LB- light brown, LP- light pink, MN- medium neck, P-pink, G-green, DB-dark brown, SF-small flower, Fu-fusiform, C-cylindrical, B-brown, A-absent and LN-long neck.

After incubation, 2.5 ml of 5 M potassium acetate was added and mixed well by inversion and incubated on ice for 20 min. The sample was centrifuged at 10,000 rpm for 10 min at 4°C. After centrifugation the supernatant was recovered and isopropanol was added to 2/3 of the previous volume by inverting slowly until the DNA precipitated. The precipitated DNA was centrifuged for 10 min at 10,000 rpm at 4 °C. The supernatant was discarded and 1 ml of TE (10 mM Tris HCl and 1 mM EDTA; pH 8) was added and the nucleic acid was gently resuspended. Then 10 µl of RNase (Bangalore Genie, Bangalore, India) was added at 10 mg/ml per sample and incubated at 37°C for 1 h. Thereafter, 100 µl of 3 M sodium acetate and 2 ml of 95% ethanol was added to precipitate DNA and mixed by inversion, then centrifuged for 10 min at 10,000 rpm at 4°C.

To the DNA pellet, 500 µl of 70% ethanol was added. After centrifugation, the DNA was resuspended in 1 ml TE. Between 500 µg and 1 mg of high quality DNA was obtained from each extraction and quantified by UV absorption at 260 nm using a Shimadzu UV-260 spectrophotometer. DNA was also quantified by 0.8% agarose gel electrophoresis after staining with ethidium bromide (EtBr).

PCR assay and gel analysis

A set of 36 SSR markers developed at CIAT (Chavariagga-Aguirre et al., 1998; Mba et al., 2001) were used for the genetic variability study. The SSR markers used in the present study are listed in Table 2. The reaction mixture (25 µl) consisted of 10X buffer, 100 mM each of dNTPs, 600 mM $MgCl_2$, 600 pM of each forward and reverse primer (all from Banglore Genei), 0.5 U Taq polymerase (Finnzymes, Finland) and 25 ng of template DNA. PCR was carried out in a thermal cycler (MJ Research PTC-100, USA), under the following conditions: an initial denaturation at 94°C for 4 min followed by 40 cycles of 94°C for 1 min each, 35°C for 1 min and 72°C for 2 min and a final extension at 72°C for 5 min. The amplified DNA fragments were separated by agarose gel electrophoresis. Approximately 10 µl of the amplified products and a 1-kb molecular ruler were run for 2 h at 80 V on a 3% (w/v) agarose gel. PCR products from DNA bulks of the different accessions were each loaded into one lane. The different accessions were adjacent on each gel to enable the identification of different alleles, even in closely related accessions.

Table 2. Sequence of SSR primers used for amplification.

No	Left primers sequence	Right primers sequence	Product size
1	GGTAGATCTGGATCGAGGAGG	CAATCGAAACCGACGATACA	NA
2	CGACAAGTCGTATATGTAGTATTCACG	GCAGAGGTGGCTAACGAGAC	194
3	ACTGTGCCAAAATAGCCAAATAGT	TCATGAGTGTGGGATGTTTTTATG	291
4	AGTGGAAATAAGCCATGTGATG	CCCATAATTGATGCCAGGTT	182
5	AACTGTCAAACCATTCTACTTGC	GCCAGCAAGGTTTGCTACAT	266
6	TGTCCAATGTCTTCCTTTCCTT	CTTTTTGCCAGTCTTCCTGC	196
7	TGTGACAATTTTCAGATAGCTTCA	CACCATCGGCATTAAACTTTG	211
8	CAACAATTGGACTAAGCAGCA	CCTGCCACAATATTGAAATGG	192
9	AGGTTGGATGCTTGAAGGAA	GGATGCAGGAGTGCTCAACT	298
10	CATTGGACTTCCTACAAATATGAAT	TGATGGAAAGTGGTTATGTCCTT	143
11	GGAAACTGCTTGCACAAAGA	CAGCAAGACCATCACCAGTTT	270
12	AGTGCCACCTTGAAAGAGCA	TTGAGTGGTGAATGCGAAAG	247
13	CGTTGATAAAGTGGAAAGAGCA	ACTCCACTCCCGATGCTCGC	158
14	CAGGCTCAGGTGAAGTAAAGG	GCGAAAGTAAGTCTACAACTTTTCTAA	226
15	AAGGAACACCTCTCCTAGAATCA	CCAGCTGTATGTTGAGTGAGC	220
16	GTACATCACCACCAACGGGC	AGAGCGGTGGGGCGAAGAGC	113
17	AAGACAATCATTTTGTGCTCCA	TCAGAATCATCTACCTTGGCA	290
18	ACCACAAACATAGGCACGAG	CACCCAATTCACCAATTACCA	268
19	AACGTAGGCCCTAACTAACCC	ACAGCTCTAAAAACTGCAGCC	100
20	TCGAGTGGCTTCTGGTCTTC	CAAACATCTGCACTTTTGGC	225
21	TCAAACAAGAATTAGCAGAACTGG	TGAGATTTCGTAATATTCATTTCACTT	187
22	GCAATGCAGTGAACCATCTTT	CGTTTGTCCTTTCTGATGTTC	158
23	GGCTGTTCGTGATCCTTATTAAC	GTAGTTGAGAAAACTTTGCATGAG	122
24	ATAGAGCAGAAGTGCAGGCG	CTAACGCACACGACTACGGA	287
25	TCTCCTGTGAAAAGTGCATGA	TGTAAGGCATTCCAAGAATTATCA	214
26	CATGCCACATAGTTCGTGCT	ACGCTATGATGTCCAAAGGC	203
27	ACAATTCATCATGAGTCATCAACT	CCGTTATTGTTCCTGGTCCT	278
28	TTCCAGACCTGTTCCACCAT	ATTGCAGGGATTATTGCTCG	279
29	CGATCTCAGTCGATACCCAAG	CACTCCGTTGCAGGCATTA	239
30	CCAGAAACTGAAATGCATCG	AACATGTGCGACAGTGATTG	253
31	GCTGAACTGCTTTGCCAACT	CTTCGGCCTCTACAAAAGGA	130
32	TGAGAAGGAAACTGCTTGCAC	CAGCAAGACCATCACCAGTTT	272
33	TTGGCTGCTTTCACTAATGC	TTGAACACGTTGAACAACCA	179
34	CCTTGGCAGAGATGAATTAGAG	GGGGCATTCTACATGATCAATAA	163
35	ATCCTTGCCTGACATTTTGC	TTCGCAGAGTCCAATTGTTG	210
36	ACAATGTCCCAATTGGAGGA	ACCATGGATAGAGCTCACCG	NA

NA- Not available.

The gels were stained in an EtBr solution (1 mg/L) for 15 min, rinsed in double distilled water for 15 min and observed under a Gel Doc System for DNA fragment analysis (Syngene).

Genetic diversity study

Allelic frequencies of SSR markers were used to estimate the percentage of polymorphic loci (P), mean number of alleles per locus (A), effective number of alleles (A_E), and observed heterozygosity (H_E) (Hedrick, 2004) using the computational program POPGENE 32 (Yeh and Yang, 1999). DNA bands were scored for the presence (1), absence (0) or ambiguous (9) for each accession by visual inspection. To ensure accurate scoring, all markers were scored twice from two different gels. Loci were considered to be polymorphic if more than one allele was detected. Wright's fixation index (F) was estimated using the formula:

$$F = 1-(Ho/He)$$

To quantify the lack of or excess heterozygosity, out-crossing rate (t) was estimated using $t = (1-F)/(1+F)$ (Weir, 1996). The portioning of genetic diversity within and among cassava cultivars was analyzed using F-statistics (Nei, 1973) according to the equations

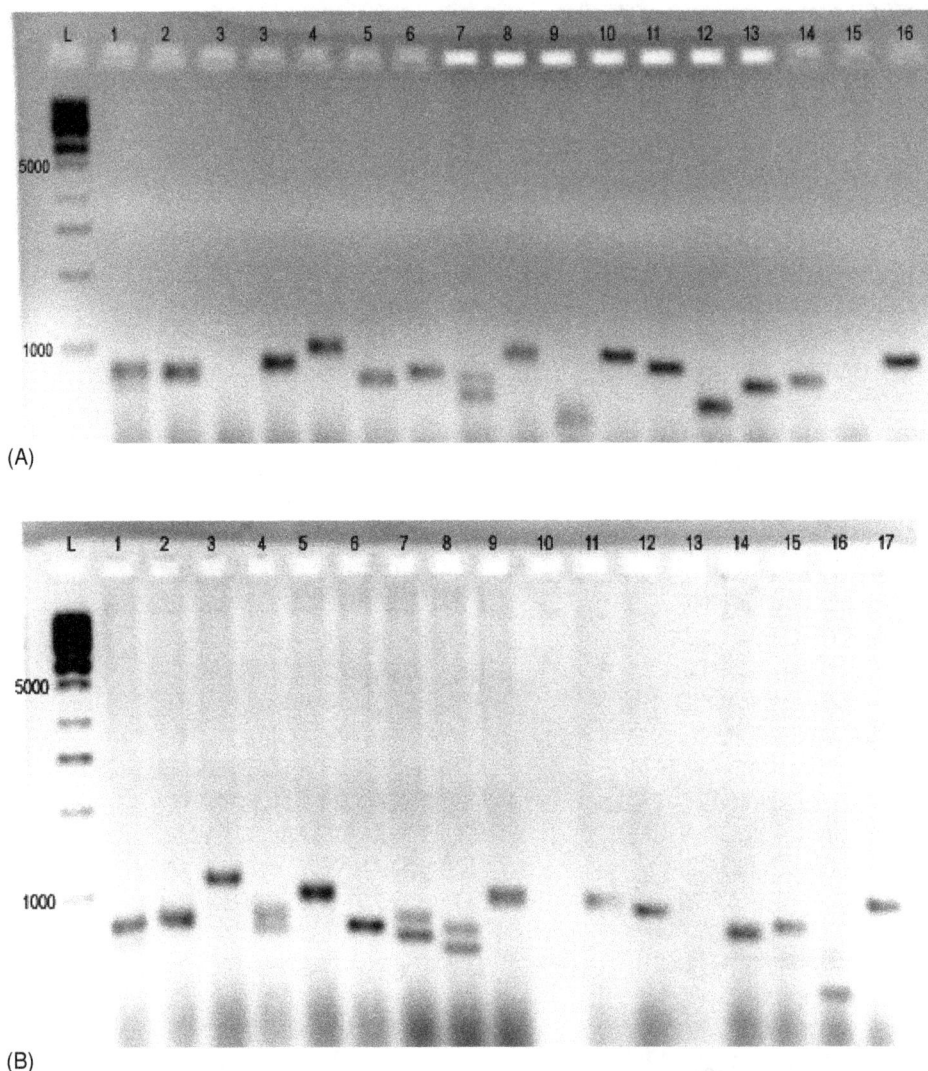

Figure 1. Representative gels showing SSR marker profile of 17 (Lanes 1 to 17) released varieties (A) or central Kerala accessions (B). Lane M: 1-kb molecular weight marker.

of Weir and Cockerham (1984). Cluster analysis of the SSR data was performed separately with the assistance of the SIMQUAL programme of NTSYS software, version 2.10 (Applied Biostatistics Inc., Setauket, NY, USA). Similarity matrices were generated using DICE and simple matching coefficients. An unweighted pair grouping by mathematical averaging (UPGMA) cluster analysis was produced from similarity matrices constructed for SSR data and resulting dendrograms were compared. Principal component analysis (PCA) was applied to identify groups of primers which contributed to the variation among the genotypes and to identify groups of lines which showed a similar response to primers. PCA removes any intercorrelation that may exist between genotypes by transforming the original variables into a few hypothetical components.

New PCAs are orthogonal to each other (Smith, 1991). Statistical analysis was done using SAS v. 8 (1999). A scatter diagram was plotted for the 36 primers using the scores obtained from first two principle components in the case of both released varieties and the central Kerala cassava collections.

RESULTS

Genetic diversity in cassava was evaluated using 12 released varieties and 24 central Kerala varieties of cassava with SSR primers. The primers utilized were highly informative. Each band produced by the primers was distinct and reproducible. The polymorphic bands produced were efficient in assessing genetic diversity among the cultivars. Band size ranged from 0.2 to 0.3 kb and the number of scorable bands per primer ranged from 1 to 2. SSR primers used in DNA amplifications resulted in scorable PCR bands or loci (Figure 1a and b).

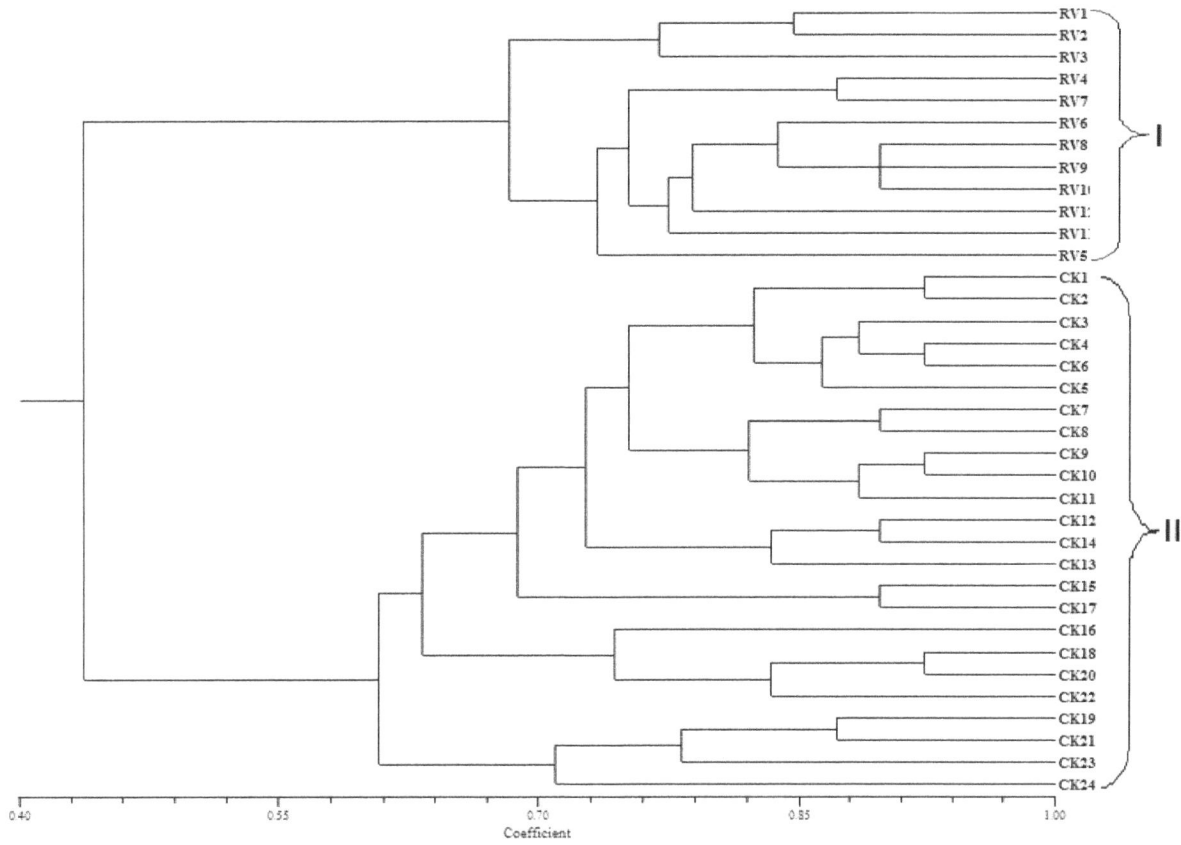

Figure 2. Unweighed Pair Group Method with Arithmatic Average (UPGMA) dendrogram of 12 accessions of released variety of cassava collections based on SSR data. The dendrogram was constructed from the matrix of Dice's similarity coefficients.

Based on SSR bands amplified by 36 primers, a total of 282 clear and scorable bands were detected for both released varieties and central Kerala collections using the 36 SSR primers and used for analysis using NTSYS software. The similarity matrix coefficient generated by the 282 SSR loci based on the NTSYS analysis ranged from 0.75 to 1.00 coefficients; the dendrogram obtained using UPGMA analysis in NTSYS software package revealed 6 distinct DNA cluster groups at 0.82 similarity coefficient units (Figure 2). Both released varieties of cassava and central Kerala varieties formed a distinct group and there was no overlapping of these two varieties. When the binary data from the 12 released varieties were treated alone in NTSYS, 10 DNA cluster groups were generated among the 12 released varieties (Figure 3) at 0.82 similarity coefficient units based on these morphological characters. Similarity index based on presence or absence of a specific band showed that the genetic similarity between varieties in this region varied from 60 to 93%. In cluster 1 vars. RV1 and RV2 were present: they have a dark green stem and light sepia leaf colour.

Cluster II includes var. RV3, which has special characters such as grey leaves. Cluster III included vars. RV4 and RV7. RV4 has a dark sepia stem colour and a fusiform-shaped tuber. RV7 has particular character such as light pink petioles and leaves. Only var. RV5 formed cluster IV; it has a long neck which is absent from accessions in other clusters. Cluster V consists of vars. RV6, RV8, RV9 and RV10, all of which have a reddish-brown stem and common characters such as a conical tuber. Var. RV12 was present in cluster VI. It is resistant to cassava mosaic disease (CMD) unlike all other varieties which are susceptible to CMD (Figure 4). Cluster VII consists of var. RV11 which is a medium height plant. Twenty four varieties collected from central Kerala were grouped into 6 clusters (Figure 5) at 0.82 similarity coefficient units based on there morphological characters; the genetic similarity between varieties in this region varied from 70 to 98%. Cluster I consisted of the major varieties, none of which had a neck. CK13 is the only variety in cluster II; it has special characters like narrow leaves. Cluster III consists of vars. CK19, CK20, CK21, CK22, CK23 and CK24, all of which have a small

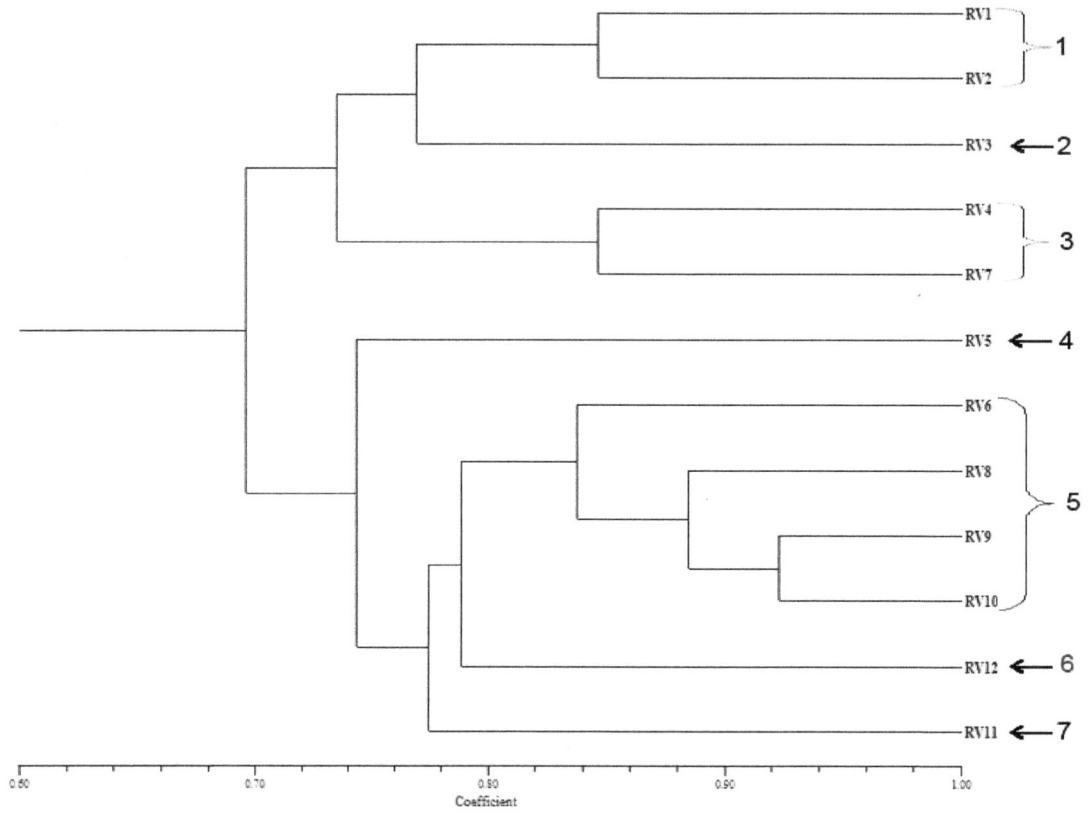

Figure 3. Unweighed Pair Group Method with Arithmatic Average (UPGMA) dendrogram of 24 accessions of cetral kerala collections based on the SSR data. The dendrogram was constructed from the matrix of Dice's similarity coefficients.

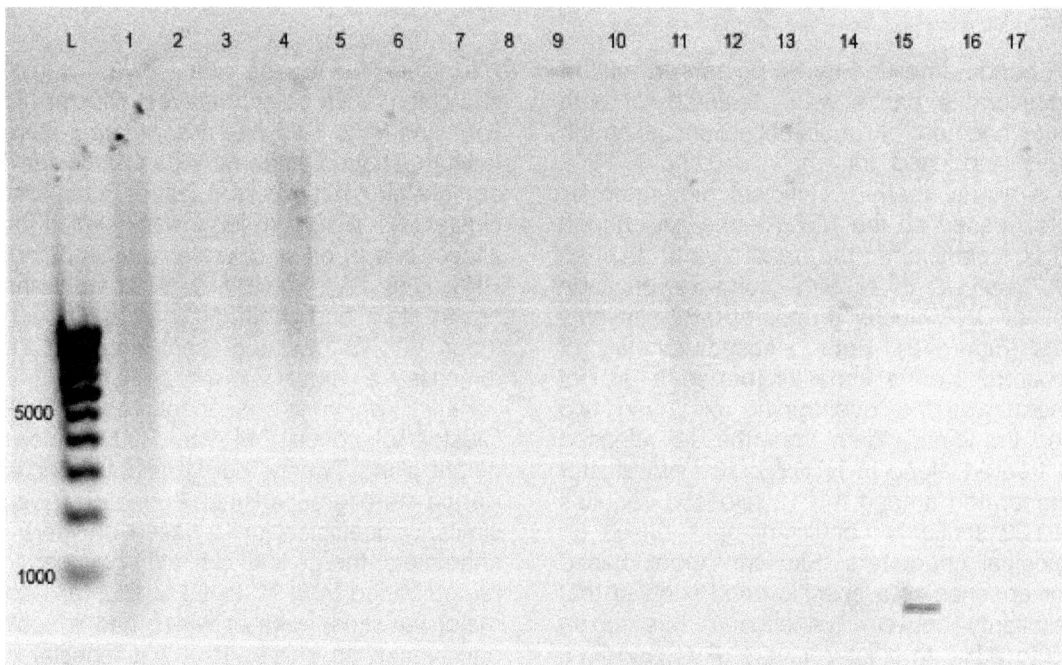

Figure 4. Representative gel showing CMD-resistant variety (Lane 15).

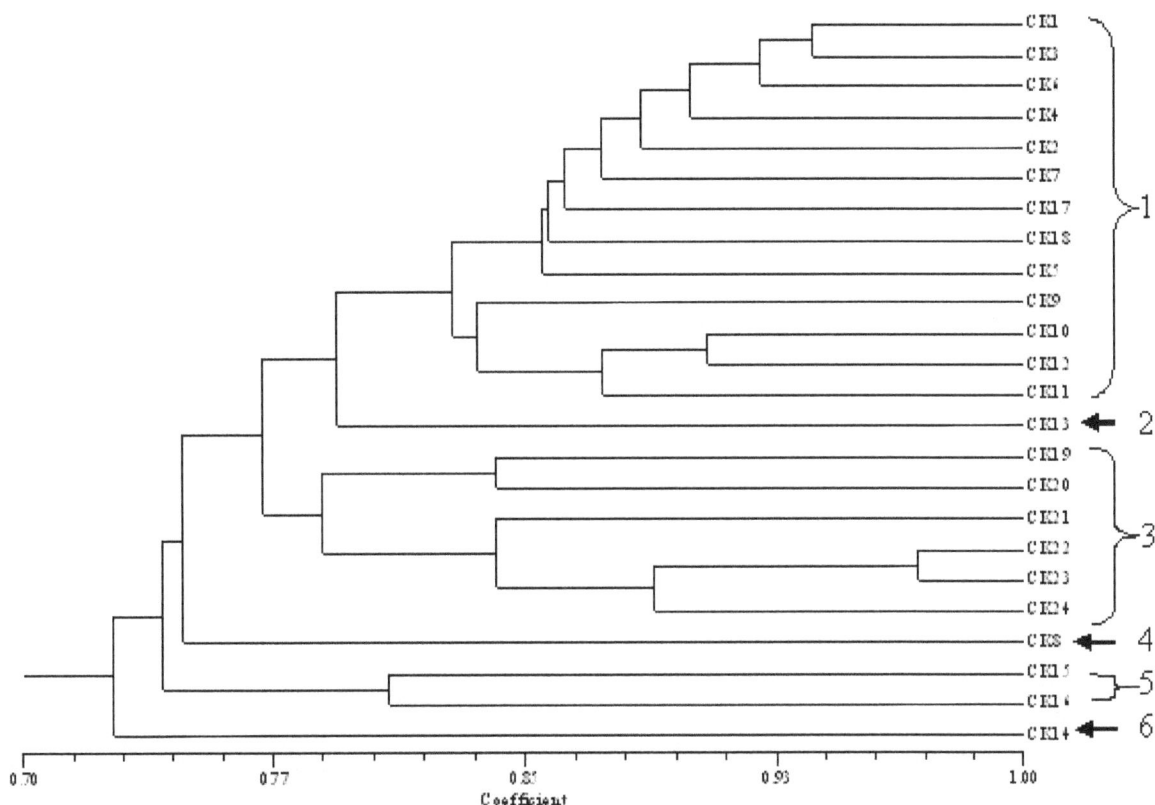

Figure 5. Unweighed Pair Group Method with Arithmetic Average (UPGMA) dendrogram of 24 accessions of central Kerala collections based on the SSR data. The dendrogram was constructed from the matrix of Dice's similarity coefficients.

or long neck. CK8 is the only variety present in cluster IV and it is an early cooking variety. CK15 and CK16 are grouped in cluster V and both have the same place of origin. Cluster VI consist of CK14 and it is an early maturing variety which matures in six months.

The binary data generated from the 36 cassava cultivars were also subjected to PCA using SAS. The first three principal components contributed 28.16, 16.76 and 8.11%, respectively of the total variation present in the data. A scatter diagram of the first two principal components (Figure 6) shows the relationship between the primers. PCA helped to identify primers which contributed much to the variation present in the population.

Population genetic analysis

Population genetic analysis in different cassava accessions was done using POPGENE software. Each band produced was treated as a locus and variations among the alleles were calculated. The SSR markers used in the study could differentiate the genetic diversity in the cassava accessions. The genetic diversity of

cassava was revealed by the percentage of polymorphic loci (P), mean number of alleles per locus (A_O), effective number of alleles (A_E), observed heterozygosity (H_O), and expected mean heterozygosity (H_E). Each band obtained by SSR was treated as a gene locus and the homozygosity and heterozygosity for each loci was determined (Table 3a and b). The genetic analysis of released varieties of cassava accessions revealed that 100% heterozygosity was present in different accessions. The number of polymorphic loci and the percentage of polymorphic loci was 39 and 100%, respectively. The A_O, A_E, H_O and H_E were 2.000, 1.3486, 0.2407 and 0.2584, respectively (Table 4a). On the other hand, the collection of central Kerala cassava accessions revealed low percentage heterozygosity in different accessions except for the homozygous gene locus which expressed only in one allele at a time. The A_O, A_E, H_O and H_E were 1.7838, 1.5120, 0.2934 and 0.3386, respectively (Table 4b).

The aforementioned data shows that new alleles are formed in a cassava population by random and natural processes of mutation and recombination while the frequency of occurrence of an allele changes regularly as a result of mutation, genetic drift and selection in released varieties of cassava.

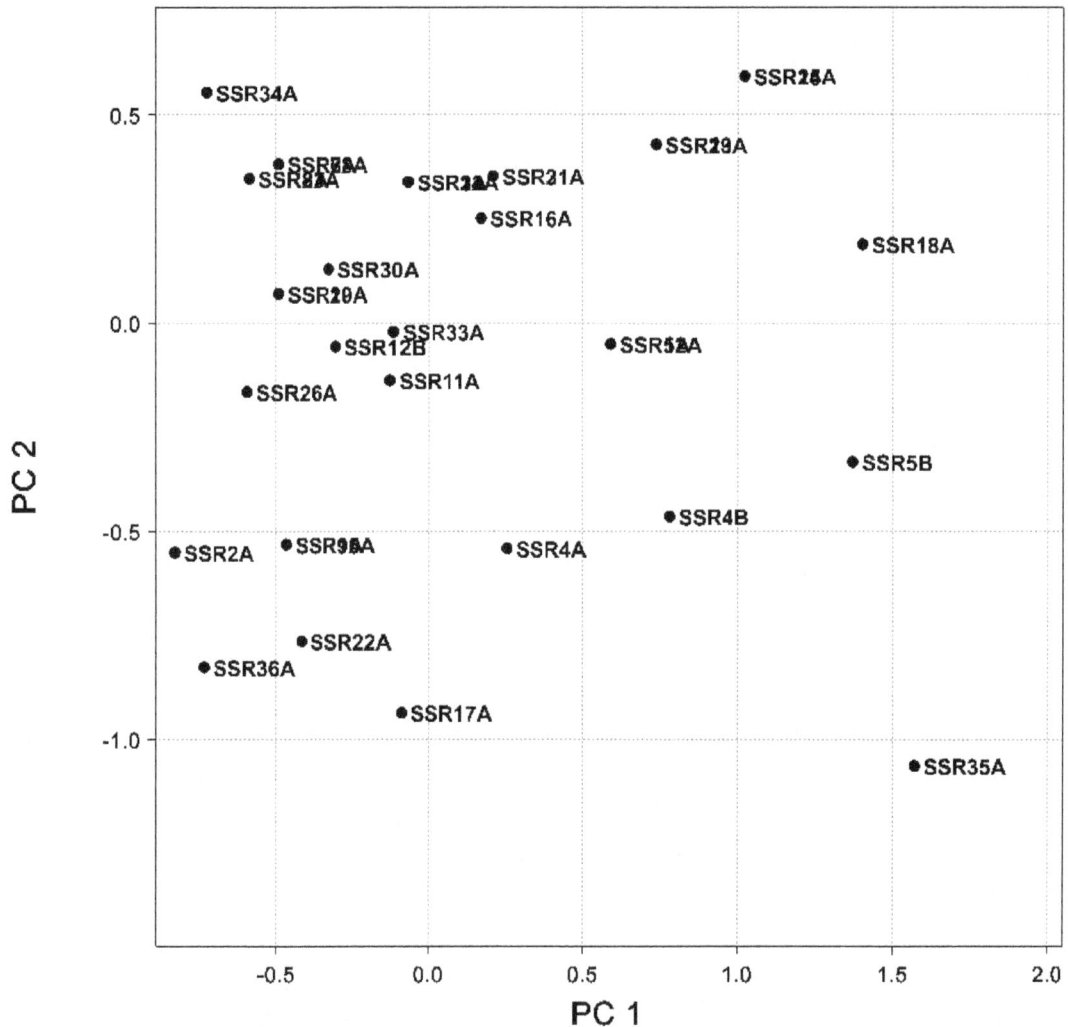

Figure 6. PCA of the studies varieties and accessions.

DISCUSSION

Understanding genetic diversity in tuber crops is important as it is the first step in harnessing their phenotypic variability for crop improvement. Morphological traits are useful tools for preliminary evaluation because they offer a fast and useful approach for assessing the extent of diversity. The estimation of descriptive statistics of 11 different morphological traits studied in the present study revealed the existence of a high level of morphological diversity among the cassava accessions, providing scope for improvement through hybridization and selection. Morphological traits have commonly been used to express genetic diversity in cassava (Lefevre and Charrier, 1993; Haysom et al., 1994; Raghu et al., 2007; Sree Lekha and Pillai., 2008, 2010), although a number of genetic marker systems have also been used for the assessment of genetic

diversity of cassava germplasm. These include isozyme markers (Sarria et al., 1992), RFLP (Angel et al., 1992), RAPD (Tonukari et al., 1997; Ugorji, 1998) and SSR (Fregene et al., 2001; Sree Lekha and Pillai., 2008, 2010) markers and low or medium genetic diversity has always been observed. In the present study there was generally high genetic diversity between released varieties and the central Kerala collections, as shown by the dendrogram. There is no relationship between these two varieties even though they are collected from same place.

The released varieties and the central Kerala varieties both form a distinct group and there is no overlapping of these two varieties observed in the dendrogram generated by NTSYS. The high differentiation between released varieties and central Kerala varieties suggests limited germplasm exchange between these two collections. SSR variation found within the released varieties of cassava was measured in terms of

Table 3a. Allelic frequencies of polymorphic loci studied in 12 cultivars of released cassava.

Locus	Allele	Allelic frequency	Locus	Allele	Allelic frequency	Locus	Allele	Allelic frequency
SSR1-A	0	0.9167	SSR12-A	0	0.8333	SSR24A	0	0.9167
	1	0.0833		1	0.1667		1	0.0833
SSR2-A	0	0.7500	SSR12-B	0	0.9167	SSR25A	0	0.7500
	1	0.2500		1	0.0833		1	0.2500
SSR3-A	0	0.9167	SSR13-A	0	0.8333	SSR26A	0	0.7500
	1	0.0833		1	0.1667		1	0.2500
SSR4-A	0	0.8333	SSR14-A	0	0.7500	SSR27A	0	0.9167
	1	0.1667		1	0.2500		1	0.0833
SSR4-B	0	0.7500	SSR15-A	0	0.9167	SSR28A	0	0.9167
	1	0.2500		1	0.0833		1	0.0833
SSR5-A	0	0.8333	SSR16-A	0	0.9167	SSR29A	0	0.8333
	1	0.1667		1	0.0833		1	0.1667
SSR5-B	0	0.5833	SSR17-A	0	0.8333	SSR30A	0	0.9167
	1	0.4167		1	0.1667		1	0.0833
SSR6-A	0	0.9167	SSR18A	0	0.3333	SSR31A	0	0.9167
	1	0.0833		1	0.6667		1	0.0833
SSR7-A	0	0.9167	SSR19A	0	0.9167	SSR32A	0	0.9167
	1	0.0833		1	0.0833		1	0.0833
SSR8-A	0	0.9167	SSR20A	0	0.9167	SSR33A	0	0.8333
	1	0.0833		1	0.0833		1	0.1667
SSR9-A	0	0.9167	SSR21A	0	0.9167	SSR34A	0	0.8333
	1	0.0833		1	0.0833		1	0.1667
SSR10-A	0	0.9167	SSR22A	0	0.8333	SSR35A	0	0.5000
	1	0.0833		1	0.1667		1	0.5000
SSR11-A	0	0.9167	SSR23A	0	0.9167	SSR36A	0	0.7500
	1	0.0833		1	0.0833		1	0.2500

percentage of polymorphic loci, alleles per locus, or genetic diversity. They are indicative of high genetic differentiation within populations. The results showed that the level of polymorphism P (78.38%) in the central Kerala collections of cassava were lower than those from released varieties of cassava (100%). These results shows high level of polymorphism when compared to studies conducted by Okogbenin et al. (2006) and

Sreelekha et al. (2010) in old and new collections of cassava collected from India. The distribution of species observed in the dendrograms (Figures 2, 3 and 5) is coherent and clearly shows that the SSR and analytical methods used in this study are powerful tools for studying the genetic diversity of *Manihot* species.

In this study the dendrograms clearly separate the released varieties from the accessions of the central

Table 3b. Allelic frequencies of polymorphic loci studied in 24 cultivars of central Kerala cassava.

Locus	Allele	Allelic frequency	Locus	Allele	Allelic frequency	Locus	Allele	Allelic frequency
SSR1-A	0	0.3913	SSR12-A	0	0.4167	SSR24-A	0	0.8333
	1	0.6087		1	0.5833		1	0.1667
SSR2-A	0	0.1667	SSR13-A	0	****	SSR25-A	0	0.0417
	1	0.8333		1	1.0000		1	0.9583
SSR3-A	0	0.7083	SSR14-A	0	0.7917	SSR26-A	0	0.7917
	1	0.2917		1	0.2083		1	0.2083
SSR4-A	0	0.4583	SSR15-A	0	0.4583	SSR27-A	0	0.1667
	1	0.5417		1	0.5417		1	0.8333
SSR5-A	0	0.2500	SSR16-A	0	****	SSR28-A	0	****
	1	0.7500		1	1.0000		1	1.0000
SSR6-A	0	****	SSR17-A	0	0.9167	SSR29-A	0	0.6250
	1	1.0000		1	0.0833		1	0.3750
SSR7-A	0	0.6522	SSR18-A	0	****	SSR30-A	0	0.3750
	1	0.3478		1	1.0000		1	0.6250
SSR7-B	0	0.5833	SSR19-A	0	0.4167	SSR31-A	0	****
	1	0.4167		1	0.5833		1	1..0000
SSR8-A	0	0.3750	SSR20-A	0	****	SSR32-A	0	0.1250
	1	0.6250		1	1.0000		1	0.8750
SSR9-A	0	0.5417	SSR21-A	0	0.7917	SSR33-A	0	0.2917
	1	0.4583		1	0.2083		1	0.7083
SSR10-A	0	0.5000	SSR22-A	0	0.7917	SSR34-A	0	0.3333
	1	0.5000		1	0.2083		1	0. 6667
SSR11-A	0	0.2917	SSR23-A	0	****	SSR35-A	0	0.2083
	1	0.7083		1	1.0000		1	0.7917
						SSR36-A	0	0.9583
							1	0.0417

Kerala collections. This clear partition into two groups is consistent with the concept that the two set of collections represent two different genetic entities. In a previous study in cassava with EST-SSR markers there was a marked separation between cultivated cassava accessions from their wild varieties (Raji et al., 2009). Studies conducted by Moyib et al. (2007) showed no differentiation between the improved varieties and Nigerian collections while Kizito (2006) showed no differentiation between the cassavas collected from different districts of Uganda. The clustering pattern shown by released varieties showed much higher diversity than the central Kerala collections. The mean fixation index (F) for released varieties was 0.0688 and

Table 4. Genetic variation parameters of both old accessions and new accessions.

	Released varieties	Central Kerala varieties
P	100	78.38
A_O	2.00	1.78
A_E	1.35	1.51
H_O	0.24	0.29
H_E	0.26	0.34
F	0.068	0.13
t	0.396	0.43

P - Percentage of polymorphic loci, A_O - mean number of allele per locus, A_E - mean effective number of alleles, H_O - mean observed heterozygosity, H_E - mean expected heterozygosity, F - Wright's fixation index, t - out crossing rate

that for central Kerala collection was 0.1337, indicating an overall conformance to Hardy-Weinberg equilibrium. The estimated F value, used to quantify an excess or deficiency of heterozygotes, was substantially higher than the mean value expected (0.05 or 5%), and positive, indicating an excess of homozygotic individuals. The excess of heterozygotes in released varieties may be the result of farmer selection during the domestication process, but an accumulation of somatic mutations can also contribute to the number of heterozygous genotypes (Birky, 1996).

The out-crossing rate (t) based on fixation indices for released varieties was 0.3964 and that for the central Kerala collection was 0.4332, which is higher than the value in released varieties. da Silva et al. (2001) reported an out crossing rate of 0.69 to 1.00 among 8 ethno-varieties of cassava from Brazil. The population genetic analysis data further provides ample evidence for the fact that recombination events that have occurred in the central Kerala accessions could be due to natural selection. Apart from maintaining a high level of genetic diversity, the formation of new varieties also serves as an insurance against crop failure due to biotic and abiotic stresses. The unique diversity suggests that the germplasm might have genes, in high frequencies, for adaptation to the area, while the high genetic diversity implies a high amount of additive genetic variance, upon which progress in plant breeding depends. The differences in allele frequencies seen among landraces in this study are probably due to genetic drift effects subsequent to mutation. The unique and broad diversity of cassava landraces found in both collections reveals an invaluable germplasm resource for cassava improvement targeted to the region. The high level of differentiation between land races from both released varieties and central Kerala collections may represent a heterotic pool and provide an opportunity for the systematic exploitation of hybrid vigor in cassava. The two collections in the present study gave different views of the amount of genetic variation and genetic relationships.

The study of population genetics is increasingly important as we struggle to maintain healthy, wild and domestic populations and ecosystems, not only for cassava. Moreover, information on the population's effective population size, heterozygosity levels and inbreeding coefficients for particular individuals can be used to design relocation or planned breeding programs which will help to maximize the genetic variation in successive generations. The current study provides a data-base for cassava breeders informed about choices in selection of parental accessions for use in a breeding program based on genetic diversity. The hierarchical clustering illustrated in a dendrogam is usually reflected in a PCA scatter plot. PCA analysis provides information about associations of accessions, which are useful to formulate better breeding strategies. It also helps to identify primers which contributed much to the variation present in the population. The results of this study, thereby, established a collection of 9 highly polymorphic SSR primers (SSRY26, SSRY11, SSRY12, SSRY10, SSRY30, SSRY16, SSRY31, SSRY22 and SSRY32) that could be readily used for genotype identification and genetic diversity studies in both released varieties and collections from the central part of Kerala. Therefore, application of few highly polymorphic SSR markers is possible for genetic variation studies in cassava and has thus great application for genetic studies on cassava in collections from around the world. This reduces the stress of applying many SSR primers for the identification of cassava cultivars in Kerala (and elsewhere) and hence, saves time and also cuts the cost of research studies for genetic diversity studies.

Cluster analysis and PCA-based scatter plots showed great similarity among Brazilian cassava landraces (Siqueira et al., 2009). Lokko et al. (2009) also reported significant diversity within clusters among African land races of cassava through PCA analysis.

ACKNOWLEDGEMENTS

A grant provided by Kerala State Council for Science, Technology and Environment, Trivandrum to carry out this research is gratefully acknowledged. The authors express deep gratitude to the Director and Head of the Division (Crop Improvement), Central Tuber Crops Research Institute, Trivandrum for providing necessary facilities. The authors are also grateful to M. Fregene, CIAT, Cali, Colombia for scientific advice. The authors also express sincere gratitude to Ajay Kumar Mishra, Kamal Sharma and Sree Kumar for their support.

REFERENCES

Angel F, Giralde F, Gomez R, Iglesias C, Tohme J, Roca WM (1992). Use of RFLPs and RAPDs in cassava Genome. In: Roca WM, Thro AM (Eds) Cassava Biotechnology Network. Proceedings of the first international scientific meeting of cassava biotechnology network held at Centro Internacional de Agriculture Tropical (CIAT) Cartagena de Indias, Colombia. 25-28 August, 1992. CIAT Working Dec., pp. 62-64.

Birky Jr. CW (1996). Heterozygosity, heteromorphy, and phylogenetic trees in asexual eukaryotes. Genetics, 144: 427-437.

Chavarriaga-Aguirre P, Maya MM, Tohme J, Duque MC, Iglesias C, Bonierbale S, Kresovich MW, Kochert G (1999). Using microsatellites, isozymes and AFLPs to evaluate genetic diversity and redundancy in the cassava core collection and to assess the usefulness of DNA-based markers to maintain germplasm collections. Mol. Breed., 5: 263-273.

Cock JH (1982). Cassava: A basic energy source in the tropics. Science, 218: 755-762.

da Silva MR, Bandel G, Martins SP (2001). Mating system in an experimental garden composed of cassava (Manihot esculenta Cranz) ethnovarieties. Euphytica, 134: 127-135.

Dellaporta SL, Word J, Hicks JB (1983). A plant DNA preparation. Plant Mol. Biol., 4: 19-21.

Fregene M, Angel F, Gómez R, Rodríguez F, Chavarriaga P, Roca WM, Tohme J, Hokanson SC, Szewe-McFadden AK, Lamboy WF, McFerson JR (1997). Microsatellite (SSR) markers reveal genetic identities, genetic diversity and relationships in a Malusx domestica Borkh, core subset collection. Theor. Appl. Genet., 97: 671-683.

Fregene M, Okogbenin E, Mba C, Angel F, Suárez MC, Janneth G, Chavarriaga P, Roca W, Bonierbale M, Tohme J (2001). Genome mapping in cassava improvement: Challenges, achievements and opportunities. Euphytica, 120(1): 159-165.

Haysom HR, Chan TLC, Hughes MA (1994). Phylogenetic relationships of Manihot species revealed by restriction fragment length polymorphism. Euphytica, 76: 227-234.

Hedrick PW (2004). Genetics of Populations (2nd Edn), Jones and Bartlett Publishers, Sudbury, MA, USA, p. 382.

Hokanson SC, Szewc-McFadden AK, Lambey WF, McFerson JR (1998). Microsatellite (SSR) markers reveal genetic identities, Genetic diversity and relationships in a Malus x domestica Borkh. Core subset collection. Theor. Appl. Genet., 97: 671-683.

Hussain AW, Bushuk H, Ramírez F, Roca WM (1987). Identification of cassava (Manihot esculenta Crantz) cultivars by electrophoretic patterns of esterases enzymes. Seed Sci. Technol., 15: 19-21.

Jones WO (1959). Manioc in Africa, Stanford University Press, p. 315.

Kizito EB (2006). Genetic and root growth studies in cassava (Manihot esculenta Crantz): Implications for breeding. PhD Thesis, Swedish University of Agricultural Sciences, Sweden, p. 127. Online:

http://dissepsilon.slu.se:8080/archive/00001220/01/E.B._Kizito's_thes is.pdf.

Lefèvre FA, Charrier S (1993). Isozyme diversity within African Manihot germplasm. Euphytica, 66: 73-80.

Lokko Y, Dixon A, Offei S, Danquah E, Fregene M (2009). Assessment of genetic diversity among African cassava Manihot esculenta Grantz accessions resistant to the cassava mosaic virus disease using SSR markers. Genet. Resour. Crop Evol., 53: 1441-1453.

Mba REC, Stephenson P, Edwards K, Melzer S, Nkumbira J, Gullberg J, Apel K, Gale M, Tohme J, Fregene M (2001). Simple sequence repeat (SSR) markers survey of the cassava (Manihot esculenta Crantz) genome: Towards an SSR- based molecular genetic map of cassava. Theor. Appl. Genet., 102: 21-31.

Moyib OK, Odunola OA, Dixon AGO (2007). SSR markers reveal genetic variation between improved cassava cultivars and landraces within a collection of Nigerian cassava germplasm. Afr. J. Biotechnol., 6(23): 2666-2674.

Nei M (1973). Analysis of gene diversity in subdivided populations. Proc. Nat. Acad. Sci. USA, 70: 3321-3323.

Okogbenin E, Marin J, Fregene M (2006). An SSR based molecular genetic map of cassava. Euphytica 147(3): 433-440.

Olsen K, Schaal B (2001). Microsatellite variation in cassava (Manihot esculenta), Euphorbiaceae and its wild relatives: Evidence for a southern Amazonian origin of domestication. Am. J. Bot., 88: 131-142.

Perera L, Rusell JR, Provan J, Powell W (2000). Use of microsatellite DNA markers to investigate the level of genetic diversity and population genetic structure of coconut (Cocos nucifera L.). Genome, 43(1): 15-21.

Pillai SV (2002). Variability and genetic diversity in cassava. Indian J. Genet., 62: 242-244.

Pillai SV, Manjusha SP, Sundaresan S (2004). Molecular diversity in the land races of cassava in India based on RAPD markers. Paper presented in the Sixth International Scientific meeting of the Cassava Biotechnology Network. CIAT, Cali, Colombia, March 8-14. p. 45 (Abstract).

Raghu D, Senthil N, Saraswathi T, Raveendran M, Gnanam R, Venkadachalam R, Shanmughasundaram P, Mohan C (2007). Morphological and simple sequence repeats (SSR)-based fingerprinting of South Indian cassava germplasm. Int. J. Integ. Biol., 1(2): 142-148.

Raji AAJ, Anderson JV, Kolade OA, Ugwu CD, Dixon AGO, Ingelbrecht IL (2009). Gene-based microsatellites for cassava (Manihot esculenta Crantz): Prevalence, polymorphisms, and cross-taxa utility. BMC Plant Biol., 9: 118.

Rogers DJ, Appan SG (1973). Manihot manihotoides (Euphorbiaceae). Flora Neotropica. Hafner Press, New York, Monograph 13: 272.

Sarria R, Ocampo C, Rodríguez H, Hershey C, Roca WM (1992). Genetics of esterase and glutamate oxaloacetate transaminases isozymes in Cassava. In: Roca WM, Thro AM (Eds) Proceedings of the First International Scientific Meeting of Cassava Biotechnology Network held at Centro Internacional de Agriculture Tropical (CIAT). Cassava Biotechnology Network, Cartagena de Indias Colombia 25-28 August, CIAT Working Doc., pp. 62-64.

Siqueira MVBM, Queiroz-Silva JR, Bressan EA, Borges A, Pereira KJC, Pinto JG, Veasey EA (2009). Genetic characterization of cassava (Manihot esculenta) landraces in Brazil assesses with simple sequence repeats. Genet. Mol. Biol., 32: 104-110.

Smith GL (1991). Principal component analysis: An introduction. Anal. Proc., 28: 150-151.

Sree Lekha S, Pillai SV (2008). SSR marker variability in a set of Indian cultivars from a typical cassava growing area. Asian Austral. J. Plant Sci. Biotechnol., 2 (2): 92-96.

Sreelekha S, Kumar S, Pillai SV (2010). Assessing genetic diversity of Indian cassava: Acomparison of old and new collection using microsatellite markers. Asian Austral. J. Plant Sci. Biotechnol., 4(1): 43-52.

Tautz D (1989). Hypervariability of simple sequences as a general source of polymorphic DNA markers. Nucl. Acid Res., 17: 6463-6471.

Tonukari NJ, Thottapilly G, Ng NQ, Mignouna HD (1997). Genetic polymorphism of cassava within the Republic of Benin detected with RAPD markers. Afr. J. Crop Sci., 5(93): 219-228.

Ugorji N (1998). Genetic characterization of cassava cultivars in Nigeria: Morphological and molecular markers. MSc Dissertation, University of Ibadan, Ibadan, Nigeria.

Weir BS, Cockerham CC (1984). Estimating F-statistics for the analysis of population structure. Evolution, 38: 1358-1370.

Weir BS (1996). Genetic Data Analysis II, Sinauer Associates, Sunderland, MA, p. 445.

Yeh FC, Yang R (1999). Microsoft Window-based Freeware for Population Genetic Analysis (POPGENE Ver. 1.31). University of Alberta, AB, Canada.

Analysis of goat production situation at Arsi Negele Woreda, Ethiopia

Gurmesa Umeta*, Feyisa Hundesa, Misgana Duguma and Merga Muleta

Adami Tulu Agricultural Research Center, P. O. Box 35, Zeway, Ethiopia.

The study was conducted at Arsi Negele District of Oromia Regional Administrative Zone with objectives of: (1) Assessing goat production situation of the area (2) identifying problems limiting goat production of the area, (3) Generating information for development practitioners working in the area to improve the situation. The sampled kebele were selected based on the potential of goat production and suitability of the area for transportation. Fifteen to twenty key informant farmers were identified with development workers for group discussion per the sampled kebele. Both female and male households were invited for group discussion. Participatory rural appraisal (PRA) techniques and methods were employed for data collection. A mix of PRA tools like group discussion, pair wise ranking, seasonal calendar and secondary data reviews were employed during data collection. The study is based on qualitative data analysis using descriptive statistics. From the current study it was realized that goat production is one of the major livelihood options for the goat keepers of the area. The study also identified that goat production plays a pivotal role in many ways for the goat keepers of the area. Its significance includes; serving as a source of milk, butter, and meat as well as income generation. In addition to this, it is considered as wealth and has contributed to social values. Furthermore, farmers also consider it as a risk mitigation strategy to cope with adverse environmental effects; this is mainly when shortage of rain occurs at the area or when scarcity of production occurs. Farmers also identified that goat production is advantageous because of having short generation intervals which give quick production for market. These huge contributions are also considered as the major reasons behind for keeping goat in the study area. Despite these benefits, goat rearing practices of the area have been constrained by many factors which can be categorized under genetic and non-genetic categories. The major non-genetic factors identified include; diseases like sheep and goat pox, diarrhea, ecto-parasite, circling disease, mastitis, anthrax, and pasteurellosis, shortage of feeds, weak extension services, and market related problems. Genetic related factors are mainly associated with lack of breed improvement interventions. Therefore, the study recommends that goat production extension package generation, development and popularization for the study area needs to be giving due attention by the stake holders working in the area.

Key words: Arsi Bale goats, Oromia, Ethiopia.

INTRODUCTION

Ethiopia has a larger livestock resource base than most countries in Africa. It is estimated that 84% of the 70

*Corresponding author. E-mail: gurme2010@yahoo.com.

Abbreviation: DA, Development agents; **EARO,** Ethiopian Agricultural Research Organization; **Oard,** office of agriculture and rural development; **MOA,** Ministry of Agriculture and Rural Development, **PA,** Peasant Association; **PRA,** participatory rural appraisal.

million people live in rural areas and depend on agriculture for their livelihoods and the sector contributes 41.4% of the Gross Domestic Product of the country (World Bank, 2006).

Mid-rift valley area is known to have a high population of sheep and goats. The environment is much more conducive for rearing of small ruminant animals. Though there is no latest and up to date information on the current small ruminants' population, Abule et al. (1998) reported that there are about 653,940 sheep and 1.8 million goats in the mid-rift valley area. Despite such

Table 1. Goat population size of the area.

Numbers	Type of ruminants	Number(heads)
1	Goats	82,211
2	Sheep	38,651
Total		120, 862

Source: Arsi Negele oARD, 2010.

huge potential, farmers in the area could not realize the expected benefit from small ruminant rearing. Poverty and food insecurity are the major phenomenon in the area. Farmers are still relying on traditional type of production system which is characterized by poor feeding, housing, breeding, and health management. As a result of this, production and productivity as well as income from sale of the animals are very low. This implies the urgent need to work toward the improvement of the conditions.

Other contributing factors also include low genetic potential; policy issues (Zinash et al., 2001) market and institutional problems and problem of credit facilities and others (Berhanu et al., 2006). Although various research and development activities have been carried out in the past, no significant increase in productivity was achieved. Therefore, improvement programs are necessary to increase productivity and sustainable development of small ruminants in different farming systems of the country innovative approach so as to meet the demands of the human population.

However, such development achievement for sheep and goats will only be successful when accompanied by a good understanding of the different farming systems that simultaneously addresses several constraints: feeding, health control, general management, cost and availability of credit as well as marketing infrastructure (Workneh, 2003). Adami Tulu Agricultural Research Center conducted Participatory rural appraisal with multidisciplinary team combined from extensioninst, economist, geneticians and nutritionists with the following objectives.

General objective

1. To assess goat production situation of the district.

Specific objectives

1. To identify problems constraining goat production of the study area;
2. To assess role of goat production for the livelihood of goat keepers;
3. To generate a piece of information on goat production situation of the area which finally contribute for the improvement of the existing situation.

METHODOLOGY

The study area and characteristics of sampled house holds

The study was conducted at the Arsi Negele district of West Arsi Zone. West Arsi zone is one of the administrative zones of Oromia region. Two kebele were selected based on the potentiality of the kebele for goat production and suitability for transportation. Based on this, Daka Dalu Harangema and Ali Wayo kebele were selected for the study purposively. Fifteen to twenty key informant farmers per the kebele were selected for group discussion. Multidisciplinary team combined from animal health, extensionists, economists, breeder and nutritionists were organized and participated on data collection. Key informant farmers who are expected to know goat production situation of the area were identified and invited with development agents. Both men and women farmers were involved during group discussion.

Method of data collection and analysis

Participatory rural appraisal (PRA) technique was employed for data collection and analysis. The term PRA refers to a series of techniques, many of them developed in India, for using local knowledge and skills to learn about local conditions, identify local development problems and plan responses to them. Using of PRA methods for research purpose has three main advantages. First, the information it provides tends to be highly accurate. This is partly because; local people's knowledge of local condition is often greater than had been supposed, as is their capacity to map, model, estimate, rank, diagram and plan. This is also because participatory approach to describing local conditions and planning allow local people discuss and cross check each other's knowledge on the spot. Secondly, plans drawn up by local people are more likely to work than plans drawn by outsiders. The third and most important of all, the participatory nature of the process is a development benefit in itself, in terms of empowering local people (Richards, 1992). The current study therefore employed PRA techniques and methods. The PRA methods employed here includes; group discussion, secondary data reviews, analytical games like pair wise ranking and scoring of results.

RESULTS AND DISCUSSION

Goat production situation of the Woreda

The goat breeds found in the study area is Arsi-Bale goat breeds. Arsi-Bale goat is distributed in the high lands of Arsi, Bale, Hararghe and mid rift valley of Ethiopia and characterized by small body size, short legs, short ears, both short and long hair as well as their glossy, wavy and gray color natue (Worknesh, 1992). Goat production system of the area is characterized by mixed farming system. Goat production system of the area is also under traditional management system. Goat production is one of the livelihood strategies for the farmers of the area. According to data taken from the districts' office of Agriculture and Rural development shows, huge numbers of goat population size are found in the district (Table1).

This somehow indicates their significance since goat production in the area is one of the livelihood strategies for the goat keepers of the area. Goat production has been serving different purposes including ensuring food

Table 2. Farmers' production objectives/ reason for keeping goats at Daka Dalu Harangema PA.

Number	Production objectives/reason for keeping goats	Rank in order of importance
1	Used as a source of milk	2nd
2	Used as a source of meat	5th
3	For income generation	4th
4	Due to having short generation interval	3rd
5	Considered as drought tolerant	1st

Source: Own PRA result.

security, considered as wealth storage and source of income for the goat keepers of the area. Despite such a huge contribution, goat production extension packages available for the goat keepers are weak which could be associated with lack of attention given for the sector by oARD compared to other sectors like crop production. The major goat production extension services available for the goat keepers of the study area is mainly veterinary services which invariably means giving less attention to other aspects of goat production extension packages like breed improvement, improved management practices like housing, feeds and feeding development and managements, marketing aspects like improved access to effective marketing information and improved fattening practices. Farmers also indicated that, the existing veterinary services available like vaccination and treatment of sick animals is not effective.

Farmers' production objectives

According to some study conducted in other parts of Ethiopia, it indicates that goat production is an important component of the livestock subsector and it is also a source of cash income, meat, milk and wool for smallholder keepers in different farming systems and agro-ecological zones of the country (Tekelye et al., 1993; EARO, 2000). The current study is also employed to see farmers' production objectives/ reasons for keeping goats in the study area. The result of the study therefore indicates that farmers have been rearing goats for different reasons/purposes. These include; they considered as a source of meat, income, milk, butter (in rare cases). On the other hand farmers indicated that, goat production is advantageous over other livestock components due to having short kidding intervals. Farmers also expressed that, goat production objective at the area is positively associated with agro- ecologies, that is, goat is considered to be drought tolerant. For social value/gathering, farmers have been using it as a gift which they call 'gegawo' during wedding. Also farmers indicated that goat production in the area is considered as wealth storage which can be served as risk mitigating strategies especially when there is shortage of rain and/ or when scarcity of food production occurs in the area

(Table 2). To analyze these production objectives, pair wise ranking methods was employed as indicated in Table 2.

Among the different reasons/objectives for producing the goat; milk production, ability to withstand drought and having of short generation interval was ranked as first, second and third respectively at this kebele. This indicates that, farmers' in the study area keeps their goats mainly for these purposes other than for other reasons. Also other study conducted by Asfaw, 1997 is in agreement with this finding. He argued that goat production is considered as investment and insurance due to their high fertility, short generation interval, adaptation in harsh environment and their ability to produce even with limited feed resources. They can tolerate drought and can feed on different leguminous trees such as acacia pods and acacia leaves. In addition, they explained that it can reach for markets within a short period of time and also, it is considered as medicinal value.

The other sampled kebele was Ali Wayo kebele. Respondents at this Kebele identified five production objectives/ reasons for keeping goats, namely; for milk consumption, considered as a source of meat, for income generation purpose, sometimes considered as a social value and used as wealth storage (Table 3).

The three production objectives namely source of milk; income generation purpose and wealth storage were ranked as first indicating that goat production plays almost equal roles with regard to the above indicated purposes. The next production objective which ranked as second was social values/gatherings. This indicates that, goat production objectives can go beyond direct benefits since it can be used for some social gatherings. Generally, goat production plays significant roles in terms of ensuring food self sufficiency and one of risk mitigation strategy for the community of the area.

Place of goat production in terms of generating income for the farmers of the study area

Farmers' livelihood strategy of the area depends on different activities. Among these, non-farm and agricultural activities are the major one. From agriculture related

Table 3. Farmers' goat production objectives at Ali Wayo PA.

Number	Production objectives/reason for keeping goats	Rank in order of importance
1	Used as source of milk	1st
2	Used as source of meat	3rd
3	For income generation	1st
4	For social value/gatherings	2nd
5	Considered as wealth storage	1st

Source: Own PRA result.

Table 4. Ranks for respondents' sources of income at both kebele.

Number	Major source of income	Rank at Daka Haregema kebele	Rank at Ali Wayo kebele
1	Crop production	2nd	1st
2	Sheep production	5th	5th
3	Poultry production	6th	6th
4	Cattle production	1st	2nd
5	Donkey cart	4th	4th
6	Selling of wood	9th	-(*)
7	Charcoal making	8th	-(*)
8	Selling of mineral soil	7th	-(*)
9	Goat production	3rd	3rd
10	Honey production	-(*)	7th
11	Horse cart	-(*)	8th

(*) Indicates the activity was not mentioned by farmers at respective kebele. Source: Own PRA result.

activities livestock production such as cattle, small ruminant, poultry rearing, bee keeping, and crop productions are the major activities identified from group discussion.

This study was therefore evaluated the place of goat production in contributing to farmers' livelihoods. Pair wise ranking was employed to identify and prioritize livelihood strategies of the farming community of the area (Table 4).

As explained by matrix ranking, cattle production was placed at first in generating of income at Daka Dalu Harangema kebele followed by crop production and goat production. This shows as that, goat production plays pivotal roles in generating income for the farmers of the study area. This contribution in more or less indicates that goat production is important and plays significant roles for livelihood improvement of the farmers of the area. Therefore, the sector need to be giving due attention by the stakeholders working in the area to develop and popularize goat production extension packages.

Problems constraining goat production in the study area

As identified by the current study, goat production of the

study area is constrained by a couple of problems. These problems are multidimensional in its scope that include; feeds related problems, disease related problems, and market related problems. Respondent farmers from Daka Dalu Haregema kebele identified different problems like shortage of feeds, occurrence of disease, predators, lack of market information, long kidding interval and shortage of water. From these problems, disease problems was found to be the most serious problems ranked first whereas shortage of labor and long kidding interval was ranked second indicating that shortage of labor and long kidding interval are the other major problems associated with goat production of the area which has almost equal negative impacts (Table 5).

As explained by Table 5, disease problem was ranked first. Group discussion was also held with farmers to identify type of diseases available in the area. During disease identification, farmers were asked to list up all type of diseases available in the area. Then, the diseases symptoms were also identified by farmers. Finally name of diseases were identified by the team of researchers based on the symptoms indicated by farmers. The major diseases identified at Daka Dalu Harengema kebele were; sheep and goat pox, Diarrhea, ecto-parasite, anthrax, circling disease and mastitis. Anthrax and diaharea was found to be the most serious disease ranked by farmers as first and second, respectively. The

Table 5. Problems constraining goat production at Daka Dalu Harangama PA.

Number	Type of problems	Rank in order of importance
1	Shortage of feeds	4
2	Lack of market information	4
3	Disease related problem	1
4	Predators	3
5	Shortage of labor	2
6	Long kidding interval	2
7	Shortage of water	7

Source: Own PRA result.

Table 6. problems constraining goat production at Ali Wayo PA.

Number	Type of Problems	Rank in order of importance
1	Shortage of feeds	2
2	Lack of market information	5
3	Disease	1
4	Predators	6
5	Shortage of labor	4
6	Long kidding interval	3
7	Shortage of water	7

Source: Own PRA result.

other problem mentioned by farmers is copper deficiency. As indicated by sampled respondents, there is few veterinary services/controlling mechanisms/available which is not effective for the treatments of copper deficiency. Farmers indicated that kids born from copper deficiency often born swaybacked, the kid stands unsteadily or cannot stand, displays muscle tumors and head shaking, and may grind its teeth. Farmers at Ali Wayo kebele stated same problems that constraining goat production and productivities but they ranked differently. Accordingly, disaease prevalence, shortage of feeds and long kidding interval was ranked first, second and third respectively (Table 6). The type of diseases mentioned by farmers at this kebele is similar with Daka Dalu Harengema kebele. Also the two diseases namely anthrax and diaharea was found to be the most serious diseases ranked first and second respectively.

Weak extension services

The fattening extension components include, purchased or farmer owned indigenous cattle and small ruminants, animal feed and feeding system, animal health, housing, selection of fattening animals, fattening period and marketing of fattened animals (MOA, 1990 E.C). This

study attempted to examine type of technologies used by farmers. According to the result of group discussion made indicates, two fattening options are available in the area; (1) improved practices and (2) fattening practices through own practices/indigenous knowledge. Under improved practices, different fattening packages like use of industrial by-products, feeding management and housing managements, and veterinary services for sick animals as well as for preventive services/vaccination services are available. The main source of the technology for improved practices was Adami Tulu Agricultural Research Center. The use of improved practices was reported only at Daka Dalu Harangema PA but farmers from Ali Wayo PA responded that there was no improved goat fattening technologies introduced at the area. For indigenous practices, farmers have been using the locally available feed supplements like beans, crop after math and acacia pods/leafs. The duration of fattening periods through farmers' own practices takes more than three months as indicated by group discussion.

These farmers' own fattening practices lacks different aspects of goat fattening extension packages including improved managements practices like housing, use of industrial by-products. In addition to these, the duration of fattening periods is longer when compared with the available improved fattening packages. Generally, goat production management systems of the area are under traditional management systems. This situation is also one of the gaps identified by this study.

Mitigation strategies used and recommended to minimize the existing problems

In participatory research like the current study, farmers can also be considered as investigators since their contribution in identifying and prioritizing the existing situation is strong in PRA principle. The current study identified different problems which have been negatively influencing goat production of the area and farmers' mitigation strategies that has been used to minimize the identified problems related with goat production. The gap existing between farmers' mitigation strategies and improved mitigation options was also identified by this study which can be an input to improve the situation. After identifying farmers' mitigation strategies, possible intervention areas were identified (Table 7). With regard to these mitigation strategies, nearly similar results were reported by Gurmesa et al. (2011) in his studies conducted at East Showa Zone (in the process for publication).

Gender participation in goat production activities

Sex and gender

Sex refers to the biological differences between men and

Table 7. Summary of mitigation strategies used by farmers and possible intervention areas identified from the study with regard to goat production.

Types of problem associated with goat production	Farmers' mitigation strategies used to stand with the existing problem	Gaps identified and intervention area to be made to improve the existing situation
Shortage of feeds	-Migration(moving to other places in searching of feeds) -Grazing in the crop/crop after math -Using house west -Reducing of flock size -Feeding of trees leafs/pods -Allocation and conservation of feeds at back yard	-Development of improved forage -Awareness creation on adoption of concentrate supplementation/agro-by product and improved feeding managements -Feed conservation -Reducing flock size -Improving the indigenous breeds by crossing/ through selection
Disease	-Using homemade medication like salt and pepper, drenching local alcohol (catikala), tobacco and other traditional healers -Medication(purchased form market) -Migrating to other places in searching of water because they perceive that if goats frequently used the available water, it will be exposed to especially Cu-deficiency -Vaccination (which could be associated with weak veterinary services, being far from veterinary site and weak awareness by farmers about the severity of disease)	-Establishing of veterinary clinic -Regular vaccination based on frequent assessment of disease occurrence -Awareness creation on disease -Monitoring of the flock
Predators	-Burning of living habitat(hole) of predators -Using bait -Shepherding	-Close supervision of the goats -Appropriate housing
Shortage of labor	-Selling -Hiring of labor -Migration(moving to other places in searching of feeds) -Tying -Keeping the goats by shifting(darabee)	- Making semi intensive by enclosing grazing area -Strengthening of group work on conservation of local forest, feeds and common managements
Water shortage	-Moving to other watering point	-Use of water harvesting
Market related problems	-Searching, waiting and selling of their goats when market price is getting better -Refusing to sell to brokers	-Improving of market information delivering systems -Linking of farmers with potential buyers - Establishing of group work(Farmers' extension groups) especially for fattening - Linking farmers with micro credit institutions -Improving infrastructure like roads
Long kidding interval	-Replacing existing breeds by well performing animal -Migration(moving to other places in searching of feeds) -Borrowing of breeding bucks	-Feed Improvement -Breed improvement -Natural resource conservation - Participatory monitoring of the flock

Source: Own PRA result.

Table 8. Gender participation in goat production activities at the study area.

Number	Type of goat production activity	House hold members			
		Husband	**Wife**	**Son**	**Daughter**
1	Feeding	√	√	√	√
2	Housing	√	√	√	√
3	Who is looking for	√	√	√	√
4	Who sells	√	√	√	√
5	Who control over cash	√	√	√	√
6	Who take to vaccine	√	√	√	√

Source: Own PRA result.

Figure 1. A sample of picture showing when group discussion was undertaken with farmers.

women and is genetically determined. Gender refers to the socially determined differences between women and men, such as roles, attitudes, behaviors and values. Gender roles are learnt and can vary across cultures over time and are therefore amenable to change. Sex is therefore universal while gender is a socially defined category that can change. The concept of gender is vital because, applied to social analysis it reveals how women's subordination (or men's domination) is socially constructed. As such, the subordination can be changed or ended. It is not biologically predetermined nor is it fixed forever (Figure 1).

Women constitute half of the world's population, they do two third of the world's work, they earn one tenth of the world's income and they own one hundredth of the world's property including land (United Nations, 1979). Women involved in different agricultural activities in the area. But most of the time their contribution is undermined/under-valued. Women's participation in livestock production activity in general and goat production activity specifically, were also assessed to identify their contribution and related issues like control over benefits that is related to goat production (Table 8).

The current study result shows that all house hold members had participated in all activities related to goat production (Table 8). But their frequency of participation varies for house hold members. For example, in most cases it is husband who took to markets and control over the benefits. Also, other family member participation is sometimes strong for some activities. This somehow indicates that, participation of household members; men and women is critical in all aspects of goat production activities. This on the other hand indicates that both men and women should be encouraged in adopting of goat production extension packages. Furthermore, the goat production extension packages including breed development activities, fattening activities, and related aspects of extension packages like capacity development through training, visits and demonstration needs to consider both men and women.

CONCLUSION AND RECOMMENDATION

Goat production is one of the major livelihood activities for the goat keepers of the study area. The contribution of goats can be viewed in multidimensional views like income generation, wealth storages, and risk mitigation strategies in case of adapting to adverse harsh environments. It can also be considered as source of meat and milk for the farmers of the area. Despite such a huge contribution, goat keepers of the area are still working under traditional production systems which will finally influence the maximum possible outputs that can be achieved from the sector. Extension services like breed improvement, promotion of improved fattening technologies, improved market services like access to market information, effective veterinary services and other management aspects is very weak. Generally, goat production packages needs to be promoted in the area to increase the production and productivity of the goat keepers of the area which can finally contribute to the overall economic development of the region as well as the country. Also, actors working in the area including research centers, oARDs and other development practitioners needs to do much to improve the existing situation related to goat production.

REFERENCES

Abule E, Amsalu S, Tesfaye AA (1998). Effect of level of substitution of lablab (Dolicos ablab) for concentrate on growth rate and efficiency in post weaning goats. In: Proceeding of the 6th Ethiopian Society of Animal Production (ESAP) conference held on 14-15 may 1998, ESAP, Addis Ababa, Ethiopia, pp. 264-269.

Asfaw W (1997). Country report: Ethiopia, Proceedings of a Seminar on Livestock Development Policies in Eastern and Southern Africa 28th July–1st August 1997, Mbabany, Organized by CTA, OAU/IBAR, The Ministry of Agriculture, Cooperative, Swaziland.

Berhanu G, Hoekstra D, Azege T (2006). Improving the Competitiveness of Agricultural input Markets in Ethiopia: Experiences since 1991. Paper presented at the Symposium on Seed-fertilizer Technology, Cereal productivity and Pro-Poor Growth in Africa: time for New thinking 26th Triennial Conference of the International Association of Agricultural Economics (IAAE), August 12 – 18, 2006, Gold Coast, Australia

EARO (Ethiopian Agricultural Research Organization) (2000). National Small Ruminants Research Strategy Document. EARO, Addis Ababa, Ethiopia.

Richards H (1992). PRA, IIED, London, 16: 13-21.

Tekelye B, Bruns E, Kasali OB, Mutiga ER (1993). The effects of endoparasites on the reproductive performance of on-farm sheep in Ethiopian highlands. Indian J. Anim. Sci., 63: 8- 12.

Workneh A (2000). Do smallholder farmers benefit more from crossbred (Somali x Anglo- Nubian) than from indigenous goats? PhD Thesis. Georg-August University of Goettingen, Goettingen, Germany. Cuvillier Verlag, Goettingen.

World Bank (2006). Africa Development Indicators 2006. Washington D.C CACC (Central Agricultural Census Commission). 2008. Ethiopian Agricultural Sample Enumeration, 2007/08. Results at country level. Statistical report on socio-economic characteristics of the population in agricultural household, land use, and area and production of crops. Part I. (December 2008) ddis Ababa, Ethiopia.

Worknesh A (1992). Preliminary survey of indigenous goat types and husbandry practices in southern Ethiopia.

Zinash S, Aschalew T, Alemu Y, Azage T (2001). Status of Livestock Research and Development in the Highlands of Ethiopia. In: P.C.Wall (ed.). Wheat and Weed: Food and Feed. Proceedings of Two Stockholders Workshop.

10

Effect of polythene packaging on the shelf life of mango fruits

Ilesanmi FF*, Oyebanji OA, Olagbaju AR, Oyelakin MO, Zaka KO, Ajani AO, Olorunfemi MF, Awoite TM, Ikotun IO, Lawal AO and Alimi JP

Nigerian Stored Products Research Institute, PMB 5044, Ibadan, Nigeria.

The effects of polythene packaging and washing were observed on fresh green unbruised wholesome mango fruits. Skin colour, weight changes, titratable acidity and the percentage of fruit that were spoiled were monitored. The samples were treated in four different ways: (i) washed and packaged in perforated polythene bags, (ii) unwashed and packaged, (iii) washed and unpackaged and (iv) the unwashed and unpackaged control. They were all placed in the fruit shed at an average ambient temperature of 31°C and pressure 70 mm/Hg. Packaging has weight loss reduction effect (by 22% in washed mangos, by 30% in the unwashed) while washing increased weight loss by 19% in the packaged mangos, but not significantly increased (7%) in unpackaged mangoes at $p=0.05$. The weight loss and spoilage were reduced by packaging.

Key words: Mangifera indica, perforated polythene bags, shelf life, spoilage, packaging.

INTRODUCTION

Mango (Mangifera indica), is one of the most important and widely cultivated fruits of the tropical world. It is a seasonal fruit believed to have originated in the sub-Himalayan plains of Indian subcontinent. Botanically, this exotic fruit belongs to the family of Anacardiaceae, which also includes numerous species of tropical fruiting trees in the flowering plants such as cashew, pistachio...etc (Whitmore,1975; Marchand, 1869). The fleshy fruit is eaten ripe or used green for pickles and other dishes and is a rich source of Vitamins A, C and D. Mangoes also contains essential vitamins and dietary minerals such as vitamins A, B, B6, C, E and K and essential nutrients such as potassium, copper and 17 amino acids in good levels. In India reported that there are over 100 varieties of mangoes, in different sizes, shapes and colours (Saleh and El-Ansari, 1975; Singh et al., 2004). Mango peel and pulp contain other phytonutrients, such as the pigment antioxidants – polyphenols and carotenoids – and omega-3 and -6 polyunsaturated fatty acids. The mango fruit is a large, fleshy drupe, containing an edible mesocarp of varying thickness. The mesocarp is resinous and highly variable with respect to shape, size, colour, presence of fibre and flavour. The mango fruit is climacteric, and increased ethylene production occurs during ripening (Mitra and Baldwin, 1997).

Mango peel contains pigments that may have antioxidant properties, including carotenoids, such as the provitamin A compound, beta-carotene, lutein and alpha-carotene, polyphenols such as quercetin, kaempferol, gallic acid, caffeic acid, catechins and tannins. People have tried various methods to extend mango shelf life or reduce post harvest losses, like irradiation (Bayers and Thomas, 1979), processing into jam (Subramanjam et al., 1975). Investigated the effect of a fungicidal wax coating on Badami (Alphonso) mangoes. The study was made of fruits dipped in aqueous solutions of fungicidal wax emulsion containing 1.7, 2.2 and 2.7% solids and 5% ortho-phenyl-phenol, and stored at 79 to 86°F and R.H. 55 to 87%. The physiological loss in weight was found to decrease with increasing quantities of solids in the wax emulsion. At the end of 20 days storage, the percentage wastage due to disease was significantly lower in wax emulsion with 2.7% solids. These treatments increased the shelf-life in non-refrigerated storage about 50% and drying into chips in order to improve the shelf-life (Thomas, 1986). Cold storage and application of skin coatings to control the ripening processes and reduce aging and water loss have been investigated in mango in India in the past two decades, to develop efficient storage practices. According to Gandhi (1955) fully mature

---

*Corresponding author. E-mail: onifunmilayo@yahoo.com

Table 1. Mean weights of stored Mango fruits given four different treatments.

Day	Average weight of stored mango fruits (g)			
	WP	WUP	UWP	UWUP
0	187.7±27.0	181.6±16.9	173.2±29.0	181.2±14.5
1	182.8±26.6	173.3±16.0	169.5±27.9	175.0±14.8
5	178.0±26.1	165.4±15.3	164.2±27.3	168.4±14.4
6	174.5±26.7	164.8±18.5	162.1±27.2	165.5±14.3
Percent weight loss on day 6 (%)	7	9.3	6.4	8.7

Alphonso mangoes could be stored at 45 to 48°F for seven weeks but below this temperature range, the fruit is injured resulting in failure to ripen properly when shifted to room temperature. Fruits kept in perforated polyethylene bags ripened very steadily at low temperature but unpackaged mangoes suffered badly from immediate rotting on removal from cold storage, due probably to chilling injury (Sundararaj et al., 2006). Several varieties, found otherwise quite suitable for cold storage showed chilling injury at 40 to 45°F. According to Singh (1990) different varieties showed a variation in the critical temperature which lies between 40 to 45°F and wastage due to chilling may be avoided by keeping the fruits above this range.

Methods of extending shelf life of mango fruits such as irradiation, waxing, processing into jam and drying into chips, are expensive and may not be readily available to local farmers. Mangoes are readily available in Nigeria during the harvest seasons, however since cold-storage for preserving quality is potentially problematic, growers are forced to resort to 'distress sales'. Continued postharvest losses are evident in markets, stores and dumpsites. There is a need to develop cheap and commonly available technology for extending the shelf life of these produce at least to manage the movement in the market chains and control the losses. This work studied the storability of mango as influenced by washing and packaging in perforated polythene bags.

METHODOLOGY

Mango fruits (Alphonso variety) were bought from a popular fruit market in Ibadan, Nigeria and were given four different treatments. Wholesome fresh matured fruits were respectively shared into 4 sub lots, each consisting of 10 fruits, marked 1- 10. The 4 sub lots were given different treatments for storage, namely: Treatment A, samples were washed in distilled water and packaged in polythene bags (WP). Treatment B, samples were hand washed in bowls and air dried to remove surface moisture but not packaged in polythene bags (WUP). Treatment C samples were left unwashed and packaged in polythene bags (UWP), while the fourth treatment was left unwashed and unpackaged (UWUP) serving as the control.

Each of the treatment was placed in perforated cartons and kept in a fruit shed at ambient (31°C, 70% r.h.). All treatments were weighed on day 0, 1, 5, and 6, of storage and changes in weight were computed. The amount of decay was assessed after storage on the basis of aggregate percentages of surface areas visibly infected by Anthracnose and cannot be consumed any longer. Percentage green skin areas were rated as : matured green (90 to 100% green),breaker (80 to 90% green), quarter ripe (70 to 80% green), half ripe (45 to 65%) and full ripe (less than 40% green).

RESULTS AND DISCUSSION

The washing effect increased the weight loss of unpackaged and packaged mangoes by (9.3 and 8.7%) respectively but were not significantly different at p=0.05 (Table 1). Visible deterioration of the mangoes was observed as rotting of fruits. Generally the weight of mangoes decreased with storage time irrespective of treatment which is attributable to continued catabolism at the ambient storage. Therefore washing will be an unnecessary expense for farmers who will be supplying a packing house. This implies that if one must package mango, one may not wash if aesthetics is not a factor. Packaging of mangoes reduced weight loss for Washed mangoes (WP) compared with Unpacked-Washed mango (WUP). The percentage weight loss in Washed-Packaged mangoes (WP) is 7% while that of Washed-Unpackaged (WUP) is 9.3%. Mango fruits possess natural waxes. This must have been removed by washing. Packaging helped to reduce weight loss of Washed mango fruits. Therefore, if mango fruits will be washed it should be packaged to limit weight loss in storage. The weight loss in the unwashed and unpackaged (UWUP) treatment was 8.7% compared with the washed unpackaged (WUP) 9.3%. Unwashed-unpacked mango had reduced weight loss in storage and this was attributable to the natural waxes on the fruits. Therefore, if mango will be Unpackaged it should be left unwashed. The percentage weight loss in Unwashed-packaged (UWP) mango is 6.4% while that of Unwashed-Unpacked (UWUP) is 8.4%. Packaging will reduce weight loss in unwashed mangoes if there are no facilities to wash. Both natural wax and packaging will helped to reduce weight loss (Tables 1 and 2).

Conclusion

Packaging mango fruits in polythene bags can extend the shelf life of the fruits and thereby minimise losses.

Table 2. Storage qualities of the mango fruits in four different treatments at the end of 7th day.

Quality	WP	WUP	UWP	UWUP
Titratable acidity	0.76±0.29	0.36± 0.04	0.87±0.41	0.66±0.13
Colour (green) (%)	20	40	60	30
Rottenness (%)	80	60	30	70

WP - Washed-packaged; WUP - washed-unpackaged; UWP - unwashed-packaged; UWUP-unwashed-unpackaged.

Future work should be tried with larger number of samples possibly this will reveal the significance.

ACKNOWLEDGEMENT

The authors appreciate the staff and management team of Nigerian Stored Products Research Institute for allowing us to use their facilities.

REFERENCES

Bayers M, Thomas P (1979). Radiation preservation of foods of plant origin III. Tropical fruits: banana, mangoes and papayas. Crit. Rev. Food Sci. Nutr., 23: 147-206.

Gandhi SR (1955). The Mango in India. Faran Bulletin 6 Indian Council for Agricultural Research, New Delhi, 64pp.

Marchand L (1869). Revision du Groupe des Anacardiacees.Bailliere, Paris.

Martinez BE,Guevera CG, Contreras, Rodriguez JR and Lavi (1997). Preservation of mango azucer variety *Mangifera indica* L. at different storage stages. Proceedings of the 5th International Mango Symposium, 2: 747-754

Mitra SK, Baldwin EA (1997). Mango. Postharvest physiology and storage of tropical and subtropical fruits.CAB International, Wallingford, UK, pp. 85-122.

Saleh NA, El-Ansari MA (1975). Polyphenolics of twenty local varieties of Mangifera indica. Planta Med., 28(2): 124-130.

Singh RN (1990) Mango. Series No.3 Indian Council of Horticulture, pp. 1169-1188.

Singh UP, Singh DP, Singh M, Maurya S, Srivastava JS, Singh RB, Singh SP (2004). Characterization of phenolic compounds in some Indian mango cultivars. Int. J. Food Sci. Nutr., 55(2): 163-169.

Subramanjam H, Krishnamurthy S, Parpia HA (1975). Physiology and biochemistry of mango fruit. Adv. Food Res., 21: 223-305.

Sundararaj JS, Muthuswamy S, Sadasivam R (2006). Effect of chilling injury on texture and fungal rot of mangoes American J. Food Technol., 1(1): 52-58.

Thompson, AK (1971). The storage of mango fruits. Trop. Agr. (Trindad), 48(1): 63-70.

Whitmore TC (1975).Tropical Rain Forests of the Far East Clarendon, Oxford, UK.

11

Assessment of the quality of crude palm oil from smallholders in Cameroon

Ngando Ebongue Georges Frank[1]*, Mpondo Mpondo Emmanuel Albert[2], Dikotto Ekwe Emmanuel Laverdure[2] and Koona Paul[1]

[1]Centre spécialisé de Recherche sur le palmier à huile de La Dibamba, BP 243 Douala, Cameroun.
[2]University of Douala, Faculty of science, Department of Biochemistry, P. O. Box 24 157 Douala, Cameroon.

Oil palm is the highest oil producing plant, with an average yield of 3.5 tons of oil/ha/year. In 2006, palm oil became the world's most important edible oil with 37 million tons produced, accounting for 25% of the total oils and fats production. In Cameroon, palm oil meets 80% of total edible oil needs and it is estimated that 30% of crude palm oil (CPO) production is provided by none industrial oil mills. However, previous studies tend to demonstrate that, there was a problem with the consumption of CPO with respect to food safety. In the present study, the effect of processing methods and storage time on some physico-chemical parameters of Cameroonian CPO was assayed. Results showed that, lipid peroxidation and oil acidity significantly increased in palm oil samples from none industrial oil mills during the first four weeks of storage; thus making them unfit for human consumption. Both processes were enhanced by moisture and impurity levels of the oil at the outset above 0.1 and 0.01% respectively. Despite its many other benefic properties, this is a clear indication that, CPO from inappropriate extraction processes is becoming a real problem in sub-Saharan African countries regarding food safety.

Key words: Food safety, oil acidity, peroxide value, crude palm oil, FFA, oil palm.

INTRODUCTION

Oil palm is by far the highest oil producing plant, with an average of 3.5 tons of oil/ha/year. Extracted from the mesocarp of the fruit, crude palm oil (usually referred to as CPO) represents 95% of the total oil production of the oil palm which also provides palm kernel oil. Since 2006, palm oil has become the world's most important edible oil with about 37 million tons produced that year, representing 25% of the total oils and fats production (Oil world Ista GmbH Mielke, 2007). Unlike palm kernel oil which has wide applications in the oleochemical industry, palm oil is used mainly for edible purposes. It is reported to be the richest natural source of carotenoids in terms of retinol (provitamin A) equivalent (Vaughan, 1990; May, 1994). In some tropical countries, such as Cameroon, it contributes up to 80% of the total edible oil needs

(Hirsch, 1999). Following the drop in the early nineties of the prices of cocoa and coffee which were then the major commercial farming crops in Cameroon, many smallholders turned out planting oil palm. This fact is clearly illustrated by the amount of germinated oil palm seeds purchased by small and medium size farmers at the Centre for oil palm research of La Dibamba (Cameroon) which rose from 20% of the total production in 1996 to an average of 60% during the past ten years. From these data, it is estimated that about 5, 000 ha of oil palm were planted by small and medium size farmers each year during the last decade, making a total of about 90, 000 ha for the none industrial palm grove in Cameroon (Bakoume and Mahbob, 2006).

However, this rush to oil palm growing also raised difficulties, mainly with regard to the quality of palm oil produced by these smallholders. For those who are located in the neighbourhood of industrial oil mills, fresh fruit bunches (FFB) are delivered to the latter for processing. But in many cases, palm plantations are very far from the industrial oil mills, small and medium scale farmers have to process the FFB themselves. In industrial oil palm estates, FFB are harvested when the fruits are at optimum ripeness and handled with care to avoid bruising. The FFB are quickly sterilized, threshed and digested. CPO is extracted from the digested fruits hydraulically or using a screw press, clarified and dried. There are some subtle differences between oil extraction methods used by smallholders and the process in force in industrial oil mills. Once harvested, FFB are allowed to ferment for a time (1 to 6 days) at ambient temperature, so as to allow easy separation of fruits from the bunch. The fruits are then boiled for some hours. In the traditional method, the boiled fruits are pounded into a pulp using a mortar and pestle or trampled underfoot, and the oil is separated by adding water and skimming it off. In many modern methods, manual or motorized screw presses are used to squeeze out the oil from boiled fruits. The oil is finally heated to remove the residual water. It is estimated that none industrial oil mills contributes 30% to Cameroon's national crude palm oil production.

In this regard, and owing to the fact that it is a perishable food item, none industrial crude palm oil must fulfill the requirements of quality applicable to all oils and fats dedicated to human consumption. CPO from the traditional oil extraction method is highly sought after in local markets, due to its better sensory qualities (red color, taste, smell) which make it an irreplaceable ingredient of many local recipes. However, when some physico-chemical parameters generally used as indicators of the quality of dietary oils and fats regarding food safety were assayed, it was demonstrated by some authors that CPO samples from the traditional oil extraction methods were of lesser quality compared to the CPO from industrial oil mills (Coursey, 1966; Broadbent and Kuku, 1977; Aletor et al., 1990). In the present study, we are intending to assay through chemical analyses the quality of CPO produced by smallholders in the main oil palm growing areas of Cameroon. The effect of processing methods and storage on these chemical parameters will also be discussed.

MATERIALS AND METHODS

Collection of oil samples

Palm oil samples were collected in four oil palm growing areas of the South-west and Littoral Regions of Cameroon. This sampling was performed between January and April, what corresponds to the peak season for FFB production. For samples from group I, palm oil was extracted by the traditional method, with the boiled fruits being trampled underfoot. For samples from groups II, III and IV, palm oil was extracted by improved processes using manual or motorized screw presses. Samples from the control (C) were obtained from an industrial oil mill (SOCAPALM). The three oil extraction processes involved are summarized in the flow charts in Figure 1. All samples were collected in 0.5 L PVC screw capped bottles filled to the

maximum and closed hermetically. The samples were kept at ambient temperature, and transported to the laboratory for analyses within 1 to 3 days after collection.

Chemical analyses

For each sample, moisture and FFA content were assayed, alongside peroxide value and impurity level. The FFA content was determined by titrating the alcoholic solution of the oils with a 0.1 N solution of sodium hydroxide using phenolphthalein and alkaline blue as indicators. The FFA content was then expressed as a percent of palmitic acid, the major fatty acid in palm oil. The peroxide value was determined by titrating chloroform/glacial acetic acid/saturated KI solution of the oil with an aqueous solution of sodium thiosulphate using starch as indicator. Moisture content was determined by the gravimetric method of air-oven drying to constant weight at 105°C. For the assessment of the impurity level, oil samples were mixed with an excess of hexane then filtrated. The residue on the filter was then washed with hexane and oven dried to constant weight at 105°C. All chemical analyses were determined by methods of the Association Française de Normalisation (AFNOR, 1988).

Effect of storage on the chemical parameters of palm oil

For this purpose, two chemical parameters were chosen, namely oil acidity and peroxide value. Oil samples were assessed every two weeks, carefully locked again and stored at room temperature.

RESULTS

The FFA content, peroxide value, moisture and impurity level of palm oil samples from different areas and/or extraction processes are illustrated in Table 1. There is an important variability between samples from different groups and within groups for all parameters assessed. The highest values for FFA content and peroxide value were recorded for samples from group III. For moisture content, the highest values were recorded for samples from groups II and III, whereas group IV samples had the highest values regarding the impurity level. Except for the peroxide value, the industrial palm oil sample used as control showed the lowest value for all the parameters assessed.

The effect of storage time on FFA content is shown by Figure 2. A continuous increase of the FFA content was recorded for all groups of samples during the first 4 weeks of storage. The highest rate was attributed to samples from group I which increased by 110% within this period, moving from 6.4 to 13.4% FFA content. On the other hand, the control sample increased by only 15% within the same period. The FFA content subsequently decreased significantly between the fourth and the sixth week of storage for all groups of samples, and for group I samples the decrease even continued during the sixth and eighth weeks. No significant variations were noticed during the sixth and the tenth weeks of storage (eighth and tenth weeks respectively for samples from group I).

The effect of storage time on the peroxide value of

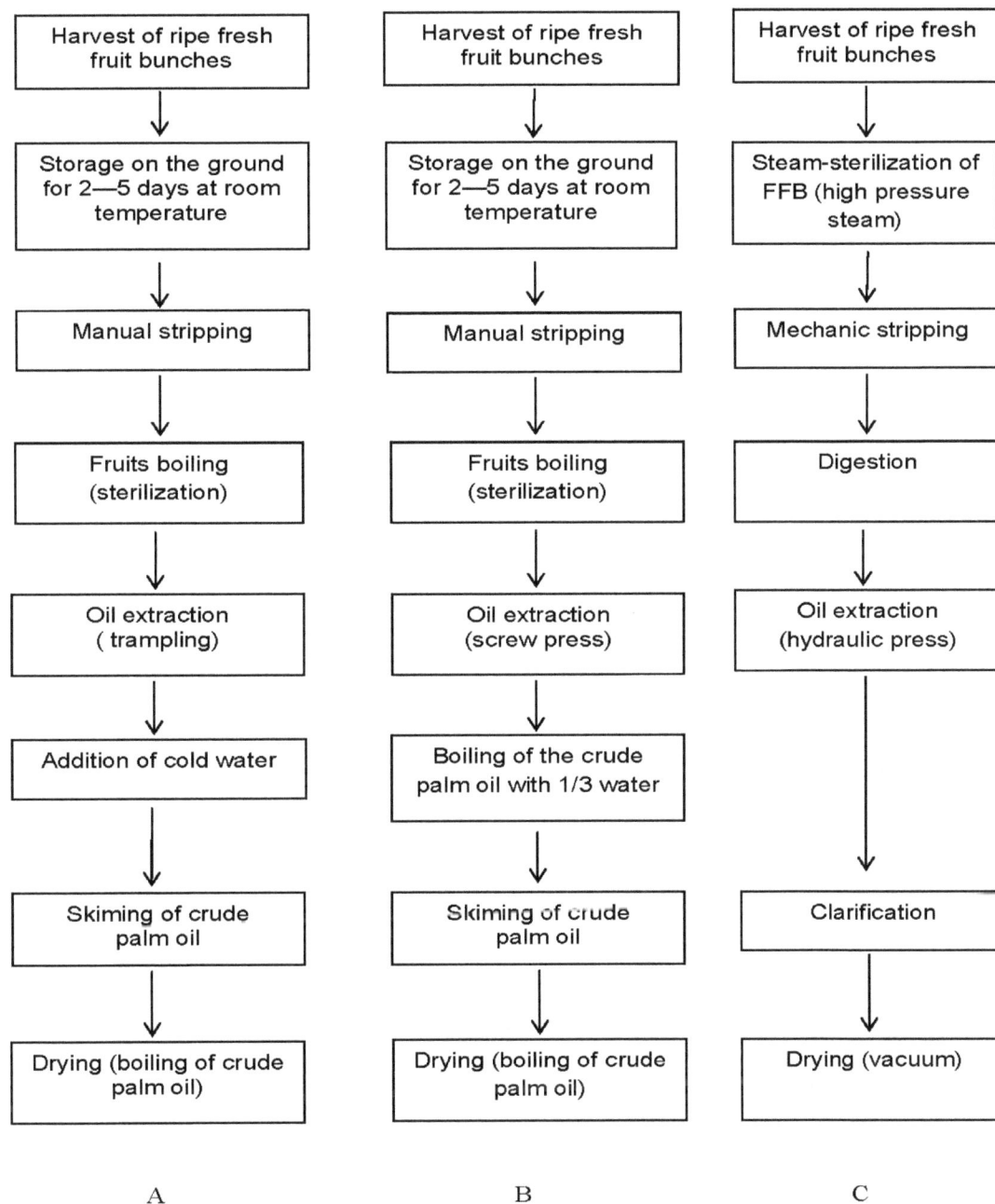

Figure 1. Flow charts of the traditional (A), semi-mechanized (B) and industrial (C) crude palm oil extraction processes from which samples of this study were obtained.

Table 1. Effects of extraction processes on chemical parameters of crude palm oil samples.

	Extraction process	Number of samples	FFA (%)	Peroxide value (Meq O_2/kg)	Moisture (%)	Impurity level (%)
Group I	Traditional	7	6.39 ± 3.20	2.07 ± 0.91	0.22 ± 0.04	0.11 ± 0.09
Group II	Semi-mechanic	4	7.72 ± 2.35	2.87 ± 0.91	0.32 ± 0.15	0.05 ± 0.03
Group III	Semi-mechanic	9	10.26 ± 4.56	5.71 ± 4.45	0.30 ± 0.05	0.08 ± 0.07
Group IV	Semi-mechanic	7	5.00 ± 1.91	1.48 ± 0.55	0.23 ± 0.05	0.31 ± 0.02
Control (C)	Industrial	1	4.71 ± 0.60	2.67 ± 0.60	0.08 ± 0.02	0.01 ± 0.00

Values are expressed as mean of a triplicate ± standard deviation.

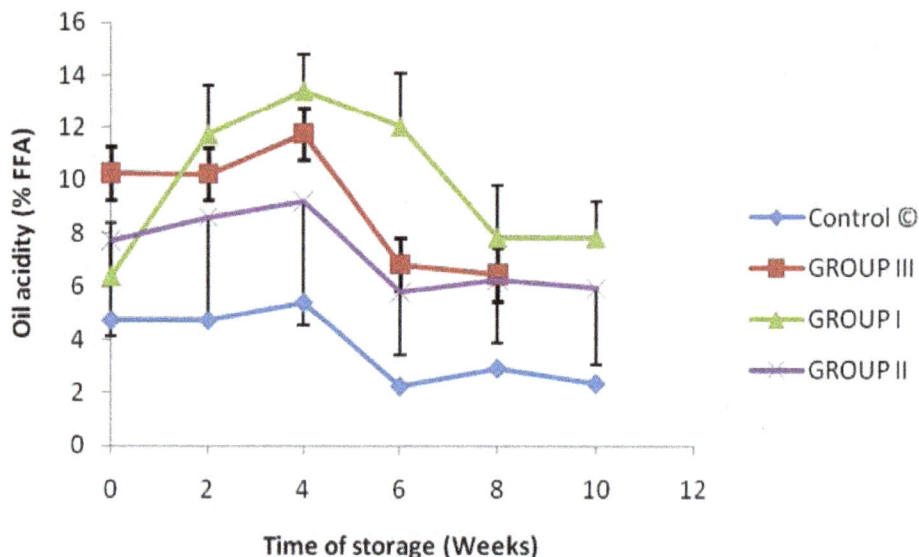

Figure 2. Changes in oil acidity during storage of crude palm oil samples from the traditional (group I), semi-mechanized (groups II and III) and industrial (Control C) oil extraction processes.

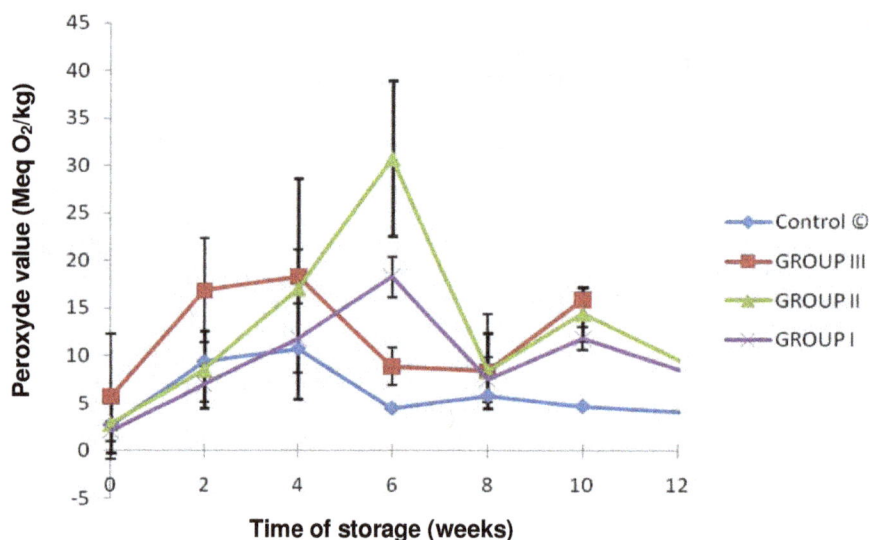

Figure 3. Changes in peroxide value during storage of crude palm oil samples from the traditional (group I), semi-mechanized (groups II and III) and industrial (Control C) oil extraction processes.

palm oil samples from different groups is shown by Figure 3. As it was the case for FFA content, a continuous increase was observed for all groups of samples during the first four weeks of storage. But unlike the FFA content where the increase was only two fold for the highest value, peroxide values increased by three to six folds within the same period for all groups of samples. For samples from groups II and III, peroxide value continued to increase between the fourth and sixth week, and final ten and eight folds increases respectively were recorded within the first six weeks of storage. A significant decrease was subsequently recorded between the fourth and the sixth weeks (sixth and eighth weeks respectively) for the control and group I samples (group II and III samples respectively). Thereafter, no significant variations were observed during the last weeks of storage.

DISCUSSION

FFA content is the most used criterion for determining the

quality of palm oil, as it must not exceed 5% (expressed as palmitic acid) according to Codex Alimentarius/ FAO/WHO norms (2005). Fatty acids are generally present in oils as part of triacylglycerol molecules. The presence of free fatty acids moieties in palm oil is an indication of the impairment of oil quality. This process is essentially attributed to an active lipase present in the mesocarp of the oil palm fruit and which is responsible for the hydrolysis of triacylglycerols (Henderson and Osborne, 1991; Sambanthamurthi et al., 1995; Ngando et al., 2006). The lipase is activated at maturity upon bruising and/or wounding of the fruit. According to Desassis (1957), 15 min are enough to hydrolyze 40% of the triacylglycerols of a bruised ripe fruit. However, this endogenous lipase activity was found to be very variable, and some lines with very low lipase activity were identified. The crude palm oil extracted from the fruits of these low lipase activity lines also showed a low FFA/oil acidity content (Ngando et al., 2008).

FFA can also be generated to some extent by contaminating lipases from microorganisms (Hiol et al., 1999; Houria et al., 2002). In order to limit lipase activity, fresh fruit bunches must be handled gently and above all, processed rapidly after harvest. For all the groups of samples, values were above the 5% limit, regarding the FFA content except for the control sample from industrial oil mill. This is not surprising, as only industrial oil mills generally process FFB shortly after harvest. The steam-sterilization of FFB at high temperature rapidly inactivates the lipase, thus limiting subsequent FFA accumulation in palm oil from industrial oil mills. For the traditional and semi-mechanized processes (Figure 1), a prior fermentation of FFB is usually carried out, in order to enable fruits easily detach from the bunch. In some cases, the harvested FFB can be kept at room temperature for one week or more before they are processed. The harmful effect of fermentation is the continuous build up of FFA in the mesocarp of the fruit under the action of the lipase. Once the fruits are processed, the lipase is no more active, but the FFA content of the resulting palm oil may also increase during storage as a result of autocatalytic hydrolysis. In that case, FFA acts as catalysts for the reaction between triacylglycerols and water to generate more FFA. Results from Figure 2 clearly illustrates this process, as the sharp increase of the FFA content of palm oil samples during the first four weeks of storage can be attributed to autocatalytic hydrolysis. Figure 2 also clearly showed that, the FFA content of the control sample from the industrial oil mill recorded the lowest increase during the first four weeks of storage, as this sample also had the lowest FFA and moisture contents at the outset. A positive correlation (0.76) was found between FFA and moisture content data from Table 1. Therefore, it is imperative to limit FFA and moisture contents of CPO prior to long term storage, as autocatalytic hydrolysis is unlikely to occur or is very limited below the 0.1%

moisture content limit recommended by Codex Alimentarius/FAO/WHO norms. Data from Figure 2 also indicate a sharp decrease of FFA content between the fourth and eighth weeks of storage. This does not necessarily express a real decrease, as unsaturated FFA may undergo subsequent chemical reactions such as peroxidation and generate secondary products which could not be detected while assaying oil acidity. FFA are more likely to undergo peroxidation reaction than fatty acids within the triacylglycerol molecule, and an increase of the FFA content may also enhance the peroxidation process.

Another important parameter used to assess the quality of palm oil is the peroxide value which is an indicator of the level of lipid peroxidation or oxidative degradation. In this process involving unsaturated fatty acids, specially reactive hydrogen atoms from methylene ($-CH_2-$) groups adjacent to double bonds undergo a chain reaction mechanism involving free radicals as intermediates and generating lipid peroxides as end products. These lipid peroxides later undergo additional chain cleavage at the level of the hydroperoxide group to form secondary oxidation products such as short chain aldehydes and products bearing ketone, epoxy or alcohol groups responsible for the rancid smell and taste of the oil. Peroxide value is used to assess the stability or rancidity of fats by measuring the amount of lipid peroxides and hydroperoxides formed during the initial stages of oxidation and thus, estimate to which extent spoilage of a dietary oil (expressed by the level of rancidity) has advanced. Beside these visible harmful effects on the sensory quality of the oil, peroxidation also makes the oil dangerous for human health, as the free radicals generated by this process are proven to be carcinogenic (Rossel, 1999). All the groups of samples in Table 1 met with the Codex Alimentarius/FAO/WHO norms which recommend a maximum peroxide value of 10 meq O_2/kg palm oil.

However, this was no longer the case after the first four weeks of storage, as sharp increases were recorded for almost all groups of samples except for the control from industrial oil mill. The means recorded at this stage for the 3 groups of samples (11.84, 17.08 and 18.35 meq O_2/kg for group I, II and III respectively) were above the Codex Alimentarius/ FAO/WHO 10 meq O_2/kg limit. Peak values obtained after 6 weeks of storage for samples from group II were even 3 fold higher than this recommended norm. Oxidation may be significantly enhanced by the impurity level, as components such as resins, hydroxyl-fatty acids, carbohydrates and oxidized fatty acids are quantified among impurities (AFNOR, 1988). In this regard, the maximum value recommended by FAO/WHO norms for insoluble impurities is 0.01% (w/w). Thus, although samples from groups I and II had almost the same peroxide value as the control (Table 1), they suffered higher increases during the first four weeks of storage, probably because of their higher impurity levels. The decrease of peroxide value observed between the fourth and eighth weeks of storage may not necessarily reflect a decrease of the amount of hydroperoxides formed, as the latter are transitory intermediates which undergo additional chain cleavage to generate secondary oxidation products as stated earlier. DeRouchey et al. (2000) showed that, primary and secondary oxidation products appeared during fat oxidation following Gaussian curves overlapping each other. In fact, rancidity is a qualitative state that is not chemically defined and is difficult to quantify. As a result, numerous methods have been developed to assess the amount of various intermediates or products of oxidation, of which the Peroxide Value assay.

However, these products are most of the time unstable, making it difficult to estimate the correct level of fat oxidation at a given period of time. Peroxide Value can only measure the amount of hydroperoxides formed but is not appropriate to quantify the amount of secondary oxidative products such as short chain aldehydes, ketones and epoxydes. It can therefore measure only current or recent oxidation. A second chemical parameter named Anisidine Value is used to provide information on oxidative history of a fat, as it is appropriate to quantify secondary oxidative products essentially made up of high molecular weight saturated and unsaturated carbonyl compounds. The correct estimate of the oxidative process of dietary oils within a long period of storage can be given by a parameter known as Total Oxidation or TOTOX value which can be calculated based on Peroxide Value (PV) and Anisidine Value (AV) using the equation:

$$TOTOX = 2PV + AV.$$

Thus, TOTOX value is a quantification of the precursor non volatile hydroperoxides present in the oil plus any further secondary oxidation compound formed during storage. However, the use of the sole peroxide value in this study is fully justified, as it was proven sufficient to monitor the significant increases above authorized limits in peroxydation levels of CPO samples, after only four weeks of storage. Data from Table 1 were generally in accordance with those obtained by Aletor et al. (1990) and Onwuka and Akaerue (2006) in similar studies on the influence of extraction processes on Nigerian crude palm oil.

Conclusion

Our results showed that, peroxide value and FFA content may significantly increase in none industrially processed crude palm oil samples from smallholders in Cameroon, during the first weeks of storage. Both processes are enhanced by high moisture and impurity levels of the oil at the outset, as it was demonstrated that oils with low moisture and impurity levels suffered very slight changes. Considering that this non-industrial palm oil may take at least one week or in some cases several months to move from the extraction site to the market and finally to the consumer, it is likely that, a significant increase of peroxide value and FFA content may occur during this period, thus making the oil unsuitable for human consumption.

In this regard, peroxide value and oil acidity are very useful indices to control dietary oils safety and quality. Given that crude palm oil is a major ingredient of many recipes in sub-Saharan Africa, this is a clear indication that palm oil from smallholders may become a real problem in these countries regarding food safety if nothing is done, despite its many other benefic properties.

REFERENCES

AFNOR (1988). Recueil des normes françaises sur les corps gras, graines oléagineuses, produits dérivés, 4e édition. Association Française de Normalisation, Paris.

Aletor VA, Ikhena GA, Egharevba V (1990). The quality of some locally processed Nigerian palm oils: An estimation of some critical processing variables. Food Chem., 36: 311-317.

Bakoume C, Mahbob BA (2006). Cameroon offers palm oil potential. Oils and fats int., 3: 25-26.

Broadbent JA, Kuku FO (1977). Studies on mould deterioration of Mid-West Nigerian palm fruits and pre-storage palm kernels at various stages of processing. Rep. Nig. Stored Prod. Res. Inst. Tech. Report, 6: 49-53.

Commission du Codex Alimentarius/FAO/OMS (2005). Normes alimentaires pour huiles et graisses. CODEX-STAN 210, FAO/OMS.

Coursey DG (1966). Biodeteriorative processes in palm oil stored in West Africa. Soc. Ehem. Indi Monograph, 23: 44-56.

DeRouchey JM, Hancock JD, Hines RH, Cao H, Maloney CA, Dean DW, Lee DJ, Park JS (2000). Effects of rancidity in choice white grease on growth performance and nutrient digestibility in weanling pigs. In: Goodbank B, Tokach M, Dritz S. 2000 swine day report of progress, Kansas state university. Manhattan, November 15-16 2000, pp. 83-86.

Desassis A (1957). Palm oil acidification. Oléagineux, 12: 525-534.

Henderson J, Osborne DJ (1991). Lipase activity in ripening and mature fruit of the oil palm. Stability in vivo and in vitro. Phytochemistry, 30: 1073-1078.

Hiol A, Comeau LC, Druet D, Jonzo MD, Rugani N, Sarda L (1999). Purification and characterization of an extracellular lipase from a thermophilic Rhizopus oryzae strain isolated from palm fruit. Enzyme Microb. Technol., 26: 421-430.

Hirsch RD (1999). La Filière Huile de Palme au Cameroun dans un Perspective de Relance. Agence Française de Developpement, Paris (France), p. 79.

Houria A, Comeau L, Deyris V, Hiol A (2002). Isolation and characterization of an extra-cellular lipase from Mucor sp. strain isolated from palm fruit. Enzyme Microb. Technol., 31: 968-975.

May YC (1994). Palm oil carotenoids. United nation university press. Food Nutr. Bull., p. 15.

Ngando EGF, Dhouib R, Carriere F, Amvam Zollo PH, Arondel V (2006). Assaying lipase activity from oil palm fruit (Elaeis guineensis Jacq.) mesocarp. Plant Physiol. Biochem., 44: 611-617.

Ngando EGF, Koona P, Nouy B, Carriere F, Amvam ZPH, Arondel V (2008). Identification of oil palm breeding lines producing oils with low acid value. Eur. J. Lipid Sci. Technol., 110: 505-509.

Oil world Ista Mielke GmbH (2007). Oil world annual 2006. Hamburg (Germany).

Onwuka GI, Akaerue BI (2006). Evaluation of the quality of palm oil produced by different methods of processing. Research J. Biol. Sci., 1: 16-19.

Rossel JB (1999). Measurement of rancidity. In Rancidity in foods. Ed By Allen JC and Hamilton RJ, UK. Aspen publishers, pp. 22-51.

Sambanthamurthi R, Oo KC, Parman SH (1995). Factors affecting lipase activity in *Elaeis guineensis* mesocarp. Plant Physiol. Biochem., 33: 353-359.

Vaughan JG (1990). The structure and utilization of oil seeds. Chapman and Hall ltd. London, p. 187.

Fracture resistance of palm kernel seed to compressive loading

Ozumba Isaac C* and Obiakor Sylvester I

Processing and Storage Engineering Department, National Centre for Agricultural Mechanization (NCAM), P. M. B. 1525, Ilorin, Nigeria.

Quasi-static compression tests were carried out on treated single palm kernel seeds to study the effects of temperature, moisture content and loading position on the rupture force, deformation and toughness of the kernels. The levels of moisture content and temperature considered were 5, 7 and 10% w.b and 70, 90 and 110°C each at the horizontal and vertical loading position. Twenty palm kernel seeds were tested at each moisture and temperature level in both horizontal and vertical loading positions making a total of 480 kernels that were individually measured and tested. The average compressive force required to rupture a palm kernel seed under compressive loading decreased as the moisture content of the seed increased from 7 to 10% (wb), while the corresponding deformation increased from 1.18 to 1.43 mm but decreases from 1.18 to 1.03 mm as the temperature increases. Maximum toughness occurred at 7% moisture content and 70°C temperature respectively, indicating the optimum moisture content and temperature for absorbing compressive energy. The value of toughness is an important indicator of the ability of the palm kernel to resist mechanical damage during loading. The loading position has significant effect on the rupture force and should be considered when palm kernel seeds are being loaded.

Key words: Resistance, compression loading, rupture, palm kernel, determination, toughness.

INTRODUCTION

During processes like planting, harvesting storage, processing and transportation, agricultural products are often subjected to mechanical forces. These forces most of the time results to deformation of the crops. Deformation may either sometimes be enough to result to cutting, pressing, crushing, tearing or just very small especially during harvesting and threshing to avoid a more severe damage. Thus, mechanical strength like compression tensile strength and shear strength of agricultural products plays an important role in harvesting, storage and processing of the crop. Sitkei (1986) reported that most agricultural products are visco-elastic in nature, they respond differently to tensile or compressive forces and also behave differently when they are subjected to vibration. Therefore, a fundamental knowledge of agricultural product behaviour under

mechanical forces is essential in determining the power requirement for different operations like cutting, crushing, pressing, milling etc. It is generally agreed that the oil palm (*Elaeis guineensis*) originated in the tropical main forest of West Africa. The main belt runs through the southern latitudes of Cameroon, Cote d' Ivoire, Ghana, Liberia, Nigeria, Sierra Leone, Togo and into the equatorial region of Angola and the Congo (FAO, 1993). Because of its economic importance as a high-yielding source of edible and technical oils, the oil palm is now grown as a plantation crop in countries with high rainfall in tropical climates within 10° of the equator. The palm tree bears its fruits in bunches (Figure 1) which vary in weight from 10 to 40 kg. The individual fruit ranging from 6 to 70 g, is made up of an outer skin (the exocarp), a pulp (mesocarp) containing the palm oil in a fibrous matrix, a central nut consisting of a shell (endocarp) and the kernel, which itself contains an oil, quite different to palm oil, resembling coconut oil (FAO, 1993) (Figure 2).

Extraction of oil from palm kernels is generally different

Figure 1. Fresh palm fruits in bunch.

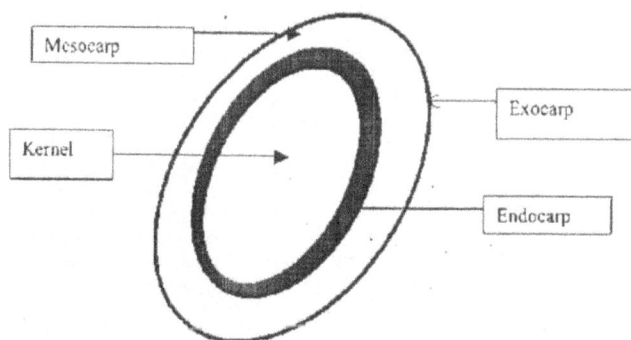

Figure 2. Cross –section of an individual palm fruit showing the main parts.

seed expeller or petroleum derived solvent (FAO, 1993). Several researchers have investigated the physical and mechanical properties of some crops considered relevant to the design of suitable machines and equipment for their production and processing (Adebayo, 2004) carried out a compression test on Dura varieties of the palm nut in order to determine the force required for cracking the palm nuts. Dev et al. (1982) investigated the size and shape of sorghum as essential properties for the analysis of the behaviours of grains during handling, storage and processing. Paulsen (1978) studied the average compressive strength, deformation and toughness of soybean seed coat rupture under quasi-static loading. Ezeaku et al. (1998) studied the measurement of the resistance of Bambara groundnut seed to compressive loading, while Makanjuola (1972) carried out a study on some of the physical properties of melon seeds.

Such similar work appears not to have been done on palm kernel seeds. A similar information on some of its mechanical properties would be essential in the design of equipment and system for the loading and processing of palm kernel into palm kernel oil. Sequel to the aforementioned, a study of the resistance of palm kernel seeds to compressive loading was undertaken. The objectives of this study are:

i) To determine the average compressive force, deformation and toughness at palm kernel seed rupture under compressive loading; and
ii) To determine the effect of loading position, moisture content and temperature on the compressive force, deformation and toughness at palm kernel seed rupture.

MATERIALS AND METHODS

Samples of Dura variety of palm kernel at moisture content of 4.7% w.b was obtained from an open market at Okuku in Osun State,

from palm oil extraction, and is often carried out in mills. The stages in this process comprise size reduction (that is grinding, the kernels into small particles), heating (cooking), and expression/extracting the oil using an oil

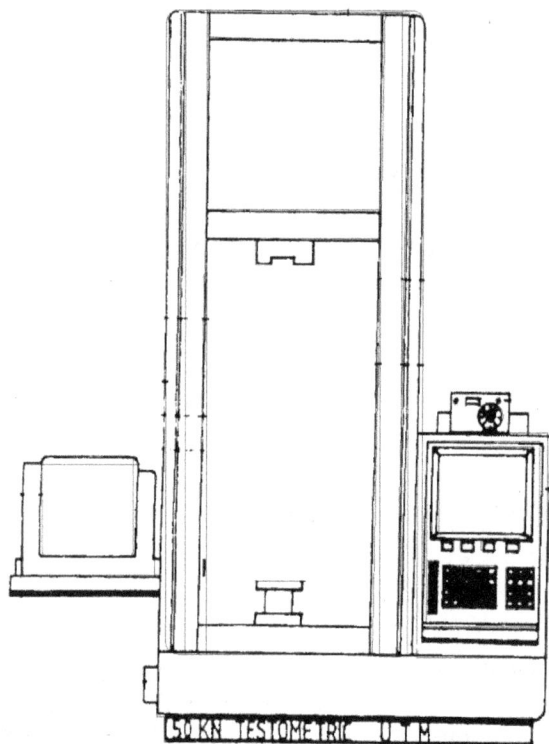

Figure 3. The instron universal testing machine (UTM).

Nigeria. The kernel was manually cleaned to remove dirt and other foreign materials in compliance with Nigeria Industrial Standard. Moisture contents were determined by oven drying respective ground samples (100 g) at 103°C for 6 h, as recommended for oil crops by Young et al. (1982). While the samples were conditioned to desired moisture levels 5, 7 and 10% (w.b) by adding distilled water as calculated from Equation (1) according to Hammond et al. (1997) as follows:

$$water \text{ to be added} = \left(\frac{100 - M_p}{100 - M_g} - 1 \right) x \ W_s$$

(1)

Where M_p = present moisture content, M_g = required moisture content, W_s = weight of samples in grammes.

After adding water, each sample was sealed in separate polyethylene bag and kept at 5°C in a refrigerator for a week to enable the moisture distribute uniformity throughout the samples. The seeds were weighted using an electronic weighing balance, while a vernier calliper was used in measuring the dimensions of the seed. The average minor diameter, intermediate diameter and major diameter of palm kernel seeds used for this investigation were 11.16, 13.4 and 21.11 mm respectively. Heating of the samples to the desired temperatures of 70, 90 and 110°C was achieved by spreading the kernel seeds in a closed container in a preset temperature controlled Gallenkamp or 440 oven at the different temperatures for 30 min in line with Olaniyan (2006). All ranges of temperature and moisture content were selected based on findings from literature review and preliminary tests.

All laboratory quasi-static compression tests were performed using the Instron Universal Testing Machine (UTM) equipped with a 50 kN load beam available at the National Centre for Agricultural mechanization Ilorin, Nigeria (Figure 3). The loading rate adopted throughout the test was 2.5 mm/min as recommended by Olaniyan

(2006) for oil-bearing seeds. Each palm kernel seed was loaded by hand between two circular seated nosepieces by chuck attachment pins. The force at bio-yield point and the corresponding deformation for each seed sample were read off from the force deformation curve. The effect of the palm kernel seed position was determined by loading the seed in either the horizontal or vertical loading position. Before loading, the palm kernel seeds were visually inspected. Those with visible cracks were not tested, thus results from the tests should be considered as the maximum force or toughness that a palm kernel seed sample could withstand and prior to rupture. The average room temperature throughout the duration of the test was 30°C.

A 3 x 4 x 2 factorial experiment in a randomized complete block design (RCBD) (Cox, 1952; Montgomery, 1976) was used to evaluate the effects of moisture content, temperature and loading position on rupture force, deformation and toughness of palm kernel seed under compressive loading. The factors included in the design were moisture content, temperature and loading position, with moisture content being blocked. The range of each factor was selected based on findings from literature review and preliminary investigations. Twenty palm kernel seeds were tested at each moisture level and each temperature level in each loading position, making a total of 480 seeds that were individually measured and tested. The volume of individual palm kernel was calculated from the principal dimensions earlier measured. Assuming ellipsoidal shaped material, the volume as given by Mohsenin (1978) and Sitkei (1986) is:

$$V = \frac{\wedge}{6}(abc)$$

(2)

Where a = major diameter (mm), b = intermediate diameter (mm), c = minor diameter (mm), and v = volume of palm kernel seed (mm^3).

Toughness is regarded as the energy absorbed by palm kernel up to the rupture point per unit of the kernel volume. The energy absorbed during compression was indicated by the data from the UTM as energy at break (rupture). Toughness was then calculated by using the formula according to Mohsenin (1978) as:

$$P = \frac{E}{V}$$

(3)

Where p = toughness (J/mm^3), E = energy (J), V = volume of kernel (mm^3).

The data recorded as means of twenty test samples in each test condition were subjected to analysis of variance test and Duncan's New Multiple Range Test was used to compare the means.

RESULTS AND DISCUSSION

All palm kernel seeds tested typically exhibited a force – deformation curve as shown in Figures 4 and 5. The bio-yield point in the force – deformation curves indicates the seed rupture point and this point was determined by a visual decrease in force as deformation increased. The force-deformation curves obtained in this investigation are similar to those obtained by previous researchers on different agricultural products (Ezeaku et al., 1998; Paulsen, 1978; and those reported in ASAE Standards ASAE S368.1 (1980). Tables 1 and 2 show a summary of the raw data obtained while the effects of temperature, moisture content and loading position on the average

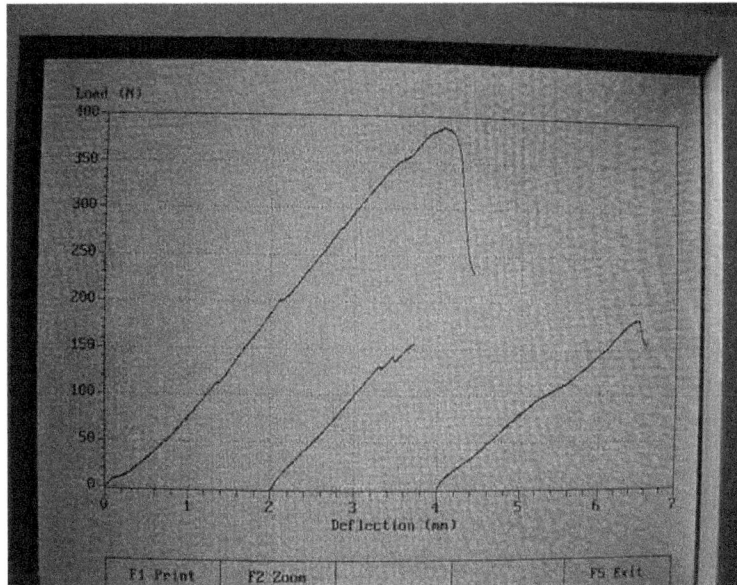

Figure 4. A typical load-deformation curve of palm kernel under horizontal compressive loading.

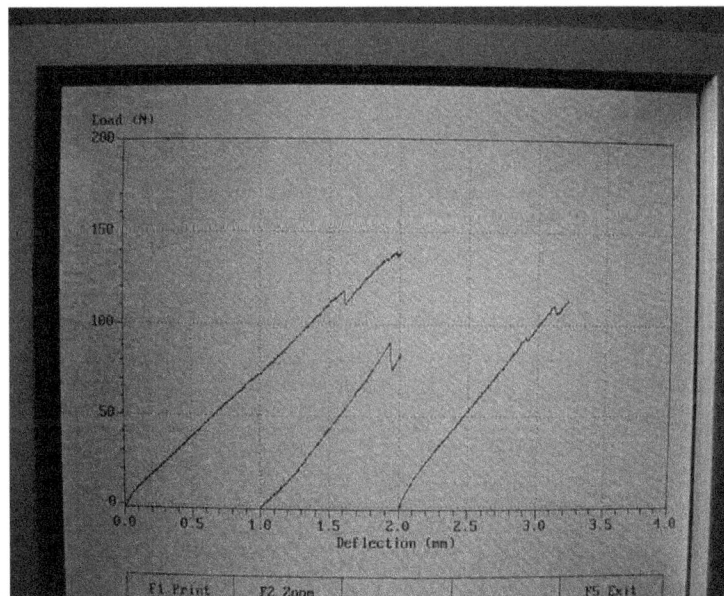

Figure 5. A typical load-deformation curve of palm kernel under vertical compressive loading.

force, deformation and toughness are shown in Tables 3, 4 and 5. These effects are thus discussed hereunder:

Effect of temperature

The effects of temperature on the average rupture force, deformation and toughness at palm kernel seed rupture are presented in Table 3. The table shows that average

rupture force and deformation at seed rupture are significantly affected by the seed temperature at P (0.05) level, while the seed temperature does not significantly affect the toughness. However, the table also shows that as the temperature increases, the deformation and toughness decreases respectively. This may be attributed to the weakening of the cell walls of the kernel as a result of high temperature, thus yielding easily to pressure.

Table 1. Summary of mechanical properties of palm kernel seed at rupture for three moisture levels and two loading positions.

Moisture content (%)	Loading position	Mechanical Properties		Toughness × 10⁻⁴ J/mm³
		Rupture force (N)	Deformation at rupture (mm)	
5	Horizontal	256.05 (38.46)	1.48 (0.38)	2.7938
	Vertical	181.23 (30.21)	1.310 (0.35)	2.4290
7	Horizontal	475.79 (78.12) [80]	1.24) (0.32)	4.2409
	Vertical	192.52 (31.64)	1.12 (0.30)	2.8034
10	Horizontal	323.56 (58.44)	1.23 (0.32)	2.4678
	Vertical	208.60 (46.26)	1.63 (0.40)	1.8171

SD Values in parenthesis are standard deviations, *each value is the means of 20 test samples.

Table 2. Summary of mechanical properties of palm kernel at seed rupture for three temperature levels and two loading position.

Temperature (°C)	Loading position	Mechanical Properties		Toughness × 10⁻⁴ J/mm³
		Rupture force (N)	Deformation at rupture (mm)	
70	Horizontal	252.49	1.22	1.33
	Vertical	159.57	1.13	1.12
90	Horizontal	599.18	1.14	1.2 8
	Vertical	217.53	1.02	1.06
110	Horizontal	210.33	1.07	1.136
	Vertical	205.25	0.98	1.072

* Each value is the means of 20 test samples.

Table 3. Effect of temperature on average rupture force, deformation and toughness of palm kernel seed

Temperature (°C)	Rupture force (N)	Deformation at rupture (mm)	Toughness at rupture × 10⁻⁴ J/mm³
70	206.03(a)*	1.18(c)	1.23(b)
90	408.36(b)	1.08(c)	1.17(a)
110	207.79(a)	1.03(c)	1.10(c)

*In each column, means with the same letters are not significantly different at P (0.05) based on Duncan's new multiple range test.

Effect of moisture content

The effect of moisture content on average rupture force, deformation and toughness is presented in Table 4. The table shows that the average rupture force, deformation and toughness are significantly affected by the moisture content of the palm kernel seed at P (0.05) level. The force required to initiate seed rupture increased when moisture content increased from 5 to 7% and later begins to decrease as the moisture content approaches 10%. This may be attributed to the fact that as the seed absorbs more moisture; it softens and tends to yield easily under pressure. Also, it may be indicated, within the range of moisture contents tested that palm kernel with lower moisture contents are generally more resistant to rupture, under compressive loading than those with higher moisture content.

Effect of loading position

The effect of loading position on average rupture force,

Table 4. Effect of moisture content on average rupture force, deformation and toughness of palm kernel seed.

Moisture content	Rupture force (N)	Deformation and rupture (mm)	Toughness × 10⁻⁴ J/mm³
5	218.64(a)*	1.409	2.61(a)
7	33.4.16(b)	1.18(b)	3.52(b)
10	266.08(a)	1.43(a)	2.15(c)

*In each column, means with the same letters are not significantly different at P (0.05) based on Duncan's new multiple range test.

Table 5. The effect of loading position on the average rupture force, deformation and toughness of palm kernel seed.

Loading position	Rupture force (N)	Deformation at rupture (mm)	Toughness × 10⁻⁴ J/mm³
Horizontal	352.90(a)*	1.23(a)	2.04(a)
Vertical	194.12(b)	1.20(a)	1.55(b)

*In each column, means with the same letters are not significantly different at P(0.05) based on Duncan's new multiple range test.

deformation and toughness is presented in Table 5. The table shows that rupture force and toughness are significantly affected by the loading positions, while the deformation is not significantly affected. This suggests that when palm kernel seeds are compressed, they absorb more energy at the horizontal position than at the vertical position. Thus, consideration should be given to the manner in which palm kernel seeds are loaded with respect to compression surfaces since loading position has significant influence on the rupture force.

Conclusions

The average compressive force required to cause palm kernel seed to rupture decreases as moisture content of the seed increased from 7 to 10% (w.b). Consideration should be given to the manner in which palm kernel seeds are loaded since loading position has significant influence on the rupture force. Deformation of palm kernel seed generally decreases as temperature increased and increases as moisture content decreases, suggesting that palm kernel seeds at lower moisture content are harder and less susceptive to rupture than palm kernel at higher moisture contents. Toughness is significantly different at all temperatures, suggesting that increase in temperature weakens the cell walls of the kernel, thus making it yield easily to pressure and consequently reducing the toughness of the kernel. However, the lower the moisture content of the kernel, the higher the toughness.

REFERENCES

Adebayo AA (2004). Development and Performance Evaluation of a Motorised Palm Nut Cracker. Proceedings of the Nigerian Institution of Agricultural Engineers (NIAE), 26: 326-330.
ASAE Standards ASAE368-1 (1980). Compression Test of Food Materials of Convex Shape. Agricultural Engineering year book. American Society of Agricultural Engineers, St. Joseph, Michigan.
Cox DR (1952). Physical Planning of Experiments. John Willey, New York.
Dev DK, Satwadhar PN, Ingle UM (1982). Effects of Variety and Moisture Content on Selected Physical Properties of Sorghum Grain. J. Agric. Engr., 19(2): 43-48.
Ezeaku CA, Akubuo CO, Onwualu AP (1998). Measurement of the Resistance of Bambara Groundnut Seed to Compressive Loading. J. Agric. Engr. Technol., 6: 12-18.
FAO (1983). Agricultural Services Bulletin, No. 22.
Makanjuola GA (1972). A Study of Some of the Physical Properties of Melon Seeds. J. Agric. Eng., 17: 128-137.
Mohsenin NN (1978). Physical Properties of Plants and Animals Materials. Gordon and Breach Science Publishers. 2nd Edition, New York.
Montgomery DC (1976). Design and Analysis of Experiments. John Willey, New York.
Olaniyan AM (2006). Development of Dry Extraction Process for Recovering Shea Butter from Shea Kernel. Unpublished Ph.D thesis, Department of Agricultural Engineering, University of Ilorin, Nigeria.
Paulsen MR (1978). Fracture Resistance of Soybean to Compressive Loading. Trans. Am. Soc. Agric. Eng., 21(6): 1210-1216.
Sitkei G (1986). Mechanics of Agricultural Materials. ELSERVIER. Amsterdam, Oxford, New York, Tokyo.
Young JH, Whitaker TB, Blankenship PA, Brusewrt GH, Tweger TM, Stede TL, Reanut(jr) NK (1982). Effect of Oven Drying Time on Peanut Moisture Determination. Trans. ASAE, 25(2): 491-525.

13

Best fitted thin-layer re-wetting model for medium-grain rough rice

M. A. Basunia[1]* and M. A. Rabbani[2]

[1]Department of Soils, Water and Agricultural Engineering, Sultan Qaboos University, P. O. Box 34, Al-Khod 123, Muscat, Sultanate of Oman.
[2]Department of Farm Power and Machinery, Bangladesh Agricultural University, Mymensingh 2202, Bangladesh.

Five commonly cited thin-layer rewetting models, that is, Diffusion, Page, Exponential, Approximate form of diffusion and Polynomial were compared for their ability to fit the experimental re-wetting data of medium grain of rough rice, based on the standard error of estimate (SEE) of the measured and simulated moisture contents. The comparison shows that the Diffusion and the Page models have almost the same ability to fit the re-wetting experimental data of rough rice. The Exponential, the Approximate form of diffusion and the Polynomial models have less fitting ability than the Diffusion and the Page models for the entire period (> 4 days) of re-wetting of 25 tests at different combinations of temperatures (17.8 to 45°C) and relative humidites (56.0 to 89.3%). The Diffusion and the Page models were found to be most suitable equations, the average SEE value was less than 0.0015 (dry-basis, decimal), respectively, to describe the thin-layer re-wetting characteristics of rough rice over a typical five day re-wetting. These two models can be used for the simulation of deep-bed re-wetting of rough rice occurring during ventilated storage.

Key words: Thin-layer, rough rice, re-wetting parameters, temperature, relative humidity.

INTRODUCTION

International rice prices have been increasing in recent years. According to the crop prospects and food situation report, at the end of March in 2008, the rice prices were nearly double those of the previous year (FAO, 2008). Increased rice price forced several countries to rebuild their rice stocks. Thus, the increased price makes rice storage and improvement of rice storage techniques increasingly important because the deterioration in rice quality during storage results in considerable economic losses (Genkawa et al., 2008a). Moisture content is one of the most important factors influencing the quality of rough rice during storage and it remains at a high level, 18 to 30% wet-basis, during the harvest and must be reduced to, 14 to 15% wet-basis, with an appropriate drying process (Hacihafizoğlu et al., 2008). In addition,

low-moisture rice is likely to crack due to rapid water adsorption (Siebenmorgen and Jindal, 1986; Banaszek and Siebenmorgen, 1990), and the cracking of rice results in breakage of kernels during milling and decrease head rice yield (Cnossen et al., 2003; Iguaz et al., 2006). Hence, it is necessary for low moisture rice to be conditioned by increasing its moisture content to prevent a decline in its eating quality (Genkawa et al., 2008b).

To design a rough rice drying equipment adequately, data are required both for drying and re-wetting characteristics from low to high temperatures. Moisture adsorption occurs when the vapour pressure within kernels is lower than the vapour pressure of the surrounding air. The moisture adsorbing environments can exist in the field before harvesting and subsequently during harvesting, holding, transport, drying and storage (Kunze and Prasad, 1978). Kunze (1988) stated that when rice in the field reaches moisture content of 30%

---

*Corresponding author. E-mail: basunia@squ.edu.om.

or below, there already may be grains which are sufficiently dry to fissure when they readsorb moisture. The lower the moisture content of the rice in the field, the higher in general is the percentage of fissured grains. Kernels can fissure in the field by first drying during the day and then rapidly readsorbing moisture during the evening or night. The grain is exposed to fluctuating air temperatures and relative humidities causing drying and re-wetting cycles in low temperature drying. Grain prepared by desorption has a higher moisture content at a given relative humidity than grain prepared by adsorption. In both the near-ambient and natural drying process, ambient air is forced through the stored grain. For both processes, it is common to run the fan continuously even if it involves running the fan during periods of high ambient relative humidity which can cause re-wetting of grain. It is desirable to know how fast a grain bed would re-wet when the fans are running during high humidity periods.

Both moisture adsorption and desorption are equally important concerns when developing models simulating deep bed drying and aeration of grain. In the early stage of deep bed drying, the lower layer of grain desorbs moisture while the upper layer of grain adsorbs moisture. Thin-layer moisture transfer equations are used in simulation models for deep bed drying of grains. Most of the earlier studies on thin-layer moisture transfer relationships were concerned with thin-layer drying of cereal grains or oilseeds for a short duration and very little work was done on thin-layer re-wetting rates (Misra and Brooker, 1980; Jayas and Sokhansanj, 1986, 1988; Basunia and Abe, 1999, 2005). Sokhansanj et al. (1984) stated that the drying rate of wheat, barley and canola changes as a result of re-wetting. Limited work has been done in developing thin-layer re-wetting equations for rough rice. Much research has been conducted to develop thin-layer drying equations and to better understand the desorption phenomenon in rice (Agrawal and Singh, 1977, 1984; Wang and Singh, 1978; Noomhorn and Verma, 1986; Basunia and Abe, 1998, 2001). Far less research has been devoted to developing thin-layer adsorption equations. Genkawa et al. (2008a) developed a re-wetting method for brown rice by using film packaging technique and established a mathematical model suitable for designing the packaging. The moisture content of brown rice was increased from 10% wet-basis (w.b.) to 19% w.b. without cracking when the polymeric package was exposed to a humid environment of 25°C and 89.5% RH for 10 weeks.

Banaszek and Siebenmorgen (1990) conducted a study on re-wetting characteristics of long grain variety of rough rice for a limited range of temperatures and relative humidities. They collected the weight data manually, removing the test sample from the conditioned chamber, at different intervals, and only 12 data points for each test. Hacihafızoğlu et al. (2008) investigated the suitability of several drying models available in literature in defining the thin layer drying behaviour of long-grain rough rice by using statistical analysis and found that Midilli et al.'s is the most appropriate model for drying behaviour of thin layer rough rice. They also found that the Page model gives better fit among the two parameter models. Chen and Tsao (1994) used several models to define their experimental data for the thin layer rough rice and concluded that the two-term model gives the best fit among them. Das et al. (2004) showed that the Page model describes the experimental data adequately for drying of high moisture rough rice. Four different thin layer drying models were used by Chen and Wu (2001) to simulate thin layer drying of rough rice with high moisture content and the two term model was found as the best fit in this study. Cihan et al. (2007) found that the most accurate model is the Midilli et al. model in defining the intermittent drying process of rough rice.

Basunia and Abe (1999, 2005) conducted a study on moisture adsorption isotherms and thin-layer re-wetting characteristics of rough rice over a wide range of temperature and relative humidities. They fitted only a single thin-layer drying equation to describe the moisture re-wetting characteristics of rough rice. So, there is need to find out the best fitted equation to describe the rewetting characteristics of medium grain rough rice from low to high temperature which is commonly used in rough rice drying and re-wetting The object of this work is to determine the rate of moisture transfer in re-wetting rough rice over a range of temperature and relative humidity, and to find the most suitable thin-layer re-wetting model for rough rice which could be used in the simulation of moisture transfer during ventilated storage.

MATHEMATICAL EQUATIONS TO PREDICT THIN-LAYER REWETTING

The drying characteristics of rough rice have been examined by many researchers and various models for the prediction of the drying rate have been performed with more or less success. Mathematical modeling of drying is crucial for the optimization of operating parameters and performance improvements of the drying system. The most commonly used thin-layer rewetting or drying models of grain are Diffusion (Newman, 1931), Approximate form of diffusion (Boyce, 1965), Page (Page, 1949), Exponential (Jayas et al., 1991) and Polynomial (Wang, 1978). The following models were therefore chosen for this study to fit the observed rewetting data:

1. Simplifications of the well-known diffusion model for large drying or re-wetting times that is frequently used to

predict the drying and re-wetting of grain is given as:

$$M_R = \frac{M_t - M_e}{M_i - M_e} = C \times \exp\left(-\frac{\pi^2 D t}{R^2}\right) \qquad (1)$$

where $C = 6/\pi^2$, M_R is the moisture ratio, M_t is the moisture content at any time in dry-basis, M_e is the equilibrium moisture content in dry-basis, M_i is the initial moisture content (dry-basis), t is the drying time in hour (h), D is the diffusion coefficient in m^2/h, R is the sphere radius in m.

2. The most commonly used empirical equation to describe the thin-layer drying and re-wetting of cereals is that of Page (Page, 1949):

$$M_R = exp\,(-K \times t^N) \qquad (2)$$

Where, t is the re-wetting time in min; and K, N are the re-wetting parameters.

3. Exponential model (Lewis, 1921) can be written as:

$$M_R = exp\,(-K \times t) \qquad (3)$$

Where, t is the re-wetting time in min, K is the re-wetting parameters.

4. Approximate form of diffusion equation (Boyace, 1965) for thin layer-drying or rewetting can be written as

$$M_R = a\,exp\,(-K \times t) \qquad (4)$$

Where, a is product dependent constant, t is the re-wetting time in min; and K is the re-wetting parameter.

5. Second order polynomial equation (Thompson et al., 1968; Wang and Singh, 1978) is of the from:

$$M_R = a + b\,t + c\,t^2 \qquad (5)$$

Where, t is the re-wetting time in min, a is the product dependent constant and b, c are the re-wetting parameters.

MATERIALS AND METHODS

The range of re-wetting conditions for the experiment is presented in Table 1. The procedure to determine weight data of the sample in the thin-layer rewetting and the adsorption equilibrium moisture content were described elsewhere (Basunia and Abe, 1999, 2005). Thin-layer re-wetting characteristics of rough rice were determined at temperature ranging from 17.8 to 45°C and for relative humidities ranging from 56 to 89.3%, with initial moisture contents in the range of 10.26 to 12.71% dry-basis. The data of sample weight, and dry and

wet bulb temperatures of the re-wetting air were recorded continuously throughout the re-wetting period for each test of 25 tests. The re-wetting process was terminated when the moisture content change in 24 h was less than 0.1% dry-basis (weight change was less than 0.05 g). Normally, such an experiment lasted for 4 to 6 days. The final points were recorded as the dynamic equilibrium moisture contents. Each data file consisted of more than 300 measured points.

Re-wetting parameters of each of the models were found for each test run using linear regression. The coefficients of determination R^2 were all above 0.90. The 25 sets of values for different parameters were used in a multiple regression procedure to find expressions for each parameter of the model equations. The measured and simulated moisture contents were compared and statistically analyzed for determining the best fit equation. The standard error of estimate (SEE) indicates the fitting ability of a model to a data set. The smaller the SEE value, the better the fitting ability of an equation. For the same data set, the equation giving the smallest SEE value represents the best fitting ability (Basunia and Abe, 1999, 2001). The standard error of estimate (SEE) is expressed as:

$$SEE = \sqrt{\frac{\sum_{i=1}^{m}(M_t - M_s)^2}{df}}$$

Where M_s is the measured moisture content in dry-basis and df is the degree of freedom. For large data set, as in this experiment, it is defined as:

$$SEE = \sqrt{\frac{\sum_{i=1}^{m}(M_t - M_s)^2}{m}}$$

where m is the number of data points.

RESULTS AND DISCUSSION

Expressions for the parameter of model Equation (1)

The values of the parameters C and diffusivity D of Equation (1) were obtained by linear regression analysis. It was observed that C varies between 0.802 to 0.892 within the temperatures and relative humidities studied. Hence, for analysis and interpretations of the results, an overall average value of C from all tests was used. The average value of C for 25 tests was 0.848. This effectively assumes C to be a product-dependent constant instead of 0.608 for a perfectly spherical grain kernel as in Equation (1). Table 1 shows the values of D and the corresponding values of standard error of estimate (SEE) values of moisture content of all tests when the parameter C was fixed at this overall average of 0.848. The average SEE value of 25 tests was only 0.121% dry-basis for a fixed value of $C = 0.848$. The assumption, therefore, of taking C as a product-dependent constant seems valid for representing the re-wetting rate data of rough rice. The expression relating diffusivity, D in m^2/h, and re-wetting air temperature, T in °C, was found as:

Table 1. List of the experimental conditions, and standard errors of estimate (SEE) for each test.

Re-wetting conditions		Initial moisture content (%db)	Diffusion model (%db)	Page Model (%db)	Exponential (%db)	Approximate form of diffusion (%db)	Polynomial (%db)
$T(^0C)$	R_H						
17.8	72.1	10.30	0.164	0.241	0.708	0.967	0.366
17.8	72.3	12.71	0.157	0.143	0.414	1.093	0.386
21.9	69.1	11.08	0.108	0.216	0.204	0.140	0.312
21.9	80.6	11.04	0.127	0.143	0.184	0.386	0.328
21.9	89.3	11.04	0.129	0.293	0.204	0.382	0.391
25.8	56.0	11.31	0.090	0.118	0.062	0.097	0.390
25.8	62.8	11.38	0.072	0.127	0.108	0.066	0.363
25.8	69.5	11.60	0.074	0.162	0.169	0.063	0.394
25.8	79.4	11.24	0.126	0.112	0.292	0.173	0.326
25.8	88.7	11.31	0.125	0.126	0.475	0.234	0.311
30.2	61.0	11.11	0.031	0.070	0.094	0.046	0.438
30.2	69.0	11.02	0.102	0.192	0.189	0.106	0.565
30.2	80.0	10.96	0.112	0.278	0.238	0.106	0.311
30.2	88.3	11.01	0.213	0.143	0.216	0.175	0.420
35.5	64.7	10.54	0.128	0.157	0.165	0.098	0.309
35.5	70.8	11.79	0.040	0.048	0.148	0.056	0.150
35.5	79.1	11.90	0.068	0.137	0.079	0.146	0.158
35.5	86.9	11.73	0.112	0.170	0.157	0.150	0.321
40.2	68.6	11.31	0.055	0.055	0.151	0.120	0.192
40.2	70.2	10.69	0.065	0.117	0.116	0.470	0.149
40.2	80.0	10.26	0.146	0.157	0.233	0.196	0.409
40.2	88.2	12.08	0.222	0.168	0.143	0.318	0.312
45.0	61.5	10.32	0.094	0.078	0.033	0.049	0.099
45.0	71.7	10.30	0.128	0.059	0.212	0.212	0.453
45.0	77.4	10.40	0.062	0.045	0.173	0.207	0.378
			Average = 0.121	Average = 0.142	Average = 0.207	Average = 0.242	Average = 0.329

*SEE of estimate of predicted moisture content with more than 300 observations for each test.

$$D = 10.0599 \times \left(-\frac{6111.24}{T + 273.15} \right) \quad (6)$$

with a coefficient of determination 0.985.

The very low SEE (0.121% dry-basis) shows the accuracy of the model to predict the moisture content at any time during the re-wetting period. The SEE of individual tests is shown in Table 1. The highest SEE was 0.222%, dry-basis and the lowest was only 0.031%, dry-basis.

Expressions for the parameters of model Equation (2)

The multiple regression analysis for K as a function of temperature T in °C and relative humidity R_H in decimal, yielded:

$$K = 0.01627 + 0.000313\ T - 0.021434\ R_H \quad (7)$$

with a coefficient of determination R^2 of 0.951.

The regression analysis for N as a function of temperature T in °C and relative humidity R_H in decimal, yielded:

$$N = 0.40441 + 0.00248\ T + 0.34959\ R_H \quad (8)$$

with a coefficient of determination R^2 of 0.97.

The highest SEE was 0.293%, dry-basis and the lowest was only 0.045%, dry-basis. The average standard error of estimate between the measured and predicted values of moisture contents for the full data set was only 0.142%

dry-basis. This very low SEE (0.00142) shows the accuracy of the model to predict the moisture content at any time during the re-wetting period. The SEE of individual tests are shown in Table 1.

Expression for the parameter of model equation (3)

The multiple regression analysis for K as a function of temperature T in °C and relative humidity R_H in decimal, yielded:

$$K = -0.00267 + 0.000095 \times T + 0.00186 \times R_H$$
$$(R^2 = 0.95) \tag{9}$$

The average standard error of estimate between the measured and predicted values of moisture contents for the full data set was 0.207% dry-basis which is higher than the Page and Diffusion models. The SEE of individual tests are shown in Table 1. The highest SEE was 0.708%, dry-basis and the lowest was only 0.033%, dry-basis.

Expression for the parameter of model equation (4)

The multiple regression analysis for K as a function of temperature T in °C and relative humidity R_H in decimal, yielded:

$$K = 0.00112 - 0.00008 T + 0.00010 R_H \tag{10}$$

with a coefficient of determination R^2 of 0.91.

It was observed that a varies between 0.832 to 0.931 within the temperatures and relative humidities studied. Hence, for analysis and interpretations of the results, an overall average value of a from all tests was used. The average value of a for 25 tests was 0.866. The highest SEE was 1.093%, dry-basis and the lowest was only 0.046%, dry-basis. The average standard error of estimate between the measured and predicted values of moisture contents for the full data set was 0.242% dry-basis which is higher than the Page and the Diffusion models.

Expressions for the parameters of model equation (5)

The multiple regression analysis for b as a function of temperature T in °C and relative humidity R_H in decimal, yielded:

$$b = 0.00165 - 0.000062 \times T - 0.00088 \times R_H \tag{11}$$

with a coefficient of determination R^2 of 0.90.

The regression analysis for c as a function of temperature T in °C and relative humidity R_H in decimal, yielded:

$$c = -9.957E\text{-}07 + 3.81E\text{-}08 \times T + 2.14E\text{-}07 \times R_H \tag{12}$$

with a coefficient of determination R^2 of 0.87.

It was observed that a varies between 0.806 to 0.932 within the temperatures and relative humidities studied. Hence, for analysis and interpretations of the results, an overall average value of a from all tests was used. The average value of a for 25 tests was 0.843. The highest SEE was 0.565%, dry-basis and the lowest was 0.099%, dry-basis (Table 1). The average standard error of estimate between the measured and predicted values of moisture contents for the full data set was 0.329% dry-basis which is higher than the Page and Diffusion models. The SEE of individual test is shown in Table 1. From Table 1, it can be observed that for most of the tests, SEE was below 0.15% dry-basis both by the Diffusion and Page models. It was found that the numerical difference between the moisture contents predicted by Equation (1), and with diffusivity D calculated with Equation (6) and the observed moisture content did not exceed 0.4% dry-basis points in any test conducted at all temperature and relative humidity combination.

Similarly, it was found that the numerical difference between the moisture contents predicted by Equations (2), (7) and (8) and the observed moisture content did not exceed 0.45% dry-basis points in any test conducted at all temperature and relative humidity combination. This amount of error can be accepted for most practical purpose when working with biological products. So Equations (1) and (6) or the Equations (2), (7) and (8) can be used in a deep bed drying simulation model to predict the re-wetting under high ambient relative humidity conditions. The moisture contents simulated by Equation (1) with $C = 0.848$, and diffusivity D with Equation (6), were compared to observe moisture contents in Figures 1 and 2. The measured and predicted values were in very good agreement. Similar agreements were also observed in other re-wetting conditions. The moisture simulated by Equation (2) with K and N calculated with Equations (7) and (8), respectively, were compared to observe moisture in Figures 3 and 4. The predicted and observed values were in good agreement. Similar agreements were also observed in other re-wetting conditions.

Conclusions

The re-wetting rates of rough rice from low to high

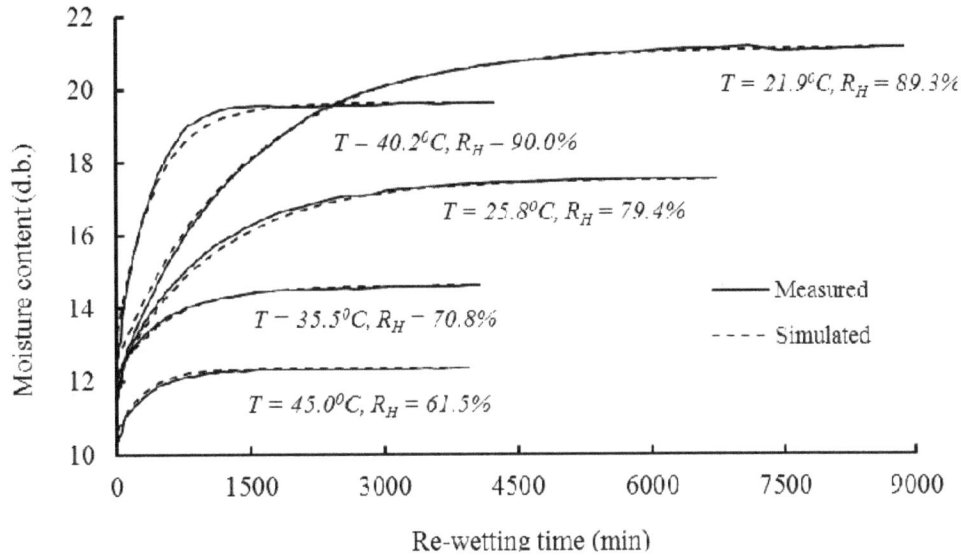

Figure 1. Comparison between the curves predicted by the diffusion model with the values of the diffusivity with Equation (6) and the experimental points at temperature (T) of 21.9, 25.8, 35.5, 40.2 and 45 °C, and various relative humidities (R_H).

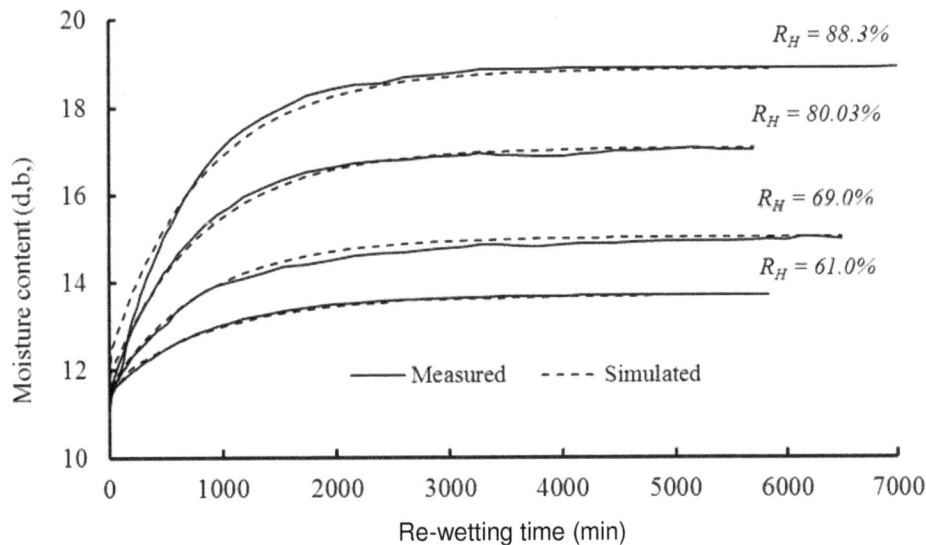

Figure 2. Comparison between the curves predicted by the diffusion model with the values of the diffusivity with Equation (6) and the experimental points at temperature (T) of 30.2 °C, and various relative humidities (R_H).

temperatures have been determined. Five models were compared based on standard error of estimate (SEE) values. The Diffusion model and the page model, based on the ratio of the difference between the initial and final moisture content and the equilibrium moisture content, fits the data well with a standard error of 0.121% dry-basis and 0.142% dry-basis, respectively. The Diffusion model and the Page model are found to be the most appropriate models for representing the rewetting characteristics of rough medium grain. Other three models, the Approximate form of diffusion, the Exponential and the Polynomial did not fit well compared to the Diffusion and the page model. The values of SEE for the Exponential, the Approximate form of diffusion

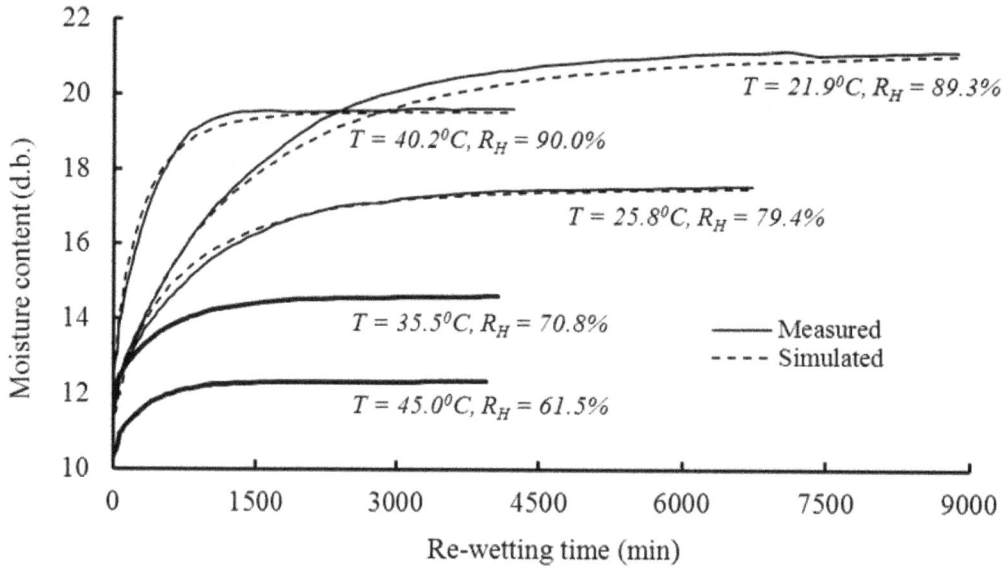

Figure 3. Comparison between the curves predicted by the Page model with the values of the re-wetting parameters with Equations (7) and (8) and experimental points at temperature (T) of 21.9, 25.8, 35.5, 40.2 and 45.0 °C, and various relative humidities (R_H).

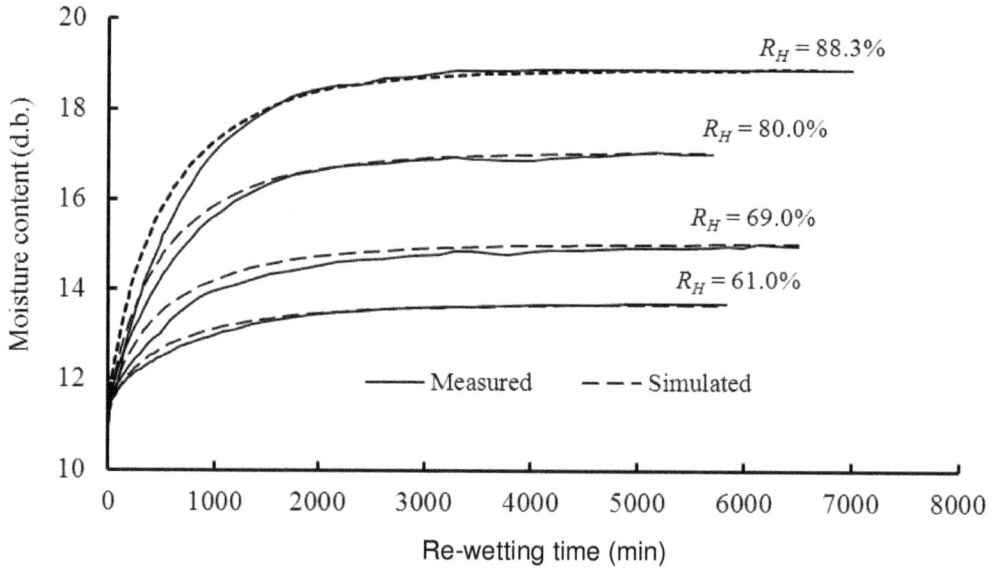

Figure 4. Comparison between the curves predicted by the Page model with the values of the re-wetting parameters with Equations (7) and (8) and experimental points at temperature (T) of 30.2 °C, and various relative humidities (R_H).

and the Polynomial models were 0.207, 0.242, and 0.329% d.b., respectively. The result presented here, over a typical 5 day re-wetting period, are useful in the longer term moisture transfer process occurring during ventilated storage.

Notation: a, Product dependent constants in Equations (4) and (5) respectively; **b, c,** re-wetting parameters in Equation (5); **C,** product dependent constant in Equation (1); **K, N,** re-wetting parameters in Equations (2) to (4); **M_e,** equilibrium moisture content of grain (d.b.); **M_i,** initial moisture content of grain (d.b.); M_t, moisture content of grain at any time, dry-basis (d.b.); M_s, simulated moisture content of grain at any time, dry-basis (d.b.); R, radius of the sphere; R^2, coefficient of determination; R_H, relative humidity; M_R, moisture ratio; t, re-wetting time (hr in Equation 1, and min in Equations (2) to (4); T, re-wetting temperature (°C).

REFERENCES

Agrawal YC, Singh RP (1977). Thin-layer drying studies on short-grain rough rice. ASAE paper No. 77-3531, ASAE, St. Joseph, Michigan 49085.

Agrawal YC, Singh RP (1984). Thin layer drying of rough rice. Journal of Agricultural Engineering. Indian Soc. Agric. Eng., 21(4): 41-47.

Banaszek MM, Siebenmorgen TJ (1990). Moisture adsorption rates of rough rice. Trans. ASAE, 33(4): 1257-1262.

Basunia MA, Abe T (1998). Thin-layer drying of characteristics of rough rice at low and high temperatures. Drying Technol., 16: 579-595

Basunia MA, Abe T (1999). Moisture adsorption isotherms of rough rice. J. Food Eng., 42: 235-242.

Basunia MA, Abe T (2001). Moisture desorption isotherms of medium-grain rough rice. J. Stored Prod. Res., 37: 205-219.

Basunia MA, Abe T (2005). Thin-layer re-wetting of rough rice at low and high temperature. J. Stored Prod. Res., 41: 163-173.

Boyace DS (1965). Grain moisture and temperature and moisture changes with position and time during trough drying. J. Agric. Eng. Res., 10(4):333-341.

Chen C, Tsao CT (1994). Study on the thin-layer drying model for rough rice. J. Chin. Agric. Res., 43(2): 208-227.

Chen C, Wu PC (2001). Thin-layer drying model for rough rice with high moisture content. J. Agric. Eng. Res., 1: 45-52.

Cihan A, Kahveci K, Hacıhafızoğlu O (2007). Modelling of intermittent drying of thin layer rough rice. J. Food Eng., 79(1): 293-298.

Cnossen AG, Jime´nez MJ, Siebenmorgen TJ (2003). Rice fissuring response to high drying and tempering temperatures. J. Food Eng., 59(1): 61-69.

Das I, Das SK, Bal S (2004). Drying performance of a batch type vibration aided infrared dryer. J. Food Eng., 64(1): 129-133.

FAO (2008). Food and Agricultural Organization of the United Nations, Food crop prospects and food situation 2. Available on: ttp://www.fao.org/docrep/010/ai465e/ai465e00.htm.

Genkawa T, Uchino T, Miyamoto S, Inoue A, Ide Y, Tanaka F, Hamanaka D (2008a). Development of mathematical model for simulationg moisture content during the re-wetting of brown rice stored in film packaging. Biosyst. Eng., 101:445-451.

Genkawa T, Uchino T, Inoue A, Tanaka F, Hamanaka D (2008b). Development of a low-moisture-content storage system for brown rice: storability at decreased moisture contents. Biosyst. Eng., 99(4): 515-522.

Hacihafizoğlu O, Cihan A, Kahveci K (2008). Mathematical modelling of drying of thin layer rough rice. Food Bioprod. Process., 86:268-275.

Iguaz A, Rodrı´guez M, Vı´rseda P (2006). Influence of handling and processing of rough rice on fissures and head rice yields. J. Food Eng., 77(4): 803-809.

Jayas DS, Sokhansanj S (1986). Thin-layer drying of wheat at low temperatures. In Drying: (Mujumdar A S ed.), Hemisphere Pub. Corp., New York, NY, pp. 844-847,

Jayas DS, Sokhansanj S (1988). Thin-layer drying of barley at low temperatures. Can. Agric. Eng., 31: 21-23.

Jayas DS, Cenkowski S, Pabis S, Muir WE (1991). Review of thin layer drying and rewetting equations. Drying Technol., 9: 551-558.

Kunze OR, Prasad S (1978). Grain fissuring potentials in harvesting and drying rice. Trans. ASAE, 21(2): 361-266

Kunze OR (1988). Simple R. H. systems that fissure rice and other grains. ASAE paper No. 88-5010, St. Joseph, Michigan, USA.

Lewis WK (1921). The rate of drying of solid materials. Introduction Eng. Chem., 13: 427-432.

Misra MK, Brooker DB (1980). Thin-layer drying and re-wetting equations for shelled yellow corn. Trans. ASAE, 23(5): 1254-1260.

Newman AB (1931). The drying of porous solid. Diffusion and surface emission equation. Trans. Am. Inst. Chem. Eng., 27: 310-333.

Noomhorn A, Verma LR (1986). Generalized single-layer rice drying models. Trans. ASAE, 29(2): 587-591.

Page C (1949). Factors influencing the maximum rate of drying shelled corn in layers. M. S. Thesis, Purdue university, W. Lafayette, IN, USA.

Siebenmorgen TJ, Jindal VK (1986). Effects of moisture adsorption on the head rice yields of long-grain rough rice. Trans. ASAE, 29(6): 1767-1771.

Sokhansanj S, Singh D, Wasserman JD (1984). Drying characteristics of wheat, barely and canola subjected to repetitive wetting and drying cycles. Trans. ASAE, 27(3): 903-914.

Thompson TL, Peart RM, Foster GH (1968). Mathematical simulation of corn drying- a new model. Trans. Am. Soc. Agric. Eng., 24: 582-586.

Wang CY, Singh RP (1978). A single layer drying equation for rough rice. ASAE paper No. 78-3001, St. Joseph, Michigan, USA.

14

Assessment of broad spectrum control potential of *Eucalyptus citriodora* oil against post harvest spoilage of *Malus pumilo* L.

Sushil Kumar Shahi[1]* and Mamta Patra Shahi[2]

[1]Bioresource Technology Laboratory, Department of Microbiology, CCS University, Meerut- 250005, India.
[2]Department of Biotechnology, Meerut Institute of Engineering and Technology, Meerut-250005, India.

In vitro Eucalyptus citriodora Hook. oil showed potent bioactivity against dominant post harvest fungal pathogens. The minimum bioactive concentrations with fungicidal action of the oil was found to be 1.0 μl ml^{-1} for *Alternaria alternata, Botrytis cinerea, Cladosporium cladosporioides, Colletotrichum capsici, Cyrtomium falcatum, Fusarium cerealis, Fusarium culmorum, Gloeosporium fructigenum, Penicillium digitatum, Penicillium expansum, Penicillium italicum, Penicillium implicatum, Penicillium minio-luteum,* 1.2 μl ml^{-1} for *Aspergillus flavus, Aspergillus fumigatus, Aspergillus niger, Aspergillus parasiticus, Curvularia lunata, Fusarium oxysporum, Fusarium udum, Penicillium variable, Helminthosporium oryzae, Helminthosporium maydis, Phoma violacea,* and 1.4 μl ml^{-1} for *Rhizopus nigricans.* The oil exhibited potency against heavy doses (30 mycelial disc, each of 5 mm in diameter) of inoculum at 2.0 μl ml^{-1} concentrations. The bioactivity of the oil was thermostable up to 100°C and lasted up to 72 months. The oil preparation did not exhibit any phytotoxic effect on the fruit skin (epicarp) of *Malus pumilo* up to 50 μl ml^{-1} concentrations. *In vivo* trials of the oil as a fungicidal spray on *M. pumilo* for checking the rotting of fruits, it showed that 30 μl ml^{-1} concentration controls 100% infection by pre-inoculation treatment, while in post-inoculation treatment, 40 μl ml^{-1} concentration of fungicidal spray were required for the 100% control of rotting. The fungicidal spray was found to be cost effective (INR 15/L) has long shelf life (72 months) and devoid of any adverse effects. Therefore, it can be used as a potential source of sustainable eco- friendly broad-spectrum herbal pesticide after successful completion of wide range trials.

Key words: *Eucalyptus citriodora* Hook., fungicidal spray, fruit rot, herbal pesticide, *Malus pumilo*.

INTRODUCTION

Edible fruits are among the most important foods of mankind as they are nutritive and indispensable for the maintenance of health. They are also high-value commodities, offering good economic return even on small area of land. Based on policy directives of the planning commission of India, on the research priority area for enhanced fruit production was identified as reducing post harvest losses (Eckert and Sommer, 1967; Harvey, 1978). They identified weak post harvest management as major constraints and quoted 50% loss from harvesting, handling, storage and marketing of fruits

according to FAO. India, being a geographically subtropical country with warm and humid climate, provides suitable environment for developing and spread of numerous plant pathogens. Harvested fruit and vegetables are attacked by microorganisms because of their high moisture content and rich nutrients (Simmonds, 1963).

Usually, synthetic pesticides are applied for the control of 'pest and disease' of the agricultural food commodities, as these are effective, dependable and economic. However, their indiscriminate use has resulted into several problems such as pest resistance to pesticides, resurgence of pests, toxic residues in food (causing health hazards to animals and human beings), water, air, soil and disruption of eco-system (Somasundaram et al.,

*Corresponding author. E-mail: shahi.sk@rediffmail.com.

1990). Natural products are an alternative to the use of these synthetic pesticides (Shahi et al., 2003). Keeping this view in mind, the present paper reports the bioactivity of the essential oil of *Eucalyptus citriodora* Hook. belonging to the family Myrtaceae also called lemon-scented gum, spotted gum, because of typical strong lemon like odour. It is grown naturally in Tamilnadu, Karnataka and Kerela, and is commercially cultivated in India. In the present investigation, the oil of *E. citriodora* were evaluated *in vitro* against dominant post harvest pathogenic fungi as well as control of rotting in apple.

MATERIALS AND METHODS

Maintenance of fungus culture

The test fungal pathogens, *Alternaria alternata* (Fr.) Keissler (MTCC 2724), *Aspergillus flavus* Link (MTCC 3396), *Aspergillus fumigatus* Fres (MTCC 2544), *Aspergillus niger* Van Tiegham (MTCC 1781), *Aspergillus parasiticus* Speare (MTCC 6768), *Botrytis cinerea* Pers. Ex. Fr. (MTCC 2104), *Cladosporium cladosporioides* (Fresenius) de Vries (MTCC 3478), *Colletotrichum capsici* (Syd) Butler and Bisby (MTCC 2071), *Cyrtomium falcatum* Went. (MTCC 2222), *Curvularia lunata* (Wakker) Boedijn (CBTC 2342), *Fusarium cerealis* (Cooke) Sacc (CBTC 2456), *Fusarium culmorum* (W.G Smith) Sacc (MTCC 2090), *Fusarium oxisporum* Schlecht.:Fr. (MTCC 2087), *Fusarium udum* (Butler) Snyder and Hansen (MTCC 2204), *Gloeosporium fructigenum* Berk (MTCC 2191), *Helminthosporium oryzae* Breda de Haan (CBTC 1256), *Helminthosporium maydis* Nisikado and Miyakel (CBTC 2314), *Penicillium digitatum* Sacc. (CBTC 1121), *Penicillium expansum* Link (MTCC 4485), *Penicillium italicum* Wehmer (CBTC 1029), *Penicillium implicatum* Biourge (CBTC 1034), *Penicillium minio-luteum* Dierckx (CBTC 1045), *Penicillium variabile* Sopp (CBTC 1046), *Phoma violacea* (Bertd) Eveleigh (CBTC 2051), *Rhizopus nigricans* Ehrenb (CBTC 2167) (Neergaard, 1977; Samson et al., 1995) were collected from Microbial type culture collection (MTCC), Chandigarh (India) and Collection of Bio-resource Type Culture (CBTC), Microbiology Department, CCS University, Meerut (India). All culture were maintained on potato dextrose agar medium (200 g scrubbed and diced potato in 1000 ml distilled water, 15 g agar, 20 g dextrose pH ± 5.6). A 7 day old culture of each fungus was used for bioactivity tests.

Isolation of active constituent(s)

The essential oil was extracted from the fresh leaves of *E. citriodora* Hook by hydro-distillation using Clevenger's apparatus (Clevenger, 1928). A clear light yellow green coloured oily layer was separated and dried with anhydrous sodium sulphate. The physiochemical properties of the oil were determined by the technique described by Langenau (1948).

In vitro studies

The minimum bioactive concentrations (MBCs) of the oil were determined following the poisoned food technique (PFT) of Grover and Moore (1962) with slight modification (Shahi et al., 1999). The requisite quantity of the oil was dissolved in 2 ml acetone and then added in 100 ml pre-sterilized potato dextrose agar (PDA) medium (pH- 5.6). In control sets, sterilized water (in place of the oil) and 2

ml acetone were used in the medium. Mycelial discs of 5 mm diameter, cut out from the periphery of 7-day old cultures of the test pathogens, were aseptically inoculated upside down on the agar surface of the medium. Inoculated Petri plates were incubated at 27 ± 1℃ and the observations were recorded on the 7th day. Percentage of mycelial growth inhibition (MGI) was calculated as follows:

MGI (%) = (dc-dt) × 100 / dc

where, dc = mycelial growth diameter in control sets, dt = mycelial growth diameter in treatment sets.

The nature of antifungal activity, fungistatic (temporary inhibition) / fungicidal (permanent inhibition) of the oil was determined by the method of Garber and Houston (1959). The inhibited fungal discs (at minimum bioactive concentrations) were reinoculated up side down on plain PDA (potato dextrose agar) medium in Petri plate. Observations were recorded on 7th day of incubation at 27±1℃. Fungal growth on 7th day indicated fungistatic action of the oil, while absence of growth indicated fungicidal action of the oil. The effect of inoculum potentiality on bioactivity of the oil (2.0 μl ml[-1]) was determined by the method of Shahi et al. (1999). Mycelial disc of 5 mm in diameter of seven day old cultures were inoculated in culture tube containing 2.0 μl ml[-1] oil in liquid medium (Potato dextrose broth) separately. In controls, sterile water were used in place of oil and run simultaneously. The number of mycelial discs in the treatment as well as control sets were increased progressively up to 30 in multiple of five. Observations were recorded after the 7th day of incubation. Absence of mycelial growth in treatment sets on the 7th day exhibited the oil's potential against heavy doses of inoculum.

Effect of temperature and duration of toxicity during storage of the oil was evaluated according to Shahi et al. (1999). Five lots of oil were kept in small vials, each containing 5 ml of oil; these were exposed at 40, 60, 80 and 100℃ in an incubator for 60 min. Residual activity was assayed by poisoned food technique of Grover and Moore (1962). Loss of toxicity of the oil was also determined by storing the oil at room temperature (30 ± 4℃) and withdrawn samples at intervals of 60 days up to 7 years and tested by poisoned food technique (Grover and Moore, 1962). All the experiments were repeated twice and each contained five replicates; the data presented mean values.

Phytotoxic investigation

Phytotoxic effect of the oil was carried out at different concentrations (ranging from 10 to 100 μl ml[-1]) on fruits skin (epicarp) of *Malus pumilo*. Two sets of 50 samples (apples) were maintained one for the treatments and another for the controls. Each sample was first washed with distilled water followed by 70% ethyl alcohol and then allowed to dry. In treatment sets, 1 ml of the different concentrations of oil was sprayed to each sample separately. In controls, sterilized water was sprayed (in place of oil). The qualitative observations (morphological changes, such as colour, odour, weight, size, changes in epicarp and taste) have been recorded at the interval of 24 h up to 3 weeks.

In vivo investigation of the oil in the form of fungicidal spray

The study was designed to see the activity of the oil in the form of fungicidal spray applied on fruit skin for the control of fruit rot of *M. pumilo* by different methods. For *in vivo* study, both pre and post inoculation treatments (fungicidal spray) were applied to the fruits. In the pre inoculation treatment, two sets were prepared, treatments

Table 1. Physico-chemical properties of the oil of *Eucalyptus citriodora* oil.

Properties studied	Observations
Plant height (m)	25-40
Oil yield (%)	0.6
Colour	Light yellow
Specific gravity at 15 °C	0.8640-0.8770
Refractive index at 20 °C	1.4511-1.4570
Optical rotation	+3 to -3°
Saponification value	8.90 to 2.0
Ester value	12-60
Solubility in 70% alcohol	1.3 to 1.5 vols
Citronellol content (%)	65-85

as well as controls. In treatment set, fruits were sprayed in known concentrations (10 to 50 µl ml^{-1}) of oil preparation in vehicle. In controls, the fruits were sprayed with distilled water in vehicle. Thereafter, the fruits were injured using a sterilized needle, and the fungal inoculum of *P. expansum, B. cinerea, P. violacea* (5 mm diameter mycelial disc of each fungus) was placed over the injured areas. All inoculated fruits were incubated at 26 ± 1 °C and the observations were recorded on the 7th day.

In post inoculation treatment, fruits were first wounded with a sterilized needle and fungal inoculum of *P. expansum, B. cinerea, P. violacea* (5 mm diameter mycelial disc of each fungus) was placed over the wounded areas. After 24 h of incubation, fruits were sprayed in different concentrations (10 to 50 µl ml^{-1}) of oil preparation. In controls, fruits were sprayed with distilled water in vehicle. Inoculated fruits were incubated at 26 ± 1 °C and the observations were recorded on the 7th day. The data were average of 5 replicates and repeated twice. Percentages of inhibition (I) were calculated as follows.

I (%) = (Ic-It) × 100 / Ic

Where: Ic = average diameter of infected area in control set, It = average diameter of infected area in treatment sets.

Statistical analysis

Analysis of variance (ANOVA) was used to determine the significance (P≥0.05) of the data obtained in all experiments. All results were determined to be within the 95% confidence level for reproducibility. The ANOVA was computed using the SPSS version 16.0 software package.

RESULTS

The leaves of *E. citriodora* on hydro-distillation yielded 0.6% essential oil. The physicochemical properties of the oil were shown in Table 1. The oil exhibited broad antifungal activity, the minimum bioactive concentrations with fungistatic action (temporary inhibition) of the oil was found to be 0 .4 µl ml^{-1} for *A. alternata,* 0.6 µl ml^{-1} for *A.*

niger, A. parasiticus, B. cinerea, C. cladosporioides, C. capsici, C. falcatum, F. cerealis, F. culmorum, G. fructigenum, P. expansum, P. digitatum, P. italicum, P. implicatum, P. minio-luteum, 0.8 µl ml^{-1} for *A. flavus, A. fumigatus, C. lunata, F. oxysporum, F. udum, H. oryzae, H. maydis, P. variable* and 1.0 µl ml^{-1} for *R. nigricans* (Table 2). The minimum bioactive concentrations with fungicidal action (permanent inhibition) of the oil was found to be 1.0 µl ml^{-1} for *A. alternata, B. cinerea, C. cladosporioides, C. capsici, C. falcatum, F. cerealis, F. culmorum, G. fructigenum, P. digitatum, P. expansum, P. italicum, P. implicatum, P. minio-luteum,* 1.2 µl ml^{-1} for *A. flavus, A. fumigatus, A. niger, A. parasiticus, C. lunata, F. oxysporum, F. udum, P. variable, H. oryzae, H. maydis, P. violacea,* and 1.4 µl ml^{-1} for *R. nigricans* (Table 2).

The oil inhibited heavy doses (30 fungal mycelial disc, each of 5 mm in diameter) of inoculum at 2.0 µl ml^{-1} concentration. The bioactivity of the oil persists up to 100 °C, and it did not expire even up to 72 months of storage.

The oil did not exhibit any phytotoxic effect up to 50 µl ml^{-1} level on fruit skin (Table 3). Formulation of the oil prepared at different concentrations (10 to 50 µl ml^{-1}) in the form of fungicidal spray. The fungicidal spray, when tested in vivo on *M. pumilo* for checking the rotting, it showed complete inhibition at 20 µl ml^{-1} concentration by pre inoculation treatment while in post inoculation treatment, 30 µl ml^{-1} concentration of spray solution was required for the 100% control of rotting (Table 4) and in Figure 1 showed. The fungicidal spray was found cost effective and free from any side effect.

DISCUSSION

Although many plants belonging to different angiospermic families have been screened for their antifungal activity, *E. citriodora* belonging to the family Myrtaceae is reported for its antifungal activity against post harvest fungal pathogens probably for the first time. Substances may inhibit the growth of fungi of either temporarily (fungistatic) or permanently (fungicidal). Essential oils obtained from the leaves of *Cymbopogon martinii* var. motia (Dikshit et al., 1980), *Hyptis suaveolens* (Pandey et al., 1982), *Melaleuca leucodendron* (Dubey et al., 1983) and the rhizome of *Alpinia galganga* (Tripathi et al, 1983) was found to have fungistatic activity. Whereas essential oils from *Cymbopogon pendulus* (Pandey et al., 1996) that have fungicidal. However, in the present investigation the oil of *E. citriodora* like those of *C. flexuosus* (Shahi et al., 2003) prove to have fungistatic activity at lower concentration and fungicidal at higher concentration. A fungicide must not be affected by extremes of temperature. Only a few workers have studied the effect of temperature on antifungal activity of the oils, but the oil of *Pepromia pellucida* was reported to

Table 2. Minimum bioactive concentrations of the oil of *Eucalyptus citriodora* against fungal pathogens.

Fungi	Percentage mycelial growth inhibition at different concentration ($\mu l\ ml^{-1}$) after 7th day of inoculation							
	0.2	0.4	0.6	0.8	1.0	1.2	1.4	1.6
Alternaria alternata	54.2	100[s]	100[s]	100[s]	100[c]	100[c]	100[c]	100[c]
Aspergillus flavus	41.2	70.6	91.2	100[s]	100[s]	100[c]	100[c]	100[c]
Aspergillus fumigatus	55.5	74.3	90.1	100[s]	100[s]	100[c]	100[c]	100[c]
Aspergillus niger	45.2	89.0	100[s]	100[s]	100[s]	100[c]	100[c]	100[c]
Aspergillus parasiticus	42.0	91.0	100[s]	100[s]	100[s]	100[c]	100[c]	100[c]
Botrytis cinerea	54.0	98.0	100[s]	100[s]	100[c]	100[c]	100[c]	100[c]
Cladosporium cladosporioides	71.0	82.2	100[s]	100[s]	100[c]	100[c]	100[c]	100[c]
Colletotrichum capsici	76.2	91.2	100[s]	100[s]	100[c]	100[c]	100[c]	100[c]
Colletotrichum falcatum	56.2	69.2	100[s]	100[s]	100[c]	100[c]	100[c]	100[c]
Curvularia lunata	76.2	81.0	90.2	100[s]	100[s]	100[c]	100[c]	100[c]
Fusarium cerealis	67.1	90.2	100[s]	100[s]	100[c]	100[c]	100[c]	100[c]
Fusarium culmorum	69.2	81.2	100[s]	100[s]	100[c]	100[c]	100[c]	100[c]
Fusarium oxysporium	70.1	89.3	100[s]	100[c]	100[c]	100[c]	100[c]	100[c]
Fusarium udum	67.2	76.0	81.2	100[s]	100[s]	100[c]	100[c]	100[c]
Gloeosporium fructigenum	45.2	76.2	100[s]	100[s]	100[c]	100[c]	100[c]	100[c]
Helmenthosporium maydis	79.2	89.2	98.9	100[s]	100[s]	100[c]	100[c]	100[c]
Helmenthosporium oryzae	75.1	92.2	95.4	100[s]	100[s]	100[c]	100[c]	100[c]
Penicillium digitatum	61.2	75.1	100[s]	100[s]	100[c]	100[c]	100[c]	100[c]
Penicillium expansum	53.1	81.2	100[s]	100[s]	100[c]	100[c]	100[c]	100[c]
Penicillium italicum	65.2	93.1	100[s]	100[s]	100[c]	100[c]	100[c]	100[c]
Penicillium implicatum	71.2	80.1	100[s]	100[s]	100[c]	100[c]	100[c]	100[c]
Penicillum minio-luteum	69.0	78.2	100[s]	100[s]	100[c]	100[c]	100[c]	100[c]
Penicillum variable	50.1	71.2	81.2	100[s]	100[s]	100[c]	100[c]	100[c]
Phoma violacea	40.2	70.1	83.1	100[s]	100[s]	100[c]	100[c]	100[c]
Rhizopus nigricans	60.2	81.2	91.0	92.8	100[s]	100[s]	100[c]	100[c]

s, Fungistatic action; c, fungicidal action.

Table 3. Phytotoxicity of oil on fruit skin (epicarp).

Concentration ($\mu l/ml$)	Phototoxic effect on different concentration					
	Colour	Odour	Weight	Size	Taste	Changes in epicarp
10	-	-	-	-	-	-
20	-	-	-	-	-	-
30	-	-	-	-	-	-
40	-	-	-	-	-	-
50	-	-	-	-	-	-
60	+	-	-	-	+	-
70	+	+	-	-	++	+
80	+	++	-	-	++	+
90	++	++	-	-	++	+
100	++	+++	-	-	++	+

-, No effect; +, mild; ++, moderate; +++, significant.

be active up to 80°C (Singh et al., 1984), in the present study the oil of *E. citriodiora* retained activity up to 100°C. A substance may be fungicidal against certain fungi yet ineffective against other pathogens.

Therefore, a clear picture of the toxicity of a fungicide comes only after it is tested against a large number of fungi. The literature shows that the essential oils have been found to exhibit a narrow or wide range of activity

Table 4. *In vivo* efficacy of the oil for the inhibition of inoculum in *Malus pumilo*.

Concentrations (µl ml⁻¹)	Percentage inoculum growth inhibition at different treatments	
	Pre-inoculation treatment	**Post-inoculation treatment**
10	76.2	59.2
20	100	68.2
30	100	100
40	100	100
50	100	100
Variance	113.288	405.432
Std. Dev.	10.6437	20.1353
Std. Err.	4.76	9.0048

ANOVA summary					
Source	**SS**	**df**	**MS**	**F**	**P**
Treatment (between groups)	238.144	1	238.144	0.92	0.365562
Error	2074.88	8	259.36		
Total	2313.024	9			

This test will be performed only if K>2 and the analysis of variance yields a significant F-ratio.

■ Post-inoculation treatment   ■ Pre-inoculation treatment

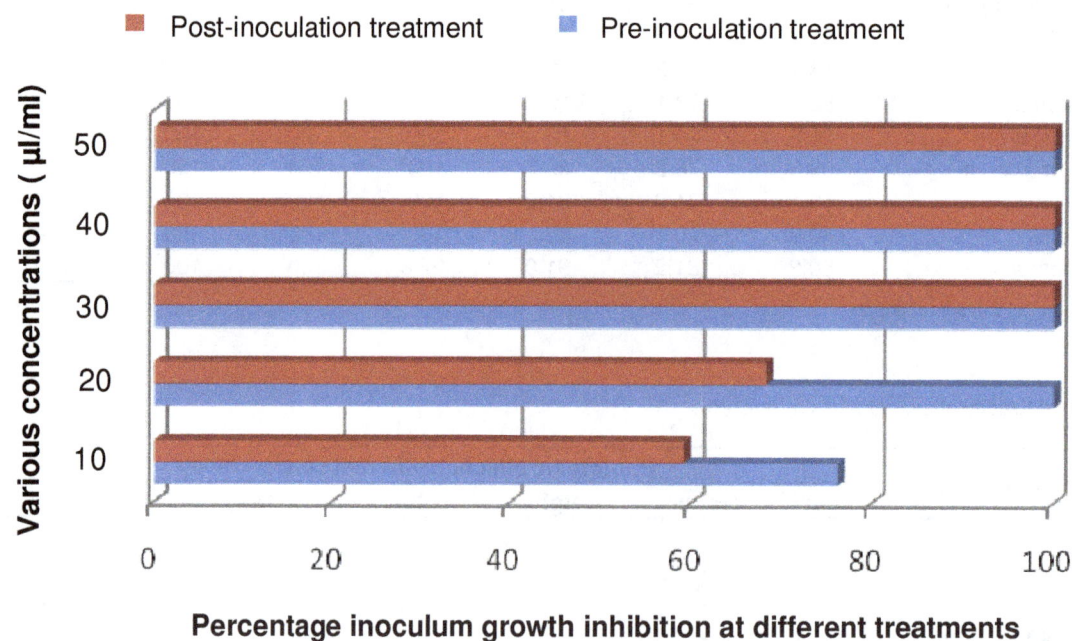

Percentage inoculum growth inhibition at different treatments

Figure 1. Bar graph showing percent fruit loss protected by the application of formulated fungicidal spray.

(Singh et al., 1980; Pandey et al., 1982; Dubey et al., 1983), but in the present study the *E. citriodora* oil exhibited a broad antifungal spectrum. Antifungal active oils derived from plants are generally non-phytotoxic (Pandey et al., 1982; Tripathi et al., 1983). In the present study, the oil was found to be non-phytotoxic at morphological level. Additionally, in preliminary *in vivo* trials, it has also been found effective in the control of fruit rot of *M. pumilo*. A chemical should be tested under both *in vitro* and *in vivo* conditions in order to prove its potential as promising antifungals for the control of disease. Since, detailed *in vitro* studies on the essential oil of *E. citriodora* indicate its potential as ideal antifungal compounds against post harvest spoilage fungi; it was further subjected to *in vivo* investigation, so as to confirm their efficacy as a natural product for the control of rotting in fruits. The present study clearly demonstrates that oil of *E. citriodora* holds a good promise as an antifungal against post harvest spoilage an account of their following virtues namely, strong efficacy against fungi with fungicidal action, potentiality against heavy fungal inoculum, long shelf life, thermostable, wide range of antifungal activity and absence of any phytotoxic effects and better result during *in vivo* trials. The oil in the form of fungicidal spray can be exploited commercially after undergoing successful completion of wide range of field trial and to find out their economic viability.

ACKNOWLEDGEMENTS

Thanks are due to Head, Department of Microbiology, Chaudhary Charan Singh University, Meerut for providing the facilities and for Department of Science and Technology for financial assistance.

REFERENCES

Barrera-Neeha LL, Garduno-Pizana C, Garduno-Pizana LJ (2009). *In vitro* antifungal activity of essential oils and their compounds on mycelial Growth of *Fusarium oxysporium F.* sp. *Gladioli* (Massey) Snyder and Hansen. Plant Pathol. J., 8(1): 17-21.

Clevenger JF (1928). Apparatus for the determination of volatile oil. J. Ame. Pharmacol. Assoc., 17: 346.

Dikshit A, Singh AK, Dixit SN (1980). Antifungal properties of *Palmarosa* oil. Ann. Appl. Biol., 97(supp): 34-35.

Dubey NK, Kishore N, Singh SN (1983). Antifungal properties of the volatile fraction of *Melaleuca leucodendron*. Trop. Agric. (Trinidad), 60: 227-228.

Eckert JW, Sommer NF (1967). Control of diseases of fruits and vegetables by postharvest treatment. Ann. Rev. Plant Pathol., 5: 391-432.

Garber RH, Houston BR (1959). An inhibitor of *Verticillium albo-atrum* in cotton seed. Phytopathology, 49: 449-450.

Grover RK, Moore JD (1962) Toxicometric studies of fungicides against brown rot organisms *Sclerotinia fructicola* and *S. laxa*. Phytopathology, 52: 876-880.

Harvey JM (1978). Reduction of losses in fresh fruits and vegetables. Ann. Rev. Phytopath., 16: 321-341.

Langenau EE (1948). The examination and analysis of essential oils, synthetics and isolates. In: Guenther (ed), The Essential Oil. Robert E. Krieger Publishing Co., Huntington, New York.

Neergaard P (1977). Seed Pathology, The Mc. Million Press Ltd. London and Basingstoke.

Pandey DK, Tripathi NN, Tripathi RD, Dixit SN (1982). Fungitoxic and phytotoxic properties of the essential oil of *Caesulia axillaris* Roxb. Angew Bot., 56: 256-257.

Pandey MC, Sharma JR, Dikshit A (1996). Antifungal evaluation of the essential oil of *Cymbopogon pendulus* (Nees ex Steud.) Wats var. Parman. Flavor Fragr. J., 11: 257-260.

Samson RA, Hoekstra ES, Frisvad JC, Filterborg O (1995). Introduction to food borne fungi. Ponsen and Looyen, Wageningen, The Netherlands.

Shahi SK, Patra M, Shukla AC, Dikshit A (2003). Use of essential oil as biopesticide against post harvest spoilage in fruits, *Malus pumilo*. Bio-Control, 48(2): 223-232.

Shahi SK, Shukla AC, Bajaj AK, Midgely G, Dikshit A (1999). Broad spectrum antimycotic drug for the control of fungal infection in human beings. Curr. Sci., 76: 836-839.

Simmonds JH (1963). Studies in the latent phase of *Colletotrichum* species causing ripe rots of tropical fruits. Qld. J. Agr. Sci., 20: 373-424.

Singh AK, Dikshit A, Dixit SN (1984). Antifungal studies of *Peperomia pellucida*. Beitr. Biol. Pflanzen, 58: 357-368.

Singh AK, Dikshit A, Sharma ML, Dixit SN (1980). Fungitoxic activiy of some essential oils. Econ. Bot., 34: 186-190.

Somasundaram L, Coats JR, Racke DK, Stahr HM (1990). Application of the Microtox system to assess the toxicity of pesticides and their hydrolysis metabolites. Bull. Environ. Contam. Toxicol., 44: 254-259.

Tripathi NN, Dubey NK, Dikshit A, Tripathi RD, Dixit SN (1983). Fungitoxic properties of *Alpinia galanga*. Trop. Plant Sci. Res., 1: 49-52.

15

Effect of small-scale irrigation on the income of rural farm households: The case of Laelay Maichew District, Central Tigray, Ethiopia

Kinfe Asayehegn[1]*, Chilot Yirga[2] and Sundara Rajan[3]

[1]Institute of Environment, Gender and Development Studies, Awassa College of Agriculture, Hawassa University, Ethiopia.
[2]Ethiopian Institute of Agricultural Research (EIAR), Ethiopia.
[3]Department of Rural Development and Agricultural Extension, Harapmaya University, Ethiopia.

Agricultural intensification is presumed to be a necessary pre-condition for the development of the agricultural sector in Ethiopia. To this end, various government and non-governmental organizations (NGOs), among others, initiated small-scale irrigation schemes throughout the country including the Tigray region. Despite these efforts, however, smallholder farmers particularly in the study area are found to be reluctant to participate in small-scale irrigation schemes. This study therefore, assessed the factors that affect smallholder farmers' participation in small-scale irrigation of the study area. A two-stage sampling procedure was used to first select peasant associations and then sample respondents. Descriptive statistics and binary probit estimation were used to estimate the determinants of small-scale irrigation participation. The analysis revealed that income, gender, access to market information and health condition of households were found to be important determinants for participating in small scale irrigation schemes. Hence, improving rural farm households' access to market information and health services, are likely to improve participation in irrigation schemes thereby improving of small holder farmers income.

Key words: Ethiopia, income, irrigation, rural farm households, small-scale.

INTRODUCTION

Ethiopia is an agrarian country where around 95% of the country's agricultural output is produced by smallholder farmers (MoARD, 2010). The contribution of agriculture to national GDP (50%), employment (85%), export earnings (90%), and supply of industrial raw materials (70%) has remained high (World Bank, 2010). Although the country is endowed with three main resources namely land, water and labor for production, agriculture in the country is mostly small- scale, rainfall dependent, traditional and subsistence farming with limited access to technology and institutional support services. Hence, the ability of the nation to address food and nutritional insecurity, poverty, and to stimulate and sustain national economic growth and development is highly dependent on the

performance of agriculture. Yet achieving higher and sustained agricultural productivity growth remains one of the greatest challenges facing the nation (Belay and Degnet, 2004; Spielman et al., 2010).

Irrigation contributes to livelihood improvement through increased income, food security, employment and poverty reduction. To this end, Hussain and Hanjira (2004) confirmed a strong direct and indirect linkage between irrigation and poverty. Direct linkages operate through localized and household level effects, whereas indirect linkages operate through aggregate or sub-national and national level impacts. Irrigation benefits the poor through higher production, higher yields, lower risk of crop failure, and higher and year-round farm and non-farm employment. Irrigation enables smallholders to adopt more diversified cropping patterns, and to switch from low-value staple production to high-value market-oriented production. Increased production makes food available and affordable for the poor. Since irrigation investments leads to production and supply shifts, indirect linkages operate through regional and national level and have a strong positive effect on the national economy. Similar study from Gambia revealed that irriga-tion provided smallholder farmers the chance for increasing income that was reflected on increased expenditure, investment in productive and household assets, saving and trade (Webb, 1991). In India poverty head count ranges from 18 to 53% in irrigated and 21 to 66% in rain fed areas and poverty incidence is 20 to 30% lower in most irriga-ted areas compared to rain fed areas. Incidence of chronic poverty is 5% lower for irrigated areas in Sri Lanka (Pakistan) than adjoining rain fed areas (Hussain and Hanjra, 2004).

Besides its positive effect, irrigation utilization decision comprises different determinant factors. Some of the fact-ors facilitate for utilization decision while others not yet. Hence, a study carried out by Desta (2004) and Tafesse (2007) on impact of community-managed irriga-tion on farm production efficiency and household income in Oromia National Regional State, Ethiopia found that education of the household head, livestock ownership, access to irrigation technology, amount of credit received, age of household head, distance from market, participation in extension package program, years of irrigation experience, total income of house-holds, access of the household to improved seed and farm size were the significant determinants of household decision on irri-gation utilization. This was also confirmed by Takele (2007) that in addition to the afore-mentioned factors dependency ratio, active labor force, sex of household head, insect and pest infestation, training received, and ownership of radio are found significant in determining the decision of small-scale irrigation utilize-tion.

The study area is one of the most land-degraded states of Ethiopia. Crop production in the region has failed to keep pace with population growth due to recurrent droughts, environmental degradation and wars, including the most recent conflict with Eritrea (Ersado, 2005). In response to severe environmental degradation and population-resource imbalance, the regional government of Tigray has initiated a major rural development program called Sustainable Agricultural and Environmental Reha-bilitation of Tigray (SAERT), through which several small-scale dams have been, constructed (Ersado et al., 2004). Farm households within the peasant association, which had rain-fed land, were given equal opportunity to own irrigated land. However, some farm households dis-regarded to possess parcels in the irrigable section of the peasant association at will due to different factors. Moreover, most studies in Ethiopia focus on technical aspects of irrigation schemes, and very little is known for the socio-economic factors that have implications on

irrigation participation (Van Den Burg and Ruben, 2006). Therefore, this research is aimed at primarily identifying, analyzing, and documenting the socio-economic and institutional factors affecting household level irrigation utilization that contributes its part to the existing body of knowledge. Secondly, it provides a base for policy ma-kers through the comparisons of positive and negative effect of irrigation with respect to similar areas in specific. Thirdly, it provides directions for further research, extension and development schemes that will benefit the scheme beneficiaries.

METHODOLOGY

Sample and sampling design

A two stage sampling procedure was followed to first select peasant associations and then sample households. In the first stage, three peasant associations where the three micro-dams found were selected purposively. Before selecting household heads to be included in the sample, the sampling frame was stratified into irrigation water user and nonuser households. The stratum of irrigation user consists of households who own, rented/shared in/out or gifted in land for direct utilization. The second stratum referred to hereafter as non-users is composed of households who neither owned irrigated land nor involved in irrigation farming. In the second stage, 130 farm households consisting of 65 irrigation users and 65 nonusers were selected from the identified list using simple random sampling technique taking into account probability proportional to size of the identified households in each of the three selected peasant associations.

Data collection and analysis

A structured interview schedule supported by personal observations of physical features and informal discussion with key informants was used to collect primary data. In addition to primary data, secondary data were collected from the District Offices of Irrigation Development (DOID) and District Offices of Rural and Agricultural development (DORAD)

Initially, the research had two objectives. However, for this paper the second objective is excluded and is organized and presented in other way with a similar analysis method. Descriptive statistics (mean, frequency, percentage and standard deviation) and binary probit were used to analyze the collected data. The statistical significance of the variables in the descriptive part was tested for both dummy and continuous variables using chi-square and t-test, respectively.

There are various ways of estimating the parameters of dichotomous qualitative response regression models. Thus, include LPM, binary probit and logit models. All these models have in common the fact that they are models in which the dependent variable is a discrete outcome such as Yes or No decision (Maddala, 1997). The most widely used discrete response models are probit (which is associated with cumulative normal distribution) and logit model (which is assume cumulative logistic probability function). In these models, the probabilities are bound between 1 if the household is user of irrigation and 0 otherwise and they fit well to non linear relationship between the probabilities and the explanatory variables. However, Maddala 1983 and Gujarati 1995 have noted that, in most applications the cumulative normal distributions (binary probit) and logistic function (logit) is quite similar, the difference being that the logistic function has slightly fatter tails. Hence, there is no compelling reason to choose one over the other and the choice is dependent upon personal preference and experience. Therefore, due to such circumstances, this study used binary probit model to analyze the factors affecting small-scale irrigation utilization.

RESULTS AND DISCUSSION

Socio-demographic characteristics of the households

Gender of the household heads regardless of the age group is an important variable influencing the participation decision in irrigation. The total sample of the study is composed of 20% female headed households while the portion of female headed households who are irrigation users is reduced to 12%. Discussion with sample households revealed that male-headed house-holds hardly faced labor shortage for irrigation as well as rain-fed farming due to physical, technological, socio-cultural and psychological fitness of farm instrument to males than females. Similarly, education plays a key role for household decision in technology adoption. It creates awareness and helps for better innovation and invention. The study revealed that 40% of the users and 60.8% of the nonusers of small-scale irrigation are illiterate. It is also found that the number of irrigation users who completed nine years of schooling and above is twice as compared to nonusers.

The average household size for the users and nonusers of small-scale irrigation is found to be 6.43 and 5.15, respectively (Table 1). This result is statistically significant suggesting labor availability is an important factor influencing households' decision to participate in small-scale irrigation schemes. The result also revealed, as active family labor or work force of a household in adult equivalent increases, the total income of the household increases, which in turn contributed to improved well-being, further providing an evidence for the importance of labor availability in influencing the participation decision of households in small-scale irrigation.

Irrigation labor force is the amount of labor needed for irrigation activities. Similarly, rain-fed labor is the labor required for rain-fed activities. Irrigated and rain-fed agriculture requires diverse labor force both in quantity and technical quality. Evidences from the study as stated in Table 2, demonstrated 44.6% of the users of small-scale irrigation faced labor shortage for irrigation activities while 30.9% of the users and 24.6% of the nonusers faced labor shortage for rain-fed activities. Farm households who faced labor shortage employ different

mechanisms to acquire additional labor required for accomplishing farm activities. A total of 76.9 and 23.1% of the irrigation users, which faced labor shortage, acquired additional labor through hiring and labor exchange mechanisms, respectively. Likewise, 77 and 23% of the labor deficient irrigation users used hired and exchange labor, respectively, to solve the problem of labor shortage for rain-fed farming. Similarly, a total of 24.6 and 75.4% of the labor deficient irrigation nonusers used hired and exchange labor for rain-fed farm activities. It worthy of note that 35.4% of the casual labor employed in irrigation farming were source from the nonusers of irrigation within the kebele/Woreda whereas 64.6% move toward from nearby kebele/Woreda that are very little in irrigation sources. This proves irrigation intensifies labor and is preeminent strategy of employment in countries like Ethiopia with elevated population growth rates.

Irrigation user households also compared the labor consumption ratio of irrigated farming to rain-fed farming, which accounts 12.3, 70.8, 15.4 and 1.5% as equal, two times, three times and four times respectively. The farm households replied from the point of view of their activities and economy. Equal and three or four times ratio is for the farm families specialized on cereal and vegetable crops respectively and two times is from the farm households which diversified on cereal and vegetable crops. This replies that the labor consumption for vegetable farming is double as compared with cereal crops.

Resource ownership and farm experience

Resource ownership and farm experience have a profound effect on the participation decision-making behavior of farm households. The variables experience in rain-fed farming and rain-fed land holding pertain to both users and nonusers of small-scale irrigation while the variables irrigation experience and irrigable land holding pertain to users only. Both irrigation user and non-user households of the area have an average land size of 1.1 and 0.627 ha respectively. The survey revealed that 10.8% of the users of irrigation do not own rain-fed land at all rather than irrigated land. On the other hand, of the total respondents, 4.6% of the users and 7.7% of the nonusers do not owned any parcel of land but always use sharecropping arrangements. Findings of the survey revealed that 58.5% of the users and 17% of the nonusers shared in land, while 16.9% of the users and 24.6% of the nonusers shared out their own land. This shows that irrigation users are better practice land shared in than nonusers are. The land shortage and searching for additional land is the motivating factor for shared in (Table 2).

Table 1. Distribution of respondents by demographic characteristics.

Sex	User		Nonuser		Total		x^2
	N	%	N	%	N	%	
Female	8	12.3	18	27.7	26	20	
Male	57	87.7	47	72.3	104	80	3.894**

	Mean	SD	Mean	SD	Mean	SD	T-value
Education (years)	2.26	2.917	1.49	2.646	1.88	2.801	1.575
Family size	6.43	2.038	5.15	1.946	5.79	2.086	3.653***
Family labor	3.71	1.665	2.57	1.468	3.14	1.665	4.135***

*** and ** statistically significant at less than 1and 5% probability level respectively.

Table 2. Distribution of respondents based on land ownership and farm experience.

	Users		Nonusers	
	Frequency	Percent	Frequency	Percent
Shared in	38	58.5	11	17
Shared out	11	16.99	16	24.6
Reasons for not using irrigation				
Land shortage			17	26.2
Limited information			21	32.4
Have fertile rain-fed land			27	41.4

Land holding in ha	User	Nonuser	Total	T-value
	Mean	Mean	Mean	
Total cultivated land	1.1	0.627	0.856	5.826***
Irrigable land	0.5	0.000	0.247	13.531***
Rain-fed land	0.6	0.627	0.608	0.546
Farm experience in years				
Rain-fed	33.37	29.68	31.52	1.706**
Irrigation	11.86	0.000	5.93	14.757***

*** and** statistically significant at 1 and 5% probability level respectively.

Irrigation non-user households which have equal opportunity with the users, have different reasons for rejecting irrigation utilization. Some of them are due to lack of farmland at the time of redistribution while others are due to information gap and lack of awareness on irrigation. Farmers' expectation of the rain-fed land they owned is too fertile and can produce better is the other reason that motivates the rejection of irrigation utilization. With regard to farm experience of the households, findings compared that 55.4% of the irrigation users and 35.4% of the irrigation nonusers have more than 30 years of rain-fed farm experience, respectively. Likewise, 55.4 and 1.5% of the users of small-scale irrigation have 12 and more than 30 years of irrigation experience respectively. The t-test on rain-fed experience between

users and nonusers of irrigation showed that there is a significant difference between irrigation user and non-user households at 5% level (Table 2).

Income distribution and inequalities of the households

Some of the households specialized in primarily irrigation dependent livelihoods while others base their livelihood on a diverse range of livelihood activities but out of irrigation. There are also households which diversify their livelihood as irrigation dependent and irrigation independent. Specifically for income and activities, households diversify to different sources. On-farm income (such as income from irrigated crop, rain-fed crop or livestock production/rearing), off-farm income (such as trading of agricultural products), and non-farm income (such as non-farm employment, non-farm trade), are the different income portfolios in which the households of the study area diversify their activities. The survey results found that there is a significant difference in mean total household income between irrigation user and non-user livelihoods. It is found that 10.8% of the irrigation users do not have any income from rain-fed crop production other than irrigation products. The results of the survey also compared that the ratio of mean total income of irrigation users to nonusers exceeds by 37.03% and nutritional status and standard of living of the users also increased by the same factor as income.

An entire 63.1% of the users and 67.7% of the nonusers of small-scale irrigation do not participate in any off-farm activities. Thus, households base their livelihood on non-farm and on-farm income portfolios. With regard to livestock production as an on-farm income, irrigation users gain income from livestock 13.8% larger than irrigation nonusers do. Remittance also covers 1.5 and 2.2% of the total income of the users and nonusers of small-scale irrigation respectively (Table 3). Generally, initial income received from non-farm and off-farm activities help farm households to participate in small-scale irrigation through coverage of initial costs such as costs for inputs, draught power e.t.c. However, once the farm families transformed from rain-fed to irrigation livelihoods, it directly minimize their off-farm income due to the load in the labor intensive activity of irrigation.

Social participation and access to infrastructural facilities

Irrigation intensifies input and labor. Credit either in the form of cash or kind from different sources, is an important institutional service to finance poor farmers for input purchase and ultimately to adopt new technologies. However, some farmers have access and utilization to credit while others do not, due to problems related to repayment and down payment in order to get input from formal sources. The survey indicated 78.4% of the non-users and 89.2% of the users of irrigation had utilization to credit although the access is equal to all households without any difference. Credit nonuser households reject credit utilization due to different reasons. The results contended that 7.7% of the users of irrigation, which spurn credit utilization, hardly faced any problem due to their limited need. On the other hand, 6.2 and 7.7% of the nonusers of irrigation eschew credit utilization due to their limited need and fear of failure to pay respectively. It is also found that 4.6% of the nonusers of irrigation reserved from irrigation utilization due to expectations of

high interest rates of the credit. An equal amount 3.1% of the users and nonusers of irrigation restricted themselves from credit utilization due to religion restrictions locally called haram.

Rural farm households engage in different positions of informal and formal institutions such as Mahber, Idir, water user association, peasant association and Woreda administration of their locality. The ratio of small-scale irrigation user households to nonuser households who are in different positions of the community exceeds by 47.7%. The main reason for the gigantic difference between irrigation user and nonuser households in their position in the community is due to the access and utilization of information. Information on market prices and channels is one of the important aspects for livelihood improvement of rural farm households. Although information on marketing of irrigation products and agricultural inputs is a determinant factor for producers, only 75.4% of the irrigation users have access to information. As a source of information, 7.7 and 67.7% of them use telephone (fixed or mobile) and person to person information sharing respectively. This shows even in the age of information era, people in such areas still using traditional way of information sources and means.

Probit model

As stated in Table 4, Farm households of the area have different income sources. On-farm income refers to the total income from irrigated and rain-fed crops. Similarly, off-farm income is a type of income, which is derived from sources such as trading of agricultural products. None-farm income on the other hand is a type of income resulted totally out of agriculture and agricultural products. The econometric results confirmed that there is a positive and significant relationship between on-farm income of households and irrigation participation at less than 1% significant level. The positive effect between on-farm household income and participation in irrigation farming suggests that income derived from on-farm activities enables households to pay for farm inputs required for profitable irrigation farming. The marginal effect shows that as on-farm income of households increases by 100 Birr, the probability of a household's participation in small-scale irrigation increases by 1%. However, off-farm income significantly and negatively influenced the likelihood of participation in irrigation-farming suggesting households engaged in off-farm activities are less likely to participate in irrigation. This negative relationship depicts the likelihood of participation in irrigation would be reduced by 1% for every 100 ETB earned from off-farm activities, as off-farm activities withdraw active labor from participating in irrigation.

Higher market prices of irrigation products are likely to

Table 3. Distribution of respondents based on their mean household income.

Source of income		Users	Nonusers	T-value
On-farm	Irrigation	12934.98	0.00	10.169***
	Rain-fed	5225.32	7084.61	-2.878***
	Irrigation and rain-fed	18160.31	7084.61	7.143***
	Livestock	1864.46	1010.46	3.026***
	Total	20024. 76	8091.07	7.497***
	Expense for crop production	6695.76	2184.64	7.273***
	Net income	12285.92	5878.73	6.065***
Off-farm		746.46	600.30	0.488
Non-farm		2023.07	2572.46	-0.669
Remittance		353.78	249.23	0.412
Property/income		33052.78	14318.91	2.723***
Total income		56200.87	25831.98	4.217***

*** and** statistically significant at less than 1 and 5% probability level respectively.

Table 4. Maximum likelihood estimates of the probit model.

Variable	Coefficients	T-value	Marginal effect
Constant	-4.75882	-2.32099**	-1.5668
Education level	0.012903	0.137887	0.0042
Family labor force	0.168341	0.866935	0.0554
Age of the household head	0.0335619	1.18129	0.0111
On farm income	0.000172252	2.81975***	0.0001
Off arm income	-0.000378195	-1.87574*	-0.0001
Nonfarm income	-0.000149225	-1.26935	0.0001
Remittance	-0.000193725	-0.901888	-0.0001
Property income	7.08725e-006	0.704812	0.0000
Distance from irrigation to market	0.0357116	0.612217	0.0118
Distance from irrigation to home	-0.598272	-3.01655***	-0.1970
Rain-fed land	-1.48404	-1.44643	-0.4886
Total livestock unit	-0.0461839	-0.306553	-0.0152
Sex	1.15819	1.70084*	0.3813
Market information	4.73361	4.18098***	1.5585
Access to credit	-0.460747	-0.589819	-0.1517
Health condition	1.54415	1.98631**	0.5084

Dependent variable	Irrigation participation decision
Weighting variable	ONE
Number of observations	130
Log likelihood function	-19.87096
Restricted log likelihood	-90.10913
Chi-squared	140.4763
Degrees of freedom	16
R-square	0.685043

***, **, and * indicates significant at less than 1, 5, and 10% probability level respectively.

motivate farm households to participate in small-scale irrigation schemes. The marginal effect revealed that the probability of participation in irrigation for a household, with a reasonably good access to market information would by nearly twice than households who do not have access to market information. Similarly, household's residence to water sources is found to have a significant and negative relationship to the probability of participation in small-scale irrigation. The negative sign indicates that the farther water source from a household's residence the lower the likelihood of participation in irrigation farming. Conversely, the nearer a household resides to a water source, the higher the probability of participating in irrigation scheme due to the fact that the opportunity cost of the time lost in travelling to and from an irrigation-farm for households located, a short distance from irrigation schemes would be much lower than households located much farther. Besides, the lower transaction cost households located near water sources enjoy are likely to have a better awareness of the associated agricultural technologies due to their proximity. Keeping other variables constant at their respective mean level, the probability of participating in irrigation for a household increased by 19.7% for as the distance of water source from his/her residence reduces by one kilometer.

Discussion with sample households and key informants revealed that male-headed households hardly faced labor shortage for irrigation as well as rain-fed farming due to physical, technological, socio-cultural and psychological fitness of farm instrument to males than females. Moreover, the income of male-headed households is higher, compared to female-headed households further increasing the comparative advantage of male-headed households to engage in irrigated farming than female-headed households do. The results of the econometric model proved that gender of the household is an important variable influencing the participation decision. The marginal effect of gender indicates the probability of participation in irrigation for a male-headed household increases by 38.13% compared to a female-headed household, given other variables are kept at the average level. In addition to gender, the health status of a household is an important variable influencing participation in program interventions. Disease, disabilities and extra old age affects irrigation participation through reduction of active labor for production and adding expenses for medication. The positive and significant relationship of health status of the household head with participation in small-scale irrigation indicates the probability of a household's participation in irrigation increases by 50.84% for a healthy household head compared to a household with poor health or with some type of disability.

CONCLUSION AND POLICY IMPLICATION

Irrigation intensifies input and labor throughout the year.

It motivates self-employment offsetting fulltime and part time off-farm or non-farm employment, due to efficient utilization of labor. This indicates off-farm income inspires to withdraw active labor force from irrigation activities and placing to off-farm income driving activities reduces irrigation participation of farm households. Farm households that have access to market information are able to compare, the net income from rain-fed and irrigation farming. Moreover, it assists purchasing of the right input at the right time from the right enterprise and supplying of the products to the right customer with a reasonable intermediary cost. However, the gender difference of household heads in irrigation participation indicated female-headed households face shortage of labor and market information, made them rent/shared out their land. Networking of rural farm households with their customers through information sources such as mobile and telephone service is a determinant factor. Accessing of labor saving technologies easily managed by women solves the workforce problem of female-headed house-holds. Special attention for female-headed households considering their gender mainly in criteria of accessing irrigable land facilitates women participation in irrigation. Insuring property ownership of female-headed households through credits and self-help groups is the other mechanism of increasing female-headed house-hold's participation in irrigation.

Access of farm households to irrigated land enables them to diversify their income sources, including cash and food crops, and to make savings. Livestock serve as a source of income for irrigation input purchase and draft power. Wealth status of households also determined by the livestock, they owned mainly oxen. Crop failure risk is minimized if the household owned livestock due to expectations of compensating failed crop through sale of their livestock. Credit is an important institutional service to finance poor farmers for input purchase, able to access draft power and ultimately to adopt new technologies. Saving livestock from sale and land from rent out or shared out, at uncertain seasons is feasible due to credit utilization and double season production. Although increasing the total land size is unfeasible, replacement of the rain-fed land by irrigable land through development of new dams and applying different irrigation technologies is crucial. Due attention to livestock production through introduction of zero grazing systems to make livestock production is friendly with environment, irrigable land and ecology conservation in general is vital. Microfinance institutions are better to provide credit, at reasonable interest rate, and at the right time credit be demanded at places where farm households can access easily.

Household members, who are free of disease, and disabilities, have productive labor for irrigation. The burden of caring and treating sick, disabled or extra old age reduces the active labor for irrigation not only labor of the diseased or disabled individual, but also labor of the other members of the household that leads to dual sentence.

Provision of social services such as health services and road at village level is essential in changing the life and active labor of the farm households. Informal education on health aspects, nutrition, hygiene and sanitation also play role on preventing and curing of disease that leads to better utilization of irrigation.

REFERENCES

Belay K, Degnet A (2004). Challenges Facing Agricultural Extension Agents: A Case Study from South-western Ethiopia. African Development Bank, Blackwell Publishing Ltd, Oxford, UK.

Ersado L, Amacher GS, Alwang J (2004). Productivity and land enhancing technologies in northern Ethiopia: Health, public investments, and sequential adoption. Am. J. Agric. Econ., 86(2): 321-331.

Ersado L (2005). Small Scale Irrigation Dams, Agriculture Production, and Health: Theory and Evidence from Ethiopia. World Bank Policy Research Working Paper 3494. The World Bank: Washington DC.

Gujarati DN (1995). Basic econometrics. 3rd edition, McGraw Hill, Inc., New York.

Hussain I, Hanjra M (2004) Irrigation poverty alleviation: review of the empirical evidence. International Water Management Institute, Colombo, Sri Lanka.

Maddala GS (1983). Limited Dependent and Qualitative Variables in Econometrics. Cambridge University Press, Cambridge.

Maddala GS (1997). Limited Dependent and Quantitative Variables in Econometrics. Cambridge University Press: Cambridge.

MoARD (Ministry of Agriculture and Rural Development) (2010). Ethiopia's Agriculture Sector Policy and Investment Framework: Ten Year Road Map (2010-2020).

Spielman D, Byerlee D, Avid J, Alemu D, Kelemework D (2010). Policies to promote cereal intensification in Ethiopia: The search for appropriate public and private roles, Food Policy, 35: 185-194.

Van Den Burg M, Ruben R (2006). Small-Scale Irrigation and Income Distribution in Ethiopia. J. Develop. Studies: Wageningen University, The Netherlands, 42(5): 868-880.

Webb P (1991). When projects collapse: Irrigation failure in the Gambia from a household perspective. J. Intern. Dev. Inst., Vol. 3(4), Washington D.C.

World Bank (2010). Ethiopian Agricultural Growth Project. Project appraisal document. Addis Ababa: Ethiopia.

16

Comparison of three methods of weight loss determination on maize stored in two farmer environments under natural infestation

Ngatia, C. M.* and Kimondo, M.

KARI – Kabete, Post harvest research, NARL. P. O. Box 14733-00800 Nairobi, Kenya.

Common methods of weight loss assessment in stored grain include the standard volume weight (SVW), count and weigh (C&W), the thousand grain mass (TGM) and the indirect with a conversion factor (CF) which have been used in varying storage environments. Apart from accuracy and reliability, practical application may limit their use in rural areas. Three of the methods: (SVW) or Bulk density (BD), C&W and CF were evaluated on maize stored in two farmer environments exposed to natural infestation. Baseline damage parameters: bulk density, grain moisture, sieved dust, weevil damage and insect pests per kilogram were established and again after 24 weeks. Weight loss was calculated using Equations 1 to 3. Percent weight loss varied by wide margins between treated and untreated maize: 4.4 to 12.3% (in Crib) and 0.3 to 9.9% (in house) for BD; 2.3 to 5% (in Crib) and 2.2 to 13.4% (in house) for C&W and 2.5 to 6.6% (in Crib) and 2 to 7% (in house) for the CF method. Generally, the house environment favoured pest establishment resulting to higher sample and cumulative weight loss in untreated maize. All the three methods had closely related weight loss figures in the same storage environment suggesting the need for careful selection of the preferred method based on practical application. C&W and CF provided the lowest results for the crib storage, but the ease in BD made it the preferred method in both environments.

Key words: Grain loss, assessment methods, bulk density, rural conditions.

INTRODUCTION

The need to reduce post harvest food loss in developing countries was first debated by the 7th Special Session of the United Nations General Assembly of 1975 (Harris and Lindblad, 1978). However, a sub-committee on methods observed that "There was no agreed methodology of loss assessment ...." Showing how hard it was to come up with a single figure for an area, country, region or global. It appears that the most important consideration is for the loss assessment method to yield realistic results which can justify loss reduction methods envisaged. But which method would work best under rural farm conditions? The main methods used for determining storage losses include the standard volume weight (SVW) (Golob, 1981)

and thousand grain mass (TGM) (Proctor and Rowley, 1983). Harris and Lindblad (1978) have given detailed accounts of the count and weigh (CW) and % damage and conversion factor (CF) in addition to the SVW. Others like Irshad and Javed (1990) have used derivatives giving rise to the multiple thousand grain mass (MTGM), multiple count and weight (MCW), indirect by weight (Ind. Wt.) and indirect by number (Ind. No.). For simplicity, Tiongson (1992) grouped the methods into: SVW; C&W; indirect (CF) and thousand grain mass (TGM). However, using one or a combination of the aforementioned methods, variable weight loss results have been reported but the main concern has been the rising trend from about 5% (De Lima, 1979) to over 30%, Golob (1981a), Muhihu and Kibata (1985). To give the monetary worth, a study on the impact of *Prostephanus truncatus* on stored grain found30% weight loss where the pest was endemic and 20% where it was not, a

---
*Corresponding author. E-mail: chrisngatia@yahoo.com or chrisngatia@gmail.com.

difference worth over Kshs 2.8 billion at the then market price of Kshs 1000 per 90 kg bag (Mutambuki and Ngatia, 2006).

Few comparisons between the loss assessment methods have been made. Golob (1981) compared the SVW and the C&W and found the former to give higher weight loss estimates. Irshad and Javed (1990) evaluated seven methods against the standard weighing (STD) but found most to be tedious and time consuming. Alonso-Amelot and Avila-Nunez (2011) found the modified standard volume/dry weight ratio and % damaged grains converted to weight loss were the most practical for wheat and barley under rural conditions. The afore-mentioned studies appear to point at the need for further refining of the methods, if farmers and traders can be expected to work with them. Farmers in rural areas know the damage caused by the maize weevil *Sitophilus zeamais,* Angoumois grain moth *Sitotroga cerealella* and even the larger grain borer *P. truncatus* but their knowledge on losses is limited. Lack of understanding robs them of the bargaining power on prices and the situation provides a rich ground for exploitation by the middlemen who buy grain in kilograms while farmers are used to trade in volume. A simplified method on loss assessment could be all the rural farmers need to understand the relationship between volume and weight before they can be expected to institute loss reduction measures.

Of the documented procedures, a few can be adapted to suit the level of understanding of the rural populations. Batch weight loss is common in central storage system where grain is weighed at the entry and again as it is disbursed. Any discrepancy is taken as the weight loss. The weight of a standard volume measure, preferably using containers commonly used in grain trade is one method that can be used. A difference in weight between original and final weight after storage could be regarded as the weight loss (Neto et al., 2006). Farmers view damaged grain in terms of "the inability to use it", and the C&W which takes into account the weevil eaten portion could be another appropriate method. Compton and Sherington (1999) adopted farmers' view to classify weevil damage on cobs using a 1 to 5 categorical scale from slightly damage (10 to 20%) to heavy damage (90 to 100%) and then applied C&W method for weight loss. The indirect method which uses the percent damage multiplied by a conversion (CF) is also suitable when the cause of damage is either *S. zeamais,* or *S. cerealella.* These appear to fit farmers' rural storage environment.

Surveys have established that maize in rural districts in Kenya is stored both as cobs in traditional outside cribs and as shelled grain in bags in the living house (Mutambuki et al., 2010). While the crib provides continuous ventilation thereby aiding drying and quality maintenance, the warm conditions in the living house could in a way, favour rapid proliferation of storage insect pests. To verify whether the conditions in the house lead to more damage and hence weight loss, a simulation trial on farmers' storage practices was set up where maize was stored in exactly the same way farmers did, with and without any chemical protection. Samples were taken on a monthly basis over a six month period and subjected to the identified three methods of weight loss assessment and the results compared with the baseline.

MATERIALS AND METHODS

Storage structures

The common maize storage practice in Bungoma were the outside crib for cobs and in-house stores after shelling and treating with dilute chemical dusts. Three traditional outside cribs were constructed in the selected homestead, one for cobs and the other two for shelled grain in bags. The farmer at the homestead also provided space in the house where experimental bags were laid on wooden logs.

Maize treatments

Eight bags of 90 kg of shelled grain were locally purchased. Two bags were treated with 1.6% pirimiphos methyl + 0.3% permethrin at the recommended rate of 50 g / 90 kg bag. These were placed upright in one of the outside cribs while the other two bags were stored untreated in the next crib. The same treatment was repeated for the in-house trial but treated and untreated bags were separated by 1 m space. The 60 kg of untreated cobs were loaded into the third crib to monitor infestation development. Farmer maize stocks were also monitored in six homesteads close to the site for comparison. Farmers' uses were not controlled and about half exhausted their stocks before end of trial, so data was pooled and the average used in the subsequent comparison.

Infestation

Infestation was left to set in naturally just as it happens in farmer stores. Pest identification was done to confirm the most prevalent storage insect pests and whether the newly introduced larger grain borer *P. truncatus* Horn had set in.

Sampling and analysis

Farmer's consumption pattern was calculated from the quantity cooked daily which translated to between 12.3 and 14.5 kg for the six farmsteads, average 13.4 kg per month. Because of handling logistics, a compromise 10 kg sample was close to farmers' consumption rate and convenient for our purpose. The 10 kg sample was reduced through conning and quartering and the 1 kg taken to designated site for analysis. After sieving and grain moisture determination, the grain was sorted into weevil damage, dust, broken pieces and mould infection.

Subsequently, regular sampling and analysis were done at 4-weekly intervals until the final at 24 weeks. In the cob store, between 9 and 17 cobs (depending on size) were picked along the 8 compass directions and shelled to give about 1 kg of grain. For subsequent sampling, the depth of cobs was visually assessed to determine the number of cob layers to be removed every month, so as to sample on the top layer left. The final sampling was from the bottom few layers.

Percent weight loss methods

Weight of standard volume method

The 1 kg sample was first sieved to remove dust, foreign matter and free living insects which were collected for identification. Grain moisture was determined using a Dickey John moisture meter. Three test weights for a 440 ml capacity glass jar was taken and average used for bulk density calculation. The results were compared with the baseline figure at 0 weeks using the formula as follows:

$$\% \text{ Weight loss} = \frac{W1 - W2}{W1} \times 100 \tag{1}$$

Where, W1 = weight of baseline sample, W2 = subsequent sample weight at different storage intervals.

Count and weigh method

The same sample was passed through the riffle divider to reduce to ⅛ for ease of handling. Grain in three ⅛ sub-samples were sorted into damage categories: insect damaged (weevil damage), mould damage, broken pieces and undamaged grains. Because the interest was on the weight loss caused by storage insect damage, only the weevil damaged grains were compared with undamaged lot in the equation as follows:

$$\% \text{ Weight loss} = \frac{(UNd) - (DNu)}{U(Nu + Nd)} \times 100 \tag{2}$$

Where, U = weight of undamaged grain, D = weight of damaged grain, Nu = number of undamaged grains, Nd = number of damaged grains.

Conversion factor method

The method has been found useful where the infestation consists mainly of the maize weevil, *S. zeamais* and the Angoumois grain moth, *S. cerealella*. The percent weevil damaged grains in Equation 2 were multiplied with ⅛, the conversion factor given in Harris and Lindblad (1978) as shown:

$$\% \text{ Weight loss} = \frac{(Nd) \times 100}{(Nu + Nd) \times 8} \tag{3}$$

Where Nd = number of damaged grains; Nu = number of sound grains.

Statistical analysis

Data was subjected to statistical analysis using the statgraphic software and the Duncan's multiple range test (DMRT) which separated the treatment means for the crib and the in-house trials as well as the methods of weight loss assessments.

RESULTS

Baseline information

The maize used was purchased locally and it was at different levels of moisture content and insect damage. Table 1 shows no statistically significant (P>0.05) differences in the bulk density (BD) for both crib and in-house at the initial stage of trials. However, weevil damage and amount of dust in 1 kg samples were significantly higher in treated maize, a reflection of the differences in farmer storage conditions. Differences in the level of grain moisture for untreated and treated maize were significant at P>0.05 level.

Infestations build up and grain damage

Infestation build up was very slow during the first three months with about one live weevil per kg at the start, which increased gradually to 112 and 150/kg in untreated samples in the two environments (Table 2). At 24 weeks, more than half (53 to 56%) of the untreated grains were weevil damaged compared with 16 to 20% in the treated maize. Dust generated as a consequence of the weevil damage increased to 20 g/kg in untreated maize compared with 5 to 7 g/kg in treated samples. Grain moisture had little variation while BD varied from 0.7407 to 0.6734, indicating a drop from the original weights. The main storage insect pests were *S. zeamais*, *S. cerealella* and *Tribolium castaneum*.

Percent weight loss

Data in Tables 1 and 2 was applied to Equations 1, 2 and 3 for the respective assessment methods. The results are shown in Table 3. ANOVA for the storage period and assessment methods were highly significant (P=0.0000) followed by store environment (P=0.006). Weight loss varied widely from a low 0.3% for BD to 13.4% in the C&W method for the two environments. In the crib at P =0.05 level, weight loss in BD was twice that of the C&W and CF respectively. Weight loss differences between BD and C&W were not statistically significant in the in-house trial. But C&W had twice that of the CF for untreated maize. Grain treatment was more effective in in-house storage, but more weight loss occurred in untreated maize in the same trial.

Cumulative weight loss

Cumulative weight loss (CWL) show the long term effect of infestation for farmers who do not apply any protectants. Figure 1 shows that farmers were likely to loose between 23 and 27% of their harvest after six months of storage in the two environments. The benefit of

Table 1. Maize conditions (parameters ± SE) before simulation trials.

| Parameter | Simulated farmer storage environment | | | |
	Crib treated*	Crib untreated	House treated*	House untreated
Initial				
Bulk density	0.7632 ± 1.2^a	0.7677 ± 0.2^a	0.7432 ± 5.0^a	0.7652 ± 2.1^a
% Grain moisture	13.3 ± 0.0^b	14.0 ± 0.0^d	13.0 ± 0.0^a	13.8 ± 0.0^c
Wt of dust (g/kg)	1.3 ± 0.0^{ab}	0.6 ± 0.4^a	2.1 ± 0.3^b	0.6 ± 0.6^a
% weevil damage	3.5 ± 0.2^b	0.9 ± 0.4^a	2.7 ± 0.3^b	0.3 ± 0.1^a
Live insect pests/kg	1.0 ± 0.0^a	0.0 ± 0.9^a	1.0 ± 1.0^a	0.0 ± 0.5

Each datum is a mean of three readings, row means followed by the same letter are not significantly different (P<0.05).
* = Analysed before chemical application.

Table 2. Maize conditions (parameters ±SE) after 24 weeks exposure to natural infestation in a simulation trial.

| Parameter | Simulated farmer storage environments | | | |
	Crib treated	Crib untreated	House treated	House untreated
Final				
BD	0.73 ± 2.9^a	0.6734 ± 3.0^b	0.7407 ± 1.9^a	0.6893 ± 3.0^b
% Grain moisture	12.7 ± 0.1^c	12.6 ± 0.0^{bc}	12.4 ± 0.2^{ab}	12.2 ± 0.0^a
Wt of dust (g/kg)	6.5 ± 0.8^a	22.0 ± 2.7^b	5.1 ± 1.7^a	19.8 ± 2.7^b
% weevil damage	19.9 ± 6.0^a	52.9 ± 3.6^b	15.7 ± 4.5^a	55.8 ± 5.0^b
Live insect pests/kg	12.6 ± 6.5^a	112.5 ± 34.5^b	15.0 ± 1.0^a	150.5 ± 5.5^b

Each datum is a mean of three readings, Row means followed by the same letter are not significantly different (P<0.05).

Table 3. Calculated percent weight loss in treated and untreated maize after 24 weeks of simulation as assessed by the three methods.

| Equation | Method | Simulated farmer storage environments | | | |
		Crib treated	Crib untreated	House treated	House untreated
1	BD	4.4^b	12.3^b	0.3^a	9.9^{ab}
2	Count & weigh	2.3^a	5.0^a	2.2^a	13.4^b
3	CF x % damage	2.5^a	6.6^a	2.0^a	7.0^a

Column means followed by same letter are not significantly different (P>0.05) DMRT. CF x % damage = Conversion Factor x % damage.

treating maize was a reduction in cumulative loss to between 10 and 13% in the in-house and the crib trials respectively. Figure 2 shows the influence of the storage environment on the cumulative weight loss by methods of assessment on untreated maize. BD had the highest (>34%) in both storage environments. C&W and CF had the lowest (15 to 20%) for untreated maize in the crib but recorded between 23 and 27% for untreated maize in the in-house trial. On cob maize storage (Figure 3), BD recorded higher cumulative weight loss (34%) while CF had 9%, the least. On farmer stored maize (Figure 4), BD and C&W methods had higher (18%, 20%) and similar pattern while CF had the lowest cumulative loss at 11%.

DISCUSSION

The simulation trial was carried out to determine the level of weight loss on farm stored maize, with and without any protection. Weight loss is defined as the difference in food stocks between two successive storage periods. The comparison between baseline and subsequent weights after 4-week intervals for 24 week storage period was the preferred approach. Weight loss determination could be influenced by the availability of the requisite equipment, methods used and the storage environment. Tiongson (1992) has outlined requisite equipment for SVW, C&W and CF. However, the listed equipment could

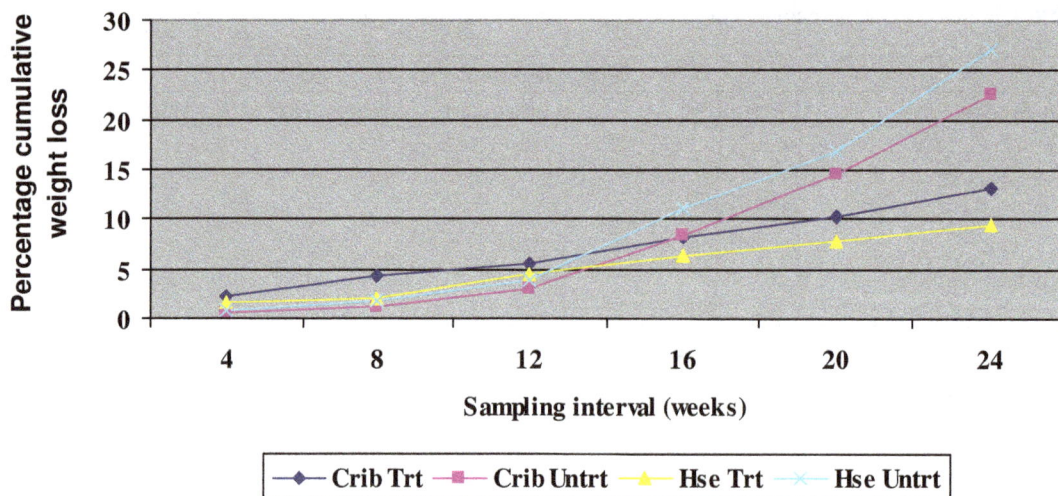

Figure 1. Influence of the two storage environments on cumulative weight loss in treated and untreated maize.

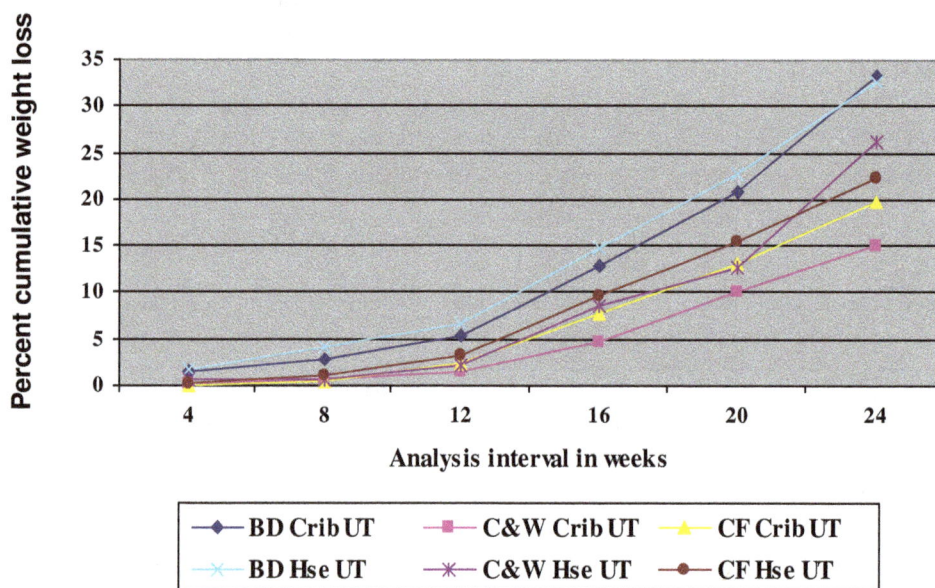

Figure 2. Mean cumulative weight loss in untreated maize in two storage environments as assessed by three methods.

be hard to find under many rural situations, making it necessary to consider the minimal that would enable the application of three methods. These could be a weighing balance for the C&W and a standard volume container for the BD methods. In this trial, the CF method utilised the results from the C&W method. All the three methods evaluated produced results in the range of 0.3 to 13.4%, which agrees with Hodges' (undated) weight loss range of 0.3 to 13.3% in maize and (Alonso-Amelot and Avila-Nunez, 2011) weight loss of 2.2 to 14.5% in wheat. The storage environments appear to have some influence

with the average 8.0% weight loss in untreated maize in the crib, compared to 10.1% for the same in the in-house storage.

The results thus confirm the house environment to have favoured insect pest damage which translated to a higher cumulative weight loss if no control was done. However, with intervention, the scenario was different as cumulative weight loss in farmer stored maize was markedly lower than in both cob and simulation trials, as a result of different interventions used. Therefore, the aforementioned losses can be regarded as being within

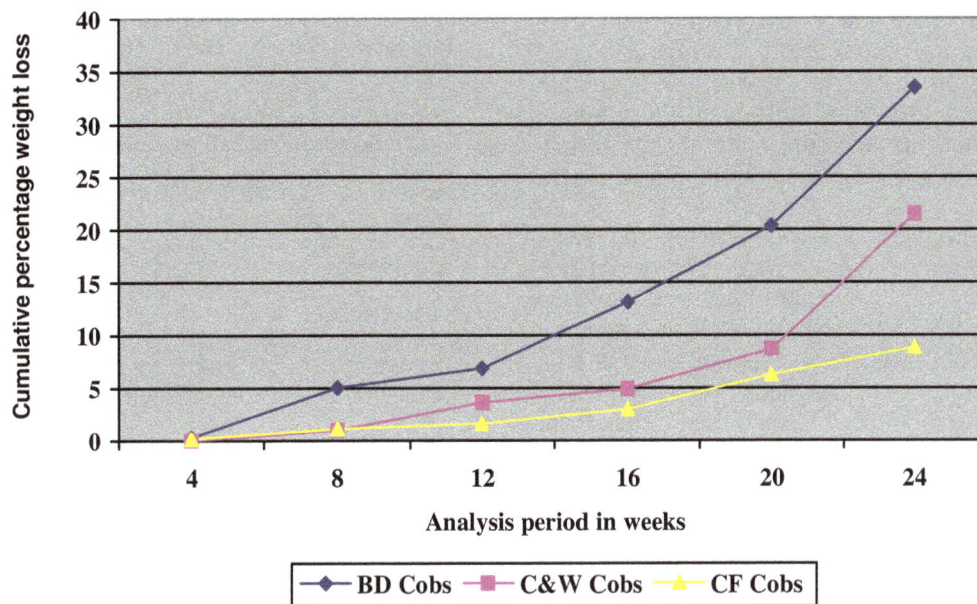

Figure 3. Cumulative weight loss on cob stored maize as assessed by three methods.

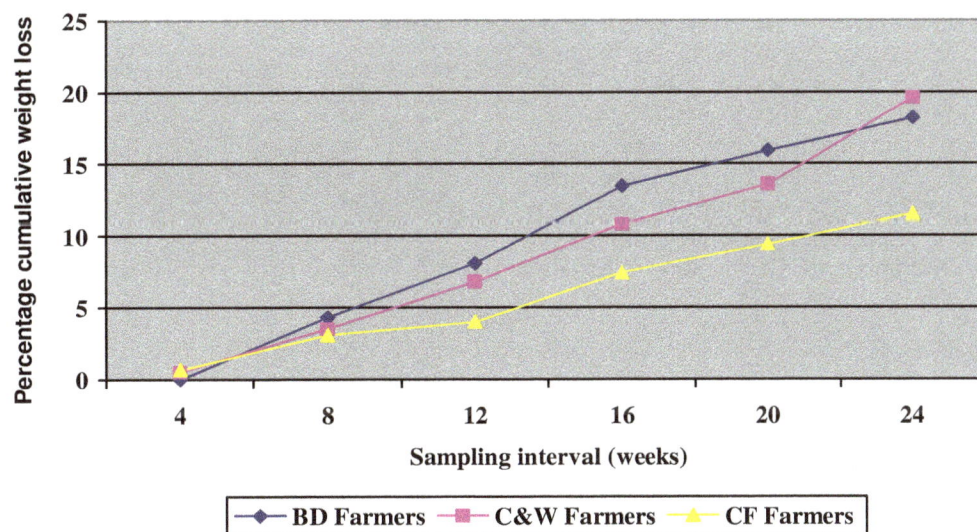

Figure 4. Cumulative weight loss in farmer stored maize as assessed by three methods.

the acceptable level for a storage period of 24 weeks. Apart from the storage environment, variation in weight loss could be due to the method of analysis. Harris and Lindblad (1978) noted that acceptable post harvest grain loss assessment methods should yield realistic results. In both the crib and in-house trials, CF results were somewhere between the BD and C&W, with the former consistently giving a higher percent sample and cumulative losses (Table 3). BD also maintained higher cumulative weight loss on cob and farmer stored maize, a fact Golob (1981) described as weight loss exaggeration.

On the other hand, the C&W method gave consistently low weight losses compared to the other methods confirming Alonse-Amelot and Avila-Nunez (2011) notion that the method grossly under estimated the losses when compared with thousand grains mass (TGM). The discrepancies between methods are not uncommon and must therefore be accepted. Miguel and Jorge (2011) reported percent weight loss of 2.3% for C&W compared with 14.5% by visual method. Braga-Caneppele et al. (2003) found 2.3, 4.7 and 21% as percent losses following three methods in Harris and Lindblad (1978).

Grain size and hidden infestation are some of the factors causing variation of the results and sorting by size could overcome the problem but can frustrate rural population because it is tedious and time consuming. One method could act as the check for another. In the simulation trial, when CF was applied to the C&W data with 13.4% weight loss, the resultant figure was 6.4% lower than in the preferred method confirming Alonse-Amelot and Avila-Nunez (2011) observation that CF was a more practical and an expedite means to evaluate losses in individual farms. The fact that BD involved only weighing makes it more attractive for both traders and farmers interested in ascertaining weight loss, but like visual inspection, BD is likely to unfairly keep prices artificially low (Miguel and Jorge, 2011). This could be one of the trade-offs between accuracy and speed.

Conclusion

The importance of weight loss at farm level cannot be ignored any more. Farmers know the causes and even the benefits of treating grain. Treated maize is of good quality and attracts premium prices. The house environment has a role to play in aggravating losses as both sample and cumulative losses showed. Simple weight loss assessment methods, like the ones evaluated were all acceptable based on level of losses found. This puts the farmers at dilemma on which to choose among them. A look at the influence of the storage place does not appear to be helpful, leaving farmers with the requisite equipment as the criterion for choice rather than weight loss levels. On this alone, BD appears to be the method of choice.

REFERENCES

Alonso A, Avila N (2011). Comparison of seven methods for stored cereal losses to insects for their application in rural conditions. J. Stored Prod. Res., 47:82-87.

Braga-Caneppele MA, Caneppele C, Lazzari FA, Noemberg-Lazzari, SM (2003). Correlation between infestation level of Sitophilus zeamais Motschulsky, 1855 (Coleoptera, Curculionidae) and the quality factors of stored corn, Zea mays L. (Poaceae). Rev. Bras. Entomol., 47: 625-630.

Compton JAF, Sherington J (1999). Rapid loss assessment methods for stored maize cobs: weight loss due to insect pests. J. Stored Prod. Res., 35: 77-87.

De Lima CPF (1979). The assessment of losses due to insects and rodents in maize stored for subsistence in Kenya. Trop. Prod. Inf., 38: 21-26.

Golob P (1981). A practical assessment of food losses sustained during storage by smallholder farmers in the Shire Valley Agricultural Development Project area of Malawi 1978/79. Report by the Tropical Products Institute, G154, vi-47pp.

Golob P (1981a). A practical appraisal of on-farm storage losses and loss assessment methods in the Shire Valley of Malawi. Trop. Stored Prod. Inf., 40: 5-13.

Harris KL, Lindblad CJ (1978). Post harvest grain loss assessment methods. A manual of methods for the evaluation of post harvest losses. American association of Cereal Chemists.

Hodges RJ (undated). Post harvest weight loss estimates for cereal supply calculations in East and Southern Africa.

Irshad M, Javed I (1990). Accuracy of different loss assessment methods for maize during storage. Sarhad J. Agric., 6(5): 491-494 ISSN 1016-4383; Record Number, 19910308008

Muhihu SK, Kibata GN (1985). Developing a control programme to combat an outbreak of Prostephanus truncatus (Horn) (Coleoptera: Bostrychidae) in Kenya. Trop. Sci., 25: 239-248.

Mutambuki K, Ngatia, CM (2006). Loss assessment of on-farm stored maize in semi arid areas of Kitui District. In: Lorini, I., Bacultchuk, B., Beckel, H., Deckles, D et al. Proceedings of the 9th International Working Conference on Stored Product protection Campinas, Sao Paulo, Brazil, pp. 15-23.

Mutambuki K, Ngatia CM, Mbugua JN (2010). Post-harvest technology transfer to reduce on farm grain losses in Kitui district, Kenya. Proceedings of the 10th International Working Conference on Stored Product Protection: 27 June to 2 July 2010, Estoril, Portugal 986.

Neto AP, Pimentel MAG, Faroni LRD'A, Garcia FM, de Souza AH (2006). Population growth and grain loss of Cathartus quadricollis (Guerin-Meneville) (Coleoptera: Silvanidae) in different stored grains.

Tiongson RL (1992). Standardized methods for the assessment of losses due to insect pests in storage. In: Towards integrated commodity and pest management in grain storage. A Training Manual for application in humid tropical storage systems. Edited by R.L. Semple, P.A. Hicks, J.V. Lozare, and A. Castermans. A REGNET (RAS/86/189) Publication in Collaboration with NAPHIRE, May, 1992.

17

Production and sensory evaluation of food blends from maize-plantain-soybean as infant complementary food

Opara, B. J.[1], Uchechukwu[1], N., Omodamiro, R. M.[2]* and Paul, N.[2]

[1]Michael Okpara University of Agriculture, Umudike, Abia State, Nigeria.
[2]National Root Crops Research Institute, Umudike, Abia State, Nigeria.

The study was conducted to produce and evaluate food blends from maize- plantain - soybean as infant complementary food. Treatments consist of diet formulated with toasted soya beans flour, ogi flour from yellow maize corn and firm ripe plantain was used for the production of the plantain flour. Six experimental diets were formulated from the above flours, to contain the following percentage ratios: OPBL1 - 50% maize, 25% plantain, 25% soybean; OPBL2 - 50% maize, 20% plantain, 30% soybean; OPBL3 - 50% maize, 15% plantain, 35% soybean; OPBL4 - 50% maize, 10% plantain, 40% soybean; OPBL5 - 50% maize, 5% plantain, 45% soybean; OPBL6 - 33.33% maize, 33.33% plantain, 33.33% soybean. Sensory evaluation of the maize-plantain-soybean composite flours based complementary foods was carried out using 7-point hedonic scale with 20 panelists. Generally, sample OPBL7 was most acceptable by the panelists, though not significantly different (P> 0.05) from samples OPBL6, OPBL5 and OPBL4. However, all samples were generally acceptable by the panelists.

Key words: Soya beans, plantain, maize and sensory evaluation.

INTRODUCTION

The period during which other foods or liquids are given to a young child along with breast milk is considered the period of complementary feeding and any nutrient containing foods or liquids other than breast milk provided the child during this period are defined as complementary foods (WHO, 1998).

Thus, it is essential that infants receive appropriate, adequate and safe complementary food to ensure the right transition from breastfeeding to the full use of family foods (WHO 2003).

Lack of appropriate feeding can set up risk factors for ill-health. The life–long impact may include poor school performance, reduced productivity, impaired intellectual and social development or chronic diseases (Nestel et al., 2003). In developing countries, complementary foods are mainly based on starchy tubers like cassava, cocoyam and sweet potato or on cereals like maize, rice, wheat, sorghum and millet. Small children are normally given these staples in the form of gruels that is mixed with boiled water or boiled with water. When prepared in this way, the starch structures bind large amounts of water, which results in gruels in high viscosity (Hellstrom et al., 1981).

Seed proteins especially from leguminous sources such as soybean have been put forward as potentially excellent sources of protein for the nutritionally quality upgrading of starchy roots and tubers for use in baby foods in countries which import all their milk requirement (Okaka and Okaka, 1990).

Hence the objective of this research work is to exploit the nutritive potentials of yellow maize, plantain and soybean for production of easy to prepare complementary food which is nutritive, and available to low earners in developing countries.

MATERIALS AND METHODS

Production of the experimental complementary foods

Production of maize into ogi flour

The method of yellow maize fermentation adopted in this study

---

*Corresponding author. E-mail: majekdamiro@yahoo.com.

followed the method of Baningo and Akpapunam (1999) and Omueti et al. (2009) with modification for the production of *ogi* flour. Ten (10) kg of cleaned yellow maize grains were sorted, washed and steeped in tap water for two days in a large basin. The contents were allowed to ferment at room temperature for 0 to 48 h. The steeped water was changed with fresh water after each day. The steeped water was decanted and the fermented cereal ground to slurry in a hydraulic mill. The slurries were sieved through a fine sieve (muslin cloth) with excess water. The seed coat and other coarse particles were discarded and the sediment allowed to settle and squeezed to remove excess water. The sediment was dried at 60°C for 12 h. The dried samples passed through the mill a second time and sieved to obtain fine particles. The 'ogi' flour was then stored in sealed air tight in food grade polyethylene bag for analysis (Figure 1).

Production of firm ripe plantain into flour

The method of Ogazi (1996) was used for the production of the plantain flour. Matured, firm ripe plantain which weighed 7.4 kg was washed with clean water and each fruit was cut into three pieces and blanched (hot water) for 20 min, peeled and sliced. The sliced plantain was then oven dried at 60°C for 12 h. The dried plantain was milled to powder with a milling machine and sieved with muslin cloth of 150 mm. The resulting flour was then packaged in sealed air tight food grade polyethylene bags for analysis (Figure 2).

Production of soybean into soybean flour

The method of Omueti et al. (2009) was adopted for the production of soybean flour with modification. Soybean grain was sorted, washed and blanched for 45 min. It was dehulled and toasted for 30 min. The toasted grain was then oven dried at 60°C for 15 min, milled and sieved to fine flour. The flour was packaged and sealed with food grade polyethylene bags for analysis (Figure 3).

Formulation of six complementary food products (Maize-Soybean-Plantain) - MAMUSOY complementary food (Mamusoy)

This is according to the method of Omueti et al. (2009). The complementary diets were prepared as shown in Figure 4. Putting the following into consideration, maize is carbohydrate present in constant ratio; soybean ratio is higher than that of plantain, being a baby food the protein content should be high. The plantain used was firm ripe, the plantain is to supply micro nutrient. The six experimental diets were formulated to contain the following percentage ratios:

OPBL1 - 50% maize, 25% plantain, 25% soybean
OPBL2 - 50% maize, 20% plantain, 30% soybean
OPBL3 - 50% maize, 15% plantain, 35% soybean
OPBL4 - 50% maize, 10% plantain, 40% soybean
OPBL5 - 50% maize, 5% plantain, 45% soybean
OPBL6 - 33.33% maize, 33.33% plantain, 33.33% soybean

The flow chart for the production of the six complementary food products (MAMUSOY) is shown in Figure 4.

SENSORY EVALUATION OF THE COMPLEMENTARY FOOD BLENDS

The method of Iwe (2002) was used for sensory evaluation

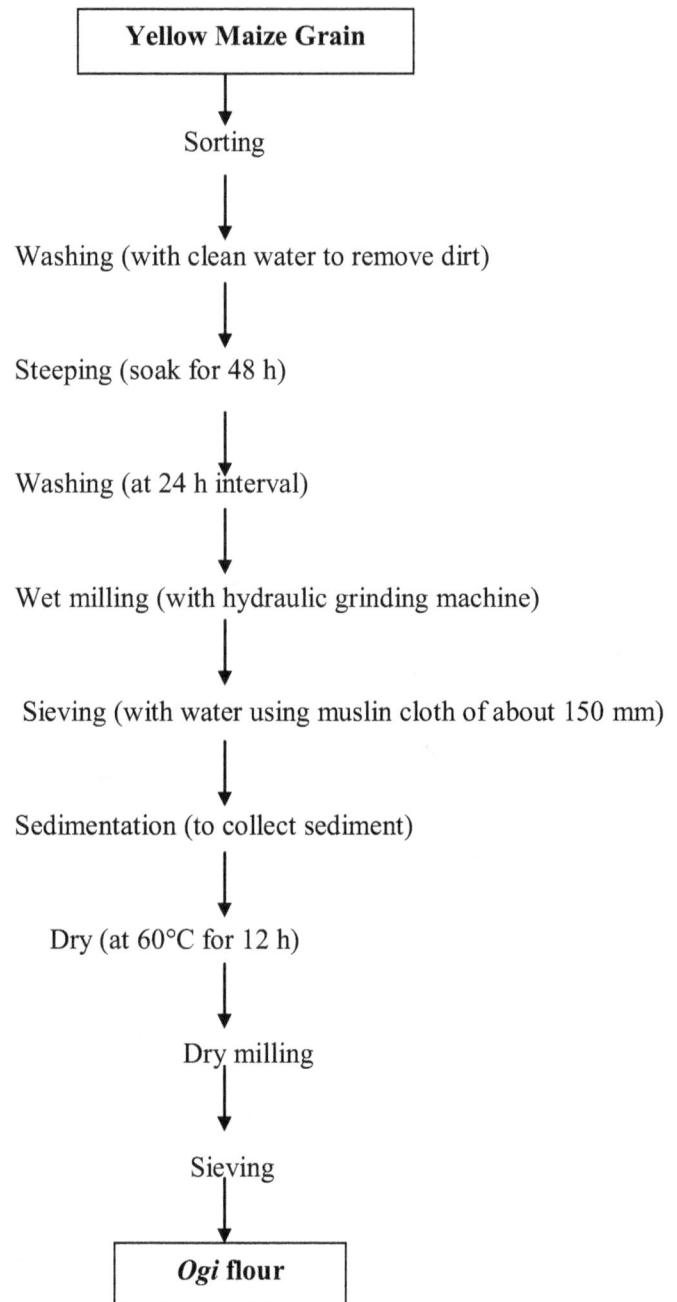

Figure 1. Flow chart for the production of *ogi* flour.

of maize-plantain-soybean composite flours based complementary foods (MAMUSOY). The effect of processing methods on the general acceptability of composite flours and the colour, aroma, taste /mouth-feel, and general acceptability of complementary foods were evaluated using a 7-point hedonic scale. The 7-hedonic scale ranged from dislike very much, through neither like nor dislike, to like very much. The complementary foods prepared to panelists for evaluation were prepared by dissolving 100 ml of clean water with 50 g complementary food (flour), after which 150 ml of boiling

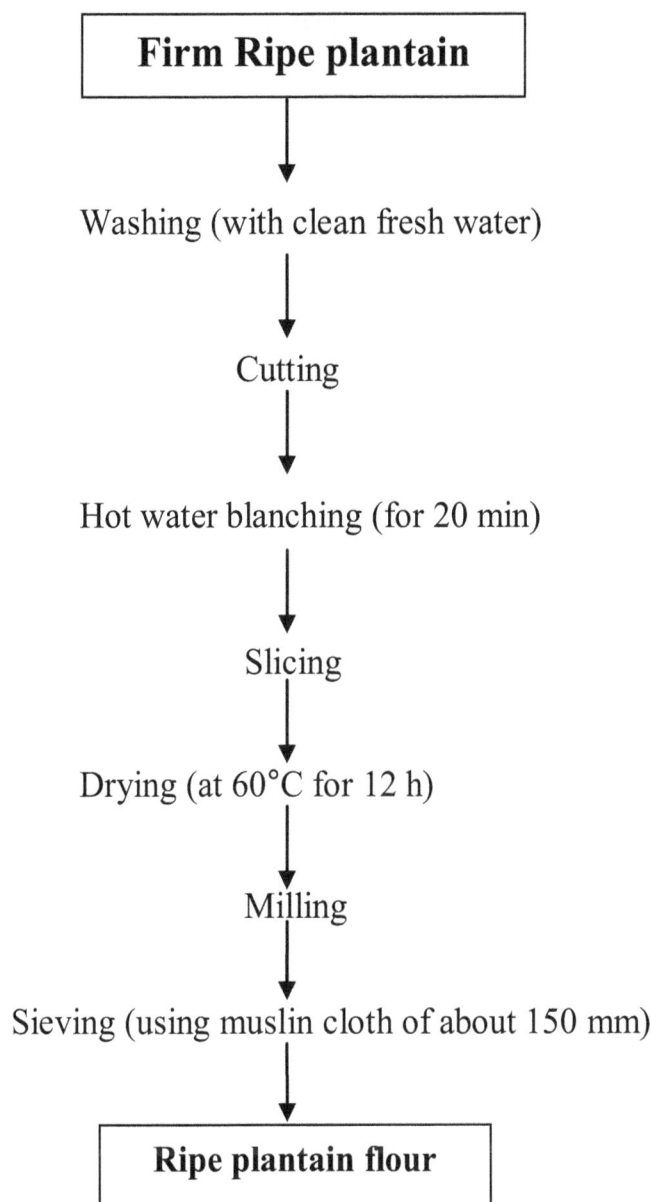

Figure 2. Flow chart for the production of plantain flour.

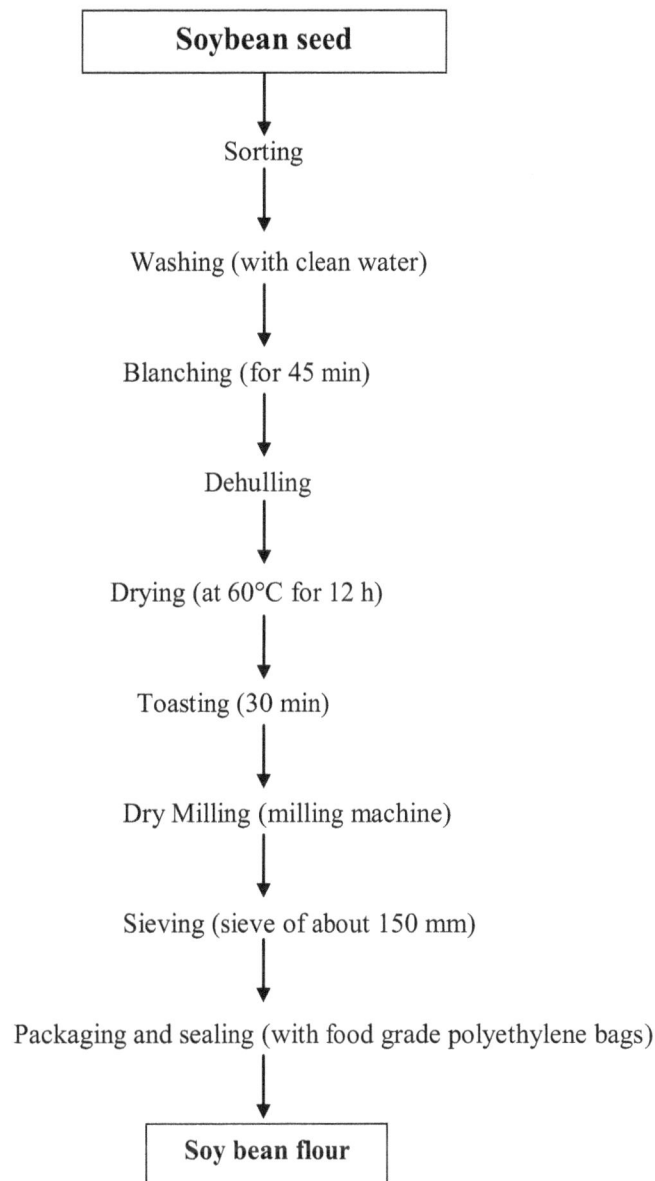

Figure 3. Flow chart for the production of soybean flour.

water (100°C) was used to reconstitute the sample of MAMUSOY which was brought to a boil for 3 to 5 min, and 2 g of sugar was added to the reconstituted samples. Panelists (trained and semi trained) were drawn from staff (males and females) of Post Harvest Technology Programme of National Root Crops Research Institute, Umudike, and students (Food Science and Technology) of Michael Okpara University of Agriculture Umudike. The samples were presented in identical sample containers coded with 4-digit random numbers, each sample having a different number. The sample order was randomized for each panelist.

The samples were presented all at once to enable the panelists evaluate the samples if desired and make comparisons between the samples. Nutrend (a maize-soya based baby food manufactured by NESTLE) was used as the reference sample. The evaluation was carried out under a conducive environment for sensory evaluation.

SENSORY EVALUATION OF COMPLEMENTARY FOOD BLENDS

In terms of colour, sample OPBL7 was most acceptable by the panelists (6.80) but was not significantly different (P> 0.05) from that of samples OPBL8, OPBL4 and OPBL2.

Preparation of the six complementary food products (maize-soybean-plantain)

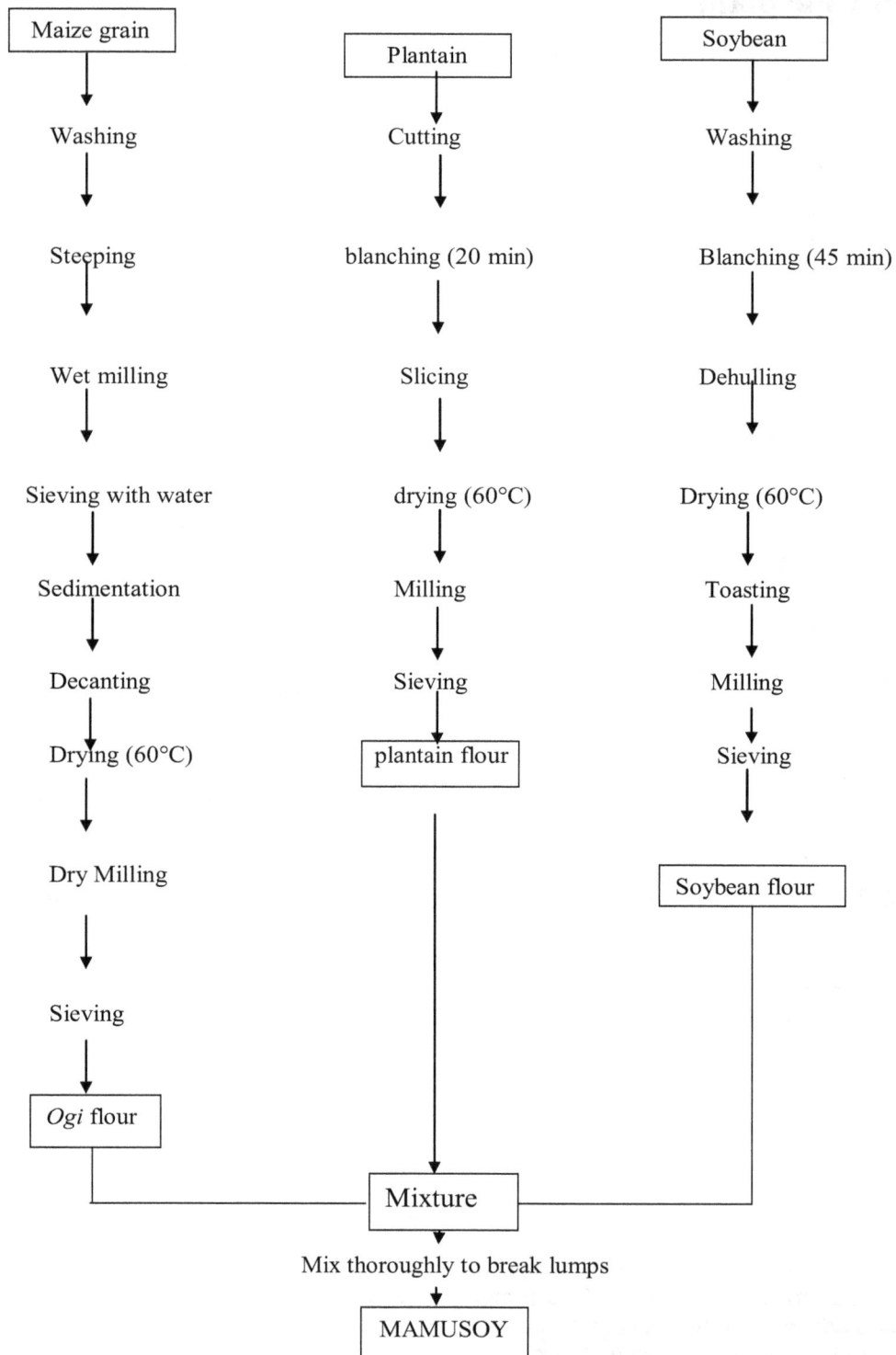

Figure 4. Flow chart for the production of six complementary food products (MAMUSOY).

However colour of all samples were acceptable by the panelists.

The taste of all the samples were general acceptable for taste and there was no significant difference (P>0.05).

The aroma was most acceptable in sample OPBL7 (6.80) and least acceptable in sample OPBL8 (5.00). Sample OPBL7 was not significantly different (P>0.05) from samples OPBL6, OPBL5, OPBL4 and OPBL2.

Table 1. Mean sensory scores of maize-plantain-soybean complementary food "MAMUSOY."

Sample	Colour	Taste	Aroma	Texture	General acceptability
OPBL1	5.80^{bc}	5.60^{a}	5.50^{bc}	5.70^{a}	5.6^{bc}
OPBL2	5.90^{abc}	5.90^{a}	5.90^{abc}	5.80^{a}	5.80^{bc}
OPBL3	5.40^{bc}	5.50^{a}	5.60^{bc}	5.60^{a}	5.30^{c}
OPBL4	5.90^{abc}	5.60^{a}	5.90^{abc}	5.70^{a}	5.90^{abc}
OPBL5	5.30^{c}	5.70^{a}	5.90^{abc}	6.00^{a}	5.90^{abc}
OPBL6	5.10^{c}	6.30^{a}	6.00^{ab}	6.10^{a}	6.20^{ab}
OPBL7	6.80^{a}	6.40^{a}	6.80^{a}	5.50^{a}	6.70^{a}
OPBL8	6.30^{ab}	5.40^{a}	5.00^{c}	5.50^{a}	5.60^{bc}
LSD (P=0.05)	0.977	1.025	0.992	1.061	0.879

Means in the same column with different superscript are significantly different at (P < 0.05) LSD = Least square difference. RPLA = Raw plantain; RSOY = raw soybean; RMAY = raw maize; MUSA = plantain flour; WINN = soybean flour; RMO = *Ogi* flour; OPBL1 (50:25:25); OPBL2 (50:20:30); OPBL3 (50:15:35); OPBL4 (50:10:40); OPBL5 (50:5:45); OPBL6 (33.33:33.33:33.33).

However, all samples were generally acceptable for aroma.

While in texture, sample OPBL6 was most acceptable by the panelist (6.10) and least acceptable in samples OPBL7 and OPBL8 (5.50). There was no significant difference (P> 0.05) in the texture of the samples and all were acceptable by the panelists.

Generally, sample OPBL7 was the most generally acceptable by the panelist (6.7), though not significantly different (P> 0.05) from samples OPBL6, OPBL5 and OPBL4. However, all samples were generally acceptable by the panelists (Table 1).

Conclusion

Sensory evaluation showed that OPBL5 (50:5:45) and OPBL6 (33.33:33.33:33.33) were most acceptable to the panelists among the six complementary foods as well as higher in nutrient. The results obtained in this study equally show that the potential exists for blends OPBL5 (50:5:45) and OPBL6 (33.33:33.33:33.33) as income generation for the rural populace if the technology is adopted.

REFERENCES

Baningo EB, Akpapunam MA (1999). Physico-chemical and Nutritional Evaluation of protein-enriched fermented maize flour. J. Manage. Technol., 1(1): 30-36.

Hellstrom A, Hermansson AM, Karlsson A, Liungquist B, Melander D, Svanberg U (1981). "Dietary bulk as a limiting factor for nutrient intake in pre-school children. II. Consistency as related to Dietary Bulk – A model study". J. Trop. Pediatr., 27: 127.

Nestel P, Briend A, de Benoist E, Decker E (2003). Complementary Food Supplements to Achieve Micro-Nutrient Adequacy for Infants and Young Children J. Pediatr. Gastroenterol. Nutr., 36: 316 – 328.

Ogazi PO (1996). Plantain: Production, Processing and Utilization. Paman and Associates Ltd, Akv Okigwe, Imo State, Nigeria. Afr. J. Biotechnol., p. 30.

Okaka JC, Okaka ANC (1990). Food: Composition Spoilage and shelf-life extension. OCJANCO Academic Publishers. Enugu.

Omueti O, Bolanle O, Olayinka J Olukayode A (2009). Functional Properties of Complementary Diets Developed from Soybean (*Glycine Max*), Groundnut (*Arachis hypogea*) and Crayfish (*Macro brachium* spp.). Elect. J. Environ. Agric. Food Chem., 8 (8): 563-573.

WHO (1998). Complementary feeding of young children in developing countries: A review of current scientific knowledge. WHO/NUT/98.1 Geneva, World Health Organisation.

WHO (2003). Feeding and Nutrition of Infants and Young Children: Guidelines for the WHO European region with emphasis on the former Soviet Union. WHO Regional Publications, European Series, No. 87, pp. 1-296.

18

Appraisal of rice production in Nigeria: A case study of north central states of Nigeria

Ajijola S.[2]*, Usman J. M.[1], Egbetokun O. A[2], Akoun J.[1] and Osalusi C. S.[1]

[1]Federal College of Forestry, P.M.B.5087, Ibadan, Nigeria.
[2]Institute of Agricultural Research and Training, PMB 5029, Moor Plantation, Ibadan, Nigeria.

The study was carried out in order to determine the level of contribution of North Central States of Nigeria in rice production to the total output of rice in Nigeria. Secondary data were collected from National Bureau of Statistics. The data include land area used for rice cultivation and output of rice from the period of 1994/1995 to 2005/2006 cropping season. For the area of land used for rice cultivation, the regression model was tried under the four basic functional forms and the double log function was chosen as the lead equation. This was based on the value of coefficient of multiple determination (R^2) of 0.625 and the significant variables in conformity with a priori expectations. For rice output, the double log function was chosen as the lead equation with R^2 value of 0.542 and explanatory variables significantly affecting total output of rice in Nigeria in consonance with a priori expectations. All explanatory variables had positive influence on rice output. It is therefore, recommended that the only sustainable and socially acceptable way forward is to enhance the competitiveness of local rice against imported rice, both in terms of quality and price. This calls for improving quality management and increasing efficiency along the entire marketing chain.

Key words: Nigeria, North Central, appraisal, rice production.

INTRODUCTION

The 20th century saw the most dramatic agricultural transformation in human history. Science-based agriculture brought about rapid changes on the farm and sped the transformation from subsistence agriculture to a more productive and profitable modern agriculture. As agricultural production improved and farmers succeeded, some began to specialize in certain crops or products. This resulted in the growth of farmer-led private enterprises and the building of non-farm private sectors in rural areas. Technological change in agriculture however, requires a constant flow of new technologies to farmers and a wide range of options (Plucknett et al., 2000). Government policies affecting rice production have been directed at protecting the local industry through tariffs and providing extension support to rice farmers. The import tariff on value-added rice was 100% in 1995, 50% in 1996 through 2000 and 85% in 2001 (Akande, 2003).

With effect from May 2008, rice imports into Nigeria were declared free from all duties and charges, including customs duty, 7% surcharge, value-added tax and levies. However, by March 2009, a 50% rice levy was instituted (Nigerian Custom Service, 2009). With this tariff level, local production of rice is expected to be expanded through increased production of paddy by farmers responding to higher paddy price. Local processors are expected to increase capacity utilization and use improved processing equipment.

The continuous increase in rice production and processing will depend on the international competitiveness and effects of policy intervention. The removal of all forms of tariffs as the government did in 2008 will change the structure of economic incentives. This, in turn, will cause major adjustments in the pattern of production, allocation of resources and trade flows. The analysis of competitiveness and comparative advantage will provide an indication of the effects of policy. Comparative advantage of a country in a commodity usually results from relative superiority in

*Corresponding author. E-mail: ajsik1967@yahoo.ca.

Table 1. Rice production trends in Nigeria (1961–2000).

Period (ha)	Average area cultivated (tons)	Average out (tons/ha)	Average yield
1961-1965	179,200	207,200	1.147
1966-1970	234,000	321,000	1.360
1971-1975	288,800	470,200	1.670
1976-1980	332,000	596,200	1.710
1981-1985	630,000	1, 300,200	2.063
1986-1990	1,06,200	2,216,064	2.090
1991-1995	1,678,000	2,979,600	1.783
1996-2000	1,742,582	3,011,028	1.733

PCU, FMARD, Nigeria (2002).

resource endowments required by the commodity. It puts the country in a vantage position to specialize in the production of the commodity. Competitive advantage is created through appropriate combination of knowledge and other critical resources to gain significant share of the world market for a particular commodity. Competitive environment and the capability of firms in the industry to innovate and improve their technologies contribute to the achievement of competitive advantage. The use of comparative advantage analysis covers not only on-farm production but incorporates downstream collection, processing and wholesaling activities as they relate to a particular commodity (Salinger, 2010).

Nigeria, Cote d'Ivoire, Zaire and Madagascar are among the biggest producers of all types of rice in Africa (Baksh, 2003). Consumers in these countries require that the domestically produced rice should satisfy minimum level of quality, health and food security standards. Generally, besides the lag in rice production, much of the food produced locally is distributed under poor marketing structure which hinders the flow of resources and virtues in the industry. Nigeria is becoming one of the major rice importers on the world market for the last five years, thus, being an important outlet for rice exporting countries. Beyond its large volume, the Nigerian rice market is even more attractive than other West African markets because Nigeria imports rice of high value (parboiled rice) against rice of lower quality in the other countries of the sub-region (WARDA, 2003). Table 1 shows the production trend of rice production in Nigeria which dictates that there is increase in the production. The macroeconomic conditions under which Nigerian rice is produced are partly responsible for the sector's lack of competitiveness. Some of the issues include high inputs costs such as cost of credit, and imported equipment and agrochemicals due to taxes (legal and illegal), tariffs and duties. There is also the problem of policy instability (ban, unban, tariffs) that makes decision-making and planning highly uncertain and puts investments at great risk. All these factors combine with discriminatory policies against agriculture to make the environment for agricultural production and agribusiness unfavorable and

uncompetitive. Other unattractive conditions include a low technology base (mechanization), decaying infrastructure, high interest rates, weak institutions (such as poorly-funded research institutes, distribution system and low rice imports increased from 1.25 for 2000/2001 to 1.8 million tons in 2001/2002 respectively (Kormawa and Akande, 2010). Therefore, the development of rice production in the country can contribute substantially to poverty alleviation, especially, for resource constrained households and can increase household food security. Encouraging the production of rice locally will lead to high reduction in dependence on imported rice. It is therefore, necessary to carry out a study of production trends of rice in the country, especially, in the North central geo-political zone of Nigeria where local production is well pronounced.

METHODOLOGY

Nigeria is divided into six geo-political zones with 36 states and the Federal Capital Territory. The six geo-political zones are South-south, South-east, South-west, North Central, North-west and North-east. The study was carried out in North central part of Nigeria. This comprises of Benue, Niger, Kwara, Kogi, Nassarawa and Plateau states. The area was purposively chosen because it was well known for rice production in Nigeria. Secondary data used for the study were obtained from National Bureau of Statistics, Abuja, via their data base on internet. The data covered the area in hectare of rice grown in the chosen area of study, and the quantities of rice harvested in tones from 1994/1995 growing season to 2005/2006. The data were subjected to statistical analysis using multiple regression models. The model is implicitly specified as follows:

$$Y = f(X_1, X_2, X_3, X_4, X_5, X_6, e)$$

Where: Y= Total land area for rice production in Nigeria (ha); X_1= land area for rice production in Benue (ha); X_2 = land area for rice production in Kogi (ha); X_3 = land area for rice production in Kwara (ha); X_4 = land area for rice production in Nassarawa (ha); X_5 = land area for rice production in Niger (ha); X_6 = land area for rice production in Plateau (ha) and e = error term.

For total output of rice production in thousand metric tons the model is implicitly specified as follows:

Table 2. Regression estimate of area of land for cultivation of rice.

Variable	Parameter	Linear	Double log	Semi log	Exponential
Constant	B_0	581.757 (536.332)	2.097 (0.621)	-2154.620 (2153.393)	2.905 (0.155)
Benue	B_1	0.198 (2.401)	0.221 (0.206)	0.220 (713.783)	0.198 (0.001)
Kogi	B_2	0.745 (7.095)	0.719 (0.189)	0.718 (655.44)	0.746 (0.002)
Kwara	B_3	0.284 (3.251)	0.422 (0.050)	0.412 (172.053)	0.290 (0.001)
Nassarawa	B_4	0.069 (2.782)	-0.002 (0.112)	-0.020 (387.836)	0.091 (0.001)
Niger	B_5	0.130 (0.462)	0.134 (0.078)	0.136 (271.800)	0.132 (0.000)
Plateau	B_6	0.082 (5.242)	0.061 (0.107)	0.069 (372.818)	0.072 (0.002)
R^2		0.563	0.625	0.613	0.573
Adjusted R^2		0.039	0.175	0.149	0.060
F-ratio		1.074	1.390	1.322	1.117
Durbin Watson		1.123	1.261	1.281	1.106

Computed from data from National Bureau of Statistics, Abuja.

$Y = f(X_1, X_2, X_3, X_4, X_5, X_6, e)$

Where: Y = total national rice production (t/ha), X_1= rice production from Benue (t/ha), X_2 = rice production from Kogi (t/ha), X_3 = rice production from Kwara (t/ha), X_4 = rice production from Nassarawa (t/ha), X_5 = rice production from Niger (t/ha), X_6 = rice production from Plateau (t/ha) and e = error term.

Four functional forms of the model namely, linear, semi-log, double-log and exponential was used out to determine the functions forms that best describe the data on the basis of econometric and statistical criteria.

RESULTS AND DISCUSSION

Land area for rice production from 1994/1995 to 2005/2006 at North central states of Nigeria

Table 2 shows the land area devoted to the cultivation of rice in the North Central States of Nigeria as regressed against total land area devoted for rice cultivation in Nigeria from 1994/1995 to 2005/2006 (PCU, 2002; FMARD, 2003; NBS, 2006). Multiple regression was used to ascertain the relationship between land area for rice cultivation in North Central and that of Nigeria as a whole. The regression model was tried under the four basic functional forms and the double log function was chosen as the lead equation. This was based on the value of coefficient of multiple determination (R^2) of 0.625 and the significant variables in conformity with *a priori* expectations. The economic estimate is presented in Table 2. The regression estimates indicated that all the variables (land area in the states) showed positive relationship. None of the states show significant contribution at P> 0.05. This coefficient of multiple determination (R^2) was 0.625, indicating that 62.5% of land area in Nigeria for rice production was from North Central states of Nigeria. The positive sign indicates that as the land area for rice cultivation in Nigeria increases that of the states in North Central also increases.

Total output of cassava production from 1994/1995 to 2005/2006

Table 3 shows the regression analysis of total rice production from 1994/1995 to 2005/2006 in North Central states of Nigeria as regressed against total output in Nigeria for the period under consideration. The double log function was chosen as the lead equation with R^2 value of 0.542 and explanatory variables significantly affecting total output of rice in Nigeria in consonance with *a priori* expectations. All explanatory variables had positive influence in rice output. This implies that the larger the output of rice in North Central states the larger the total output in Nigeria. This influence of North Central states on total output of rice in Nigeria may be due to the presence of fertile and suitable soil for rice cultivation, labour, contribution of household members in cultivating rice as well as in processing, distribution and marketing of rice products in the area.

The lead equation (double log) form was based on the normal economic, econometric and statistical criteria and was used for further analysis. The coefficient of determination of 0.542 implies that 54.2% of the variation in total rice production in Nigeria (Y) is explained by variables X_1 - X_6 (the North Central states) included in the model while the remaining 45.8% is as a result of non-inclusion of other rice cultivating states as well as, error in estimation. The explanatory variables adequately explained the model. In 2008, rice imports were 1.6 million tonnes and were estimated to remain at this level per annum up to 2010 (Childs and Baldwin, 2009). The current level of protection of the domestic rice sector in Nigeria ensures that local rice still has a significant market share, despite the lower quality and high costs. In an opening economy, the quality of the local rice will be a critical component to ensure its competitiveness and to guarantee rewards to investments in productivity improvement.

Table 3. Regression estimate of rice cultivation output in North central states on Nigeria.

Variable	Parameter	Linear	Double log	Semi log	Exponential
Constant	B_0	1834.847 (1708.088)	2.276 (1.172)	-5457.016 (7808.734)	3.321 (0.264)
Benue	B_1	0.271 (5.112)	0.270 (0.393)	0.315 (2616.718)	0.238 (0.001)
Kogi	B_2	-0.157 (10.515)	-0.129 (0.288)	-0.110 (1915.252)	-0.195 (0.002)
Kwara	B_3	0.281 (10.484)	0.265 (0.110)	0.299 (732.992)	0.240 (0.002)
Nassarawa	B_4	0.-0.554 (09.048)	-0.379 (0.246)	-0.383 (1635.253)	-0.550 (0.001)
Niger	B_5	0.856 (0.965)	0.861 (0.074)	0.801 (491.679)	0.935 (0.000)
Plateau	B_6	0.274 (12.367)	0.316 (0.222)	0.383 (1475.935)	0.210 (0.002)
R^2		0.639	0.792	0.745	0.689
Adjusted R^2		0.206	0.542	0.439	0.316
F-ratio		1.474	3.168	2.432	1.845
Durbin Watson		1.060	1.142	1.081	1.150

Computed from data from National Bureau of Statistics, Abuja.

Conclusion

There are considerable opportunities to revitalize the Nigerian rice sector. The current level of protection of the domestic rice sector provides an opportunity for such development. However, such a protection comes at a considerable social cost and therefore, should be seen as a temporary transient measure. It therefore recommended that the only sustainable and socially acceptable way forward is to enhance the competitiveness of local rice against imported rice, both in terms of quality and price. This calls for improving quality management and increasing efficiency along the entire marketing chain. The proposed strategy can be successful, but implies changing business as usual and calls for some innovative approaches and partnerships and an overall enabling environment for such an investment and adjustment to occur.

REFERENCES

Akande SO (2003). An Overview of Nigerian Rice Economy. Available at: www.unep/etu/etp/ events/agriculture/nigeria.pdf , (accessed 30 July 2011).

Baksh D (2003). The Right Way to Process Rice, J. Afr. Farming September/October Edition, 26.

Childs N, Baldwin K (2009). Rice Outlook. A Report from the Economic Research Service, US. Dept. Agric., RCF-09F. 2009. Available at www.ers.usda.gov, (accessed 11 June 2011).

Kormawa P, Akande T (2010). The Configuration of Comparative Advantage in Rice Production in West Africa: Surv. Empirical Stud.

Nigerian Custom Service (2009). Available at http://www.customs.gov.ng/Tariff/index.php (accessed 13 May 2011).

Plucknett DL, Philips TP, Kagbo RB (2000). A Global Development Strategy for Cassava: Transforming a Traditional Tropical Root Crops. Spurring Rural Industrial Development and Raising Incomes for the Rural Poor, Pp. 1-130.

Project Coordinating Unit (PCU) (2002). "Crop area yield survey (CAY)". Fed. Min. Agric. Rural Dev. Abuja, 2002.

Salinger BL (2010). Comp. Advantage Anal. World Bank. Washington. Available at http://go.worldbank.org/MDS6ZUERI0 (accessed 12 August 2011).

West Africa Rice Development Association (WARDA) – The Africa Rice Centre (2003): The Nigerian rice economy in a competitive world: Constraints, opportunities and strategic choices. Strategy for rice sector revitalization in Nigeria.

19

Influence of different soil amendments on postharvest performance of tomato cv. power (*Lycopersicon esculentum*, Mill)

Nyamah E. Y.*, Maalekuu B. K. and Oppong-skyere D.

Department of Horticulture, Kwame Nkrumah University of Science and Technology, Kumasi, Ghana.

The experiment was conducted to test the influence of different soil amendments types on postharvest performance of tomato fruits. The soil amendments were NPK 15-15-15 at 250kg/ha plus 'Asasewura' cocoa fertilizer (NPK 0-22-18 + 9CaO + 6s + 5MgO(s) at (250kg/ha), NPK 15-15-15 at 250kg/ha plus sulphate of ammonia (125kg/ha), poultry manure (1.1kg/ m²) and control (no amendment). Selected quality traits of tomato fruits harvested at the pink colour stage were evaluated after seven (7) days storage under average temperatures of 26.85°C and relative humidity of 85.75%. Fruit quality traits studied were fruit weight loss, firmness, general appearance, membrane ion leakage, pericarp thickness, decay, total soluble solids, pericarp weight, dry matter, moisture content and fruit shelf life. Significant differences among the soil amendments (P < 0.002) were observed in fruit quality traits studied. Fruits harvested from soil amended with NPK plus 'Asasewura' cocoa fertilizer recorded lower in weight loss, membrane ion leakage decay and moisture content and higher fruit firmness, general appearance, total soluble solids, dry matter and shelf life than fruits harvested from fields amended with NPK plus Sulphate of ammonia, Poultry manure and Control respectively. Significant correlations (P < 0.01 and P < 0.05) were observed among the quality traits evaluated. Fruit weight loss showed significantly but negative correlation between fruit firmness (-0.71) and shelf life (-0.71) but indicated significant but positive correlation between membrane ion leakage (0.63) and fruit decay (0.57) respectively. Fruit pericarp thickness showed significant but positive correlation between general appearance (0.69), pericarp weight (0.68) and total soluble solids (0.73).

Key words: Amendments, tomato, Asasewura cocoa fertilizer, postharvest, performance.

INTRODUCTION

Tomato (*Lycopersicon esculentum* Mill.) is one of the most widely used food crops in world vegetable economy (Chapagain and Wiesman, 2004). In Ghana, it is almost

an obligatory ingredient in the daily diets of people across all regions. Compared to other vegetables used in Ghana, tomatoes are normally used in large quantities (Ellis et al., 1998). In most tomato production areas in Ghana a range of conventional fertilizers comprising NPK 15:15:15, NPK 20:20:0, sulphate of ammonia and urea (Ellis et al., 1998) and organic fertilizer such poultry manure (FAO, 2005), is being used as part of soil amendments fertilizers. Yet fresh tomato fruits do not meet the demand of the consumers since lots of losses are counted. Among the fertilizers found in Ghana is 'Asasewura', cocoa fertilizer (NPK 0-22-18 + 9CaO + 6s + 5MgO(s), usually used to amend cocoa growing fields, NPK (15-15-15), Sulphate of and Ammonia, Poultry

---

*Corresponding author. E-mail: eddynaa@yahoo.com.

Abbreviations: CONT, Control (no amendment) field; GIPC, Ghana Investment Promotion Council; MM, millimetres; N, Newton; NPK + SA, NPK plus sulphate of ammonia amended fields; NPK+ CA, NPK pus 'Asasewura' cocoa fertilizer amended fields; °C, degree celsius; PM, poultry manure amended fields; RH, relative humidity.

(Chicken). Cultural practices such as nutrient application are claimed to be factors influencing quality of tomato before and after harvest (Watkins and Pritts, 2001). However, variations in fertilizer applications have being reported in major tomato production areas in Ghana (Adu-Dapaah and Oppong-Konadu, 2002). Post harvest losses are estimated to be about 20 to 50% (Kader, 1992) in countries like Ghana. These losses caused 650% increment in the importation of tomato paste between the years 1998 to 2003, (FAO, 2006). It is against these backgrounds that the present study is aimed at investigating the effects of two conventional and one organic fertilizer on post harvest quality tomato.

MATERIALS AND METHODS

Soil samples were randomly collected from different cores at 0 to 15 cm and 15 to 30cm for analysis before and from 0 to 30 cm after the studies (from the different soil amendment fields) for the contents of organic carbon (OC), organic matter (OM), total nitrogen (N), exchangeable potassium (K), sodium (Na), calcium (Ca), magnesium (Mg), available phosphorus and pH. Seedlings of tomato cultivar 'Power' were transplanted to randomized complete block design field in three (3) replications. Prior to application of 'Asasewura', cocoa fertilizer (NPK 0-22-18 + 9CaO + 6s + 5MgO(s) and sulphate of ammonia (21% N + 24s), NPK (15-15-15 + 2MgO + 3Zn) was applied at 5 g per plant two weeks after seedlings were transplanted. 'Asasewura', cocoa fertilizer (NPK 0-22-18 + 9CaO + 6s + 5MgO(s) and sulphate of ammonia (21% N + 24S) were applied at 5 and 3 g per plants respectively at two weeks after NPK (15-15-15 + 2MgO + 3Zn) application. Partially decomposed poultry manure (N = 2.85%, P = 2.38%, K = 24%, Ca = 15.60%, Mg = 2.50% and pH = 7.29) at 1.1 Kg/m² (applied to the soil three weeks on plots designated to be amended with poultry manure before transplanting) while control fields received no amendment.) Fruits at the pink stage were harvested (calyx attached) from plants marked on the field at three different harvests each in the morning within a month and two weeks from each plot and immediately placed under shade to maintain fruits temperature. Fruits were quickly transported to the laboratory where sorting and grading were carefully done and homogeneous colour developments were selected for further studies on qualitative parameters.

Quality parameters studied

Thirty six (36) pots containing ten (10) fruits from each plot were set up in the laboratory for each harvest in complete randomized design and stored for 7 days at ambient temperature of 26.85°C and relative humidity of 85.75%

Parameters studied

Weight loss (WL): Fruits were weighed daily for seven days and the difference in final weight loss expressed as a percentage of weight loss from the initial weight of five fruits sample.

Fruit firmness (FF): Fruit firmness was determined with a fruit tester (Effegi type Bishop FT 237). A circular portion of the peel of diameter of about 2 cm of each of the five fruits from each plot were removed before applying the plunger of the firmness tester in order to eliminate the effect due to the peel; and firmness was expressed in Newton (Batu, 1998).

General appearance (GA): Fruit general appearance was scored by overall rating that included freshness (green calyx), decay, firmness, defects, colour on a scale of 1 to 5 with: 0 to 1= Poor, 2 to 3= Good and 4 to 5= Very good.

Membrane ion leakage (MIL): Membrane ion leakage of fruit was determined using the method of Knowles et al. (2001) with some modifications. Ten (10) discs (10 mm in diameter) per cultivar were incubated in 20 ml distilled deionized at 5, 10, 15, 30 and 60 min intervals and conductivity measured and expressed as a percentage of the total electrolytes. Total electrolyte was also determined by freezing samples for 24 h after taking all readings (5 to 60 min). Samples were then thawed and dipped in boiling water for 20 min and conductivity measured. Conductivity meter (Hanna instrument) was used to measure membrane ion leakage.

Fruit decay (FD): Decay of fruit was recorded as soon as fungal mycelia appeared on the calyx or peel of the fruit. Decay was expressed as a percentage of the total initial fruit number stored.

Dry matter (DM): Dry matter content of fruits was measured by taking three (3) discs of 10 mm in diameter from the equatorial region of each of five fruits and oven dried at 105°C till constant dry weight was recorded according to AOAC (1990) and weight expressed as gram (g).

Moisture content (MC): Moisture content of fruits were determined by desiccation three (3) discs of 10 mm in diameter at the equatorial region of five fruits from each plot at 105°C for 24 h. The difference between the fresh weight and dry weight was expressed as a percentage of the initial fresh weight of the three (3) discs at the equatorial region of five fruits (AOAC, 1990).

Total soluble solids (TSS): Total soluble solid was determined in the same five fruits tested for fruit firmness, by squeezing out juice from fruits on Abbe's hand held refractometer and reflections measured in percent Brix (Anon, 1984)

Shelf life (SL): The shelf life was observed from the starts of harvesting and extended up to the start of rotting of fruits (Mondal, 2000).

Data collected were subjected to statistical analysis by using GENSTAT Discovery edition 3.0. Analytical software. Means were separated by LSD test at 5% Correlation analysis was performed at 1% and 5% using Statistical Package for Social Science Students (SPSS) edition 18.

RESULTS

Tomato fruits were harvested from fields amended with NPK plus 'Asasewura' cocoa fertilizer, NPK plus Sulphate of ammonia, Poultry manure and Control, stored for 7 days at 26.85°C and 85.75%RH to assess the influence of soil amendment types on postharvest performance in fruit. The analysis of variance indicated significant differences among soil amendment types in fruit weight loss (P = 0.002), fruits firmness (P < 0.001), general appearance (P = 0.001), membrane ion leakage (P = 0.001), fruit decay (P = 0.038), dry matter (P = 0.002), moisture content (P < 0.001) and total soluble solids (P < 0.001). Relatively low weight loss was recorded in fruits harvested from fields amended with NPK plus 'Asasewura'

Table 1. Means of fruit weight loss (WL), fruit firmness (FF), general appearance (GA), membrane ion leakage (MIL), pericarp thickness (PTK), fruit decay (FD),pericarp weight (PWT), dry matter (DM) moisture content (MC) total soluble solids (TSS), and shelf life (SL) of tomato fruits harvested from four different soil amendment after seven days storage at 26.85°C and 85.75%RH.

Amendment	WL (g)	FF (N)	GA (1-5)	MIL (%EC)	PTK (g)	FD (%)	DM (g)	MC (%)	TSS (Brix)	SL (day)
NPK+CA	2.68^b	3.31^a	3.93^a	12.52^b	3.36	6.80^b	0.33^a	90.46^b	4.04^a	9.39^a
NPK+SA	3.44^a	2.80^c	3.44^b	13.87^a	3.09	8.80^{ab}	0.32^a	91.05^b	3.91^a	7.58^b
PM	3.36^a	2.91^{bc}	3.87^a	14.61^a	3.65	9.20^a	0.29^b	92.93^a	4.02^a	8.32^b
CONT	3.07^a	3.11^{ab}	3.07^b	14.61^a	2.74	10.40^a	0.25^c	90.96^b	3.42^b	7.92^b
CV %	4.9	2.0	2.8	4.8	4.6	16.1	4.1	0.4	1.5	3.1

Figures followed by the same alphabets are not significant at 0.5%.

cocoa fertilizer (2.68 g) while high weight loss was recorded in fruits harvested from Control (3.27 g) and fields amended with NPK plus Sulphate of ammonia (3.30 g) and Poultry manure (3.36 g). No significant differences in weight loss were observed between fruits harvested from fields amended with Poultry manure, NPK plus Sulphate of ammonia and Control (no amendment) (Table 1).

The highest firmness was recorded in fruits harvested from fields amended with NPK plus 'Asasewura' cocoa fertilizer (3.31N) followed by Control (3.11N), Poultry manure (2.91N) and NPK plus Sulphate of ammonia (2.80N) amended fields. However, no significant differences were indicated between fruits harvested from fields amended with NPK + 'Asasewura' cocoa fertilizer (3.31N) and Control (3.11N) as well as NPK plus Sulphate of ammonia and Poultry manure respectively and also between Control (3.11N) and Poultry manure (2.91N) (Table1).

Fruits harvested from NPK plus 'Asasewura' cocoa fertilizer amended fields (3.93) and Poultry manure fields amended fields (3.87) showed relatively better in general appearance than fruits harvested from fields amended with NPK plus Sulphate of ammonia (3.44) and Control (3.07). However, no significant differences were observed between fruits harvested from fields amended with NPK plus 'Asasewura', cocoa fertilizer (3.93) and Poultry manure (3.87) as well as between NPK plus Sulphate of ammonia (3.44) amended fields and Control fields (3.07) (Table 1).

Comparatively, membrane ion leakage was prone in fruits harvested from fields amended with Poultry manure (14.61%), NPK plus Sulphate of ammonia (13.87%) and Control (14.61%) than fruits harvested from field amended with NPK plus 'Asasewura' cocoa fertilizer (12.52%). However, no significant differences were observed among fruits harvested from Control and fields amended with Poultry manure as well as NPK plus Sulphate of ammonia in membrane ion leakage (Table 1).

Fruits harvested from fields amended with Poultry manure (3.88 mm), NPK plus Sulphate of ammonia (3.69 mm) and NPK plus 'Asasewura' cocoa fertilizer (3.36 mm) recorded higher pericarp thickness than fruits harvested from Control fields (2.86 mm), respectively

(Table 1).

Decay was relatively high in fruits harvested from control (10.40%, poultry manure (9.20%)and NPK plus Sulphate of ammonia (8.80%) amended fields while relatively low percentage decay (6.80%) was observed in fruits from fields amended with NPK plus 'Asasewura' cocoa fertilizer. However, no significant differences were observed among fruits harvested from control poultry manure and NPK plus Sulphate of ammonia (Table 1).

Relatively high dry matter were recorded from fruits harvested from fields amended with NPK plus 'Asasewura' cocoa fertilizer (0.33 g) and NPK plus Sulphate of ammonia (0.32 g) while relatively low dry matter content (0.28 and 0.25 g) were recorded from fruits harvested from fields amended with Poultry manure (0.28 g) and Control (0.25 g) respectively. No significant differences were observed between dry matter in fruits harvested from fields amended with NPK plus 'Asasewura' cocoa fertilizer and NPK plus Sulphate of ammonia (Table 1).

Fruits harvested from Poultry manure amended fields recorded the highest moisture content (92.93%) followed by fruits harvested from NPK plus Sulphate of ammonia amended fields (91.05%), Control (90.96%) and NPK plus 'Asasewura' cocoa fertilizer amended fields (90.46%). No significant differences in fruit moisture content were observed among fruits harvested from NPK plus 'Asasewura' cocoa fertilizer, NPK plus Sulphate of ammonia amended fields and Control (Table 1).

Relatively high total soluble solids were recorded for fruits harvested from NPK plus 'Asasewura' cocoa fertilizer (4.04% Brix) and Poultry manure(4.02% Brix) amended fields while relatively low total soluble solids were recorded for fruits harvested from fields amended with NPK plus Sulphate of ammonia (3.91% Brix) and Control (3.42% Brix). No significant differences were observed between fruits harvested from NPK plus 'Asasewura' cocoa fertilizer, Poultry manure and NPK plus Sulphate of ammonia (Table 1).

The analysis of variance showed significant difference (P < 0.002) in fruits shelf life among the soil amendments. Fruits harvested from NPK plus 'Asasewura' cocoa fertilizer amended fields recorded the highest shelf life (9.39 days) followed by fruits harvested from fields

Table 2. Correlation values and P values between postharvest quality traits of tomatoes after seven days ambient (room) temperature (26.85°C) storage.

FWTL	FWTL	FF	GA	MIL	FD	DM	FMC	TSS
FF	-0.71**							
GA	-0.44**	0.39*						
MIL	0.63**	-0.57**	-0.42*					
FD	0.57**	-0.60**	-0.40*	-0.46**				
DM	-0.36*	0.55**	0.54**	0.65**	-0.33^{NS}	0.39*		
FMC	0.31^{NS}	-0.62**	-0.28	0.22^{NS}	0.23^{NS}	0.23^{NS}	-0.80**	
TSS	-0.48**	0.27^{NS}	0.60**	0.73**	-0.40**	0.63**	0.37*	0.05^{NS}
SL	-0.71**	0.73**	0.55**	0.62**	-0.63**	0.48**	-0.34*	0.67**

*, **, NS = P < 0.05, P<0.01, not significant respectively. Fruit weight loss = (FWTL), fruit firmness = (FF), general appearance = (GA,) and membrane ion leakage = (MIL), fruit decay = (FD), dry Matter = (DM), fruit moisture content (FMC), total soluble solid = (TSS), and shelf life (SL).

amended with Poultry manure (8.32 days), NPK plus Sulphate of ammonia (7.92 days) and Control (7.58 days) fields. No significant differences in fruits shelf life were observed between fruits harvested from fields amended with Poultry manure, NPK plus Sulphate of ammonia and Control (Table 1).

Correlations Fruit weight loss indicated a high correlation with fruit firmness (-0.71), shelf life (-0.71), membrane ion leakage (0.63), fruit decay (0.57), (P < 0.01) but showed no significant correlations with fruit moisture content and fruit pericarp weight (Table 2). Fruits firmness indicated high correlation (P < 0.01) with fruit decay (-0.60), fruit moisture content (-0.62) and shelf life (0.73). General appearance revealed significant positive correlation (P < 0.01) with fruit and shelf life (0.55). Membrane ion leakage indicated high correlation (P < 0.01) with fruit decay (0.61) and shelf life (-0.67) but showed no significant correlation with fruit dry matter (-0.28), moisture content (0.24) (Table 2). Fruit dry matter (0.65), total soluble solids (0.73) and shelf life (0.62) but show no significant correlation with moisture content. Fruit dry matter indicated correlation (P < 0.01) with fruit moisture content (-0.80) (Table 2).

DISSCUSION

Fruits harvested from fields amended with NPK plus 'Asasewura' cocoa fertilizer recorded significantly the lowest weight loss (2.68 g) among fruits from fields amended with NPK plus Sulphate of ammonia (3.44 g), Poultry manure (3.36 g) and Control (3.07 g) fields. This might be due to the relatively high and readily available calcium in 'Asasewura' cocoa fertilizer which is characterised by the ability to increase cell formation and reduce respiration rates. This might have contributed to the reduction of weight loss in fruits harvested from NPK plus 'Asasewura' cocoa fertilizer amended fields. Sharma et al. (1996) in their observations confirmed calcium's ability to reduce respiration, which is an indication of

weight loss (Kays, 1991).

Fruit firmness is a criterion often used to evaluate fruit quality as it is directly related to fruit storage potential. It is also related to the likelihood of bruising when fruits are subjected to impact during handling (Lesage and Destain, 1996). Relatively low nitrogen levels in NPK plus 'Asasewura' cocoa fertilizer amended fields might have caused the availability of calcium for plant utilization to increase fruit firmness (3.31N) than fruits harvested from Poultry manure (2.91N) and NPK plus Sulphate of ammonia (2.80N) amended fields where nitrogen level were relatively high. Findings of research of Siddiqi et al. (2002), Akl et al. (2003), Heeb et al. (2005b) have shown that nitrogen dominated supply may markedly increase the incidence of depression of calcium uptake which could lead to decreased firmness in fruits harvested from fields amended with poultry manure and NPK plus sulphate of ammonia. Crisosto et al. (1995) reported that, excess nitrogen during the pre-harvest stage can reduce fruit firmness.

General appearance of fruits plays an important role in making purchasing decisions (Kays, 1999). Colour, cracks, bruises, firmness, etc are factors mostly used to assess general appearance which double as important factors in the consumer preference of tomatoes. Relatively high and readily available calcium levels in 'Asasewura' cocoa fertilizer might have increased fruit firmness and also lower biosynthesis in carotenoids which are responsible for tomato fruit colour. This could have improved general appearance of fruits harvested from NPK plus 'Asasewura' cocoa fertilizers fields (3.93) more than fruits harvested from fields amended with NPK plus Sulphate of ammonia (3.44) and Control (3.07). Studies conducted by Kays (1991), indicated that insufficient calcium supply will increase the biosynthesis of carotenoids, which are responsible for tomato fruit colour (Dorais et al., 2001).

The best general appearance in fruits harvested from Poultry manure amended fields (3.87) than those from Control (3.07) and NPK plus Sulphate of ammonia fields

(3.44) could probably be due to adequate calcium and magnesium levels which has the ability to reduce defects such as shoulder cracks, increase fruit firmness and to increase overall fruit quality respectively. Hao and Papadopoulos (2004) reported that, under conditions of severe magnesium deficiency the size and overall appearance of the fruit may be reduced, unless accompanied by a commensurate increase in calcium supply.

Above and beyond wilting and dehydration of horticultural products during storage, deterioration of plant tissues is also of great concern to food and horticultural scientists. Evidence gathered supports membrane damage as the key event leading to a cascade of biochemical reactions culminating in tissues deterioration (Maalekuu et al., 2005).

Probably, the readily available calcium levels in NPK plus 'Asasewura', cocoa fertilize amended fields might have caused relatively low loss of membrane integrity resulting from membrane damage. This could have caused the lowest weight (water) loss rate which might have also led to low membrane ion leakage of fruits harvested from fields amended with NPK plus 'Asasewura', cocoa fertilizer (12.52%) than those from fields amended with Poultry manure (14.61%), NPK plus Sulphate ammonia (13.87%) and Control (14.61%) respectively. Boros-Matovina and Blakes (2001) reported that, membrane ion leakage is a measure of loss of membrane integrity resulting from membrane damage which leads to water loss and loss of other membrane-bound solute.

Among the principal causes of post harvest losses is decay (Steven and Celso, 2005). Comparatively higher nitrogen levels during the pre harvest period as characterised by poultry manure might have increased the susceptibility to decay of fruits harvested from Poultry manure (9.20) field than fruits from fields amended with NPK plus' Asasewura' cocoa fertilizers (6.80). Bechmann and Earles (2000) indicated the possibility of produce stressed by high rate of nitrogen to be susceptible to post harvest decay (diseases) in fruits.

Again, relatively high nitrogen levels available in Poultry manure amended fields might have foster calcium (a cell forming nutrient) deficiency to the plant, hence leading to week cell formation and easy degradation of cells in fruits which could increase decay incidence of fruits more than those harvested from fields amended with NPK plus 'Asasewura' cocoa fertilizers, where nitrogen levels were moderately low. Studies conducted by Siddiqi et al. (2002), Akl et al. (2003) and Heeb et al. (2005b) associated increase in fruit decay (rot) with high nitrogen levels dominated nutrients, an effect which is attributed to a depression of calcium uptake.

Attack by most organisms that cause deterioration in fruits follows physical injury or physiological breakdown. The relatively high decay recorded in fruits harvested from control (10.40%) fields than those harvested from

fields amended with NPK plus 'Asasewura' cocoa fertilizer may be due to additional calcium levels in 'Asasewura' cocoa fertilizer which is characterised to reduce the incidence of shoulder cracks and other physiological disorders that lead to deterioration of fruits. Lichter et al. (2002) who reported on the ability of calcium to reduce the incidence of shoulder check crack, and other physiological disorder that leads to deterioration in fruit quality.

High dry matter or low water content of the tomato has also been reported to affect fruit taste positively because the major components of tomato taste; sugars and acids, are more concentrated (Auerswald et al., 1999), which fits well with consumers' demand for high quality produce (El-Saeid et al., 1996). The relatively high dry matter in fruits harvested from fields amended with various fertilizers NPK plus Sulphate of ammonia (0.32 g), Poultry manure (0.29 g) and NPK plus 'Asasewura' cocoa fertilizer (0.33 g) than fruits harvested from control (0.25 g) fields may be mainly due to the additional plant nutrients for plant utilization supplied by the fertilizers used to amend fields which could improve cell formation to increase fruits dry matter.

Fruits and vegetables contain large quantities of water in proportion to their weight. Norman, (1992) indicated that, tomato fruits contain about 93% moisture.The significantly high moisture content recorded from fruits harvested from fields amended with Poultry manure (92.93%) than those harvested from fields amended with NPK plus 'Asasewura' cocoa fertilizer (90.46%), NPK plus Sulphate of ammonia (91.05%) and Control (no amendment) fields (90.96%) respectively may be due to the ability of Poultry manure (organic fertilizer) to retained more moisture (water) for plants utilization over relatively long period than fields amended with inorganic fertilizers and control. Barrett et al. (2007) who indicated that significant differences in moisture content between tomatoes produced under conventional and organic production systems where in a case, the organic tomato moisture content was higher, and in two cases, the moisture content of conventional tomatoes was higher.

Total soluble solids are known to increase fruit quality (Loboda and Chuprikova, 1999), which fits well with consumers' demand for high quality produce (El-Saeid et al., 1996). The higher total soluble solids recorded from fruits harvested from the fields amended NPK plus 'Asasewura' cocoa fertilizer (3.93) over fruits from control and NPK plus Sulphate of ammonia amended fields may partly be due to the higher dry matter content recorded from fruits harvested from NPK plus 'Asasewura' cocoa fertilizer than other amended fields. Increased in total soluble solid of fruits harvested from fields amended with Poultry manure (3.87) over fruits harvested from control (3.07) and NPK plus Sulphate of ammonia (3.44) fields may be as a result of probably higher biosynthesis or degradation of polysaccharides during storage in fruits harvested from fields amended with Poultry manure.

Artes et al. (1999) in their studies in tomato associated increment of total soluble solids to degradation of polysaccharides.

Fruits shelf life during storage is an important feature from a producer's and a distributor's point of view, allowing the determination of risks arising from the loss of commercial value of fresh fruit in trade turnover (Radajewska and Borowiak, 2002). Significantly higher fruits shelf life (9.39 days) of fruits harvested from NPK plus 'Asasewura' cocoa fertilizer amended fields than those harvested from Control (7.58 days), Poultry manure (8.32 days) and NPK plus sulphate of ammonia (7.92 days) amended fields may be due to the relatively high and readily available calcium levels (which has the ability to extend fruit shelf life) in NPK plus 'Asasewura' cocoa fertilizer amended fields for plant utilization. Sharma et al. (1996) in their research conducted supported the ability of calcium nutrients to extend shelf life of fruits. Also other desirable characteristics of calcium such as its ability to delay ripening and senescence, reduce respiration, increase firmness and reduce physiological disorders as reported by Sharma et al. (1996) could probably have affected the shelf life of tomato fruits stored possibly.

Results obtained from correlation analysis are of great concern in determining the relationship between postharvest qualities of tomato (L. esculentum Mill.) fruits after storage. Fruit weight loss indicated high significant correlation with fruit firmness (-0.71), membrane ion leakage (0.63), general appearance (-0.44), shelf life (-0.70) respectively is an indication that, increase in weight loss in fruits could directly decrease fruit firmness. Weight (water) loss is the principal cause of fruit softening (Wilson et al., 1999). Increase in fruit weight loss probably increased membrane ion leakage directly. Membrane ion leakage which is an indicator of loss of membrane integrity was found to increase in pepper genotypes susceptible to high rates of water loss (Maalekuu et al., 2005).

Fruit general appearance is negatively affected by weight (water) loss since increase in fruit water loss reduces fruit firmness, increases shriveling which are important criteria for accessing fruit general appearance qualities. Water loss is the principal cause of fruit softening and shriveling (Wilson et al., 1999). Increase in weight loss is an indication of increase respiration rate which arises from the breakdown of carbon compounds by metabolism plant products hence decrease in fruit shelf life.

Fruit firmness indicating high significant correlation with fruit shelf life (0.73) and fruit decay (-0.60) respectively is a signal that, increased in fruit firmness which may have resulted from low respiration rate which is directly proportional to increase in fruit shelf life. Increase in fruit firmness leads to harden of the dermal system of the fruits making them insusceptible to decay pathogen infestation which could lead to decrease in fruit decay.

The significant negative correlation of membrane ion leakage with fruit decay (-0.61) and shelf life (-0.67), signifying the death of fruit cells which perhaps affects the enzyme-substrate interaction thus leading to reduced starch hydrolytic activity as cause by leakage of membrane. Loss of membrane integrity resulting from membrane damage leads to water loss and loss of other membrane-bound solutes (Boros-Matovina and Blakes, 2001), an indicator for predicting fruit shelf life and decay. Fruit dry matter (0.65) indicated positive correlation with shelf life (0.63) and total soluble solids (0.73) explain why fruit with highest dry matter, total soluble solids, shelf life of fruits. Fruit dry matter indicated high correlation with fruit moisture content (-0.80), this explains why fruits with higher dry matter had lower moisture content.

Conclusion

The study revealed that, pre harvest (soil amendment and cultivar types) practices can influence post harvest performance of tomato (Lycopersicon esculentum Mill.) fruits. Fruits harvested from fields amended with NPK plus 'Asasewura' cocoa fertilizer performed best among the soil amendments on post harvest quality of tomato studied. However, fruits harvested from all fertilizer (NPK plus 'Asasewura' cocoa fertilizer, NPK plus sulphate of ammonia and poultry manure) amended fields on average performed better than fruits harvested from control (no amended) fields.

Fruit weight loss had indirect consequence on fruit firmness, general appearance and shelf life but direct effect on increased membrane ion leakage in tomato cultivar studied. Membrane ion leakage influenced fruit decay negatively in tomato cultivar studied.

Based on the study, both exotic and local types of tomato should be promoted with emphasis on the exotic types such as Royal which performed comparatively better with regard to postharvest quality. Moreover, tomato farmers are encouraged to amend their fields with NPK plus 'Asasewura' cocoa fertilizer to ensure higher yield and fruit quality.

REFERENCES

Adu-Dapaah HK, Oppong-Konadu EY (2002). Tomato production in four major tomato-growing districts in Ghana: Farming practices and production constraints. Ghana Jnl agric. Sci, 35: 11-22.

Akl IA, Savvas D, Papadantonakis N, Lydakis-Simantiris N, Kefalas P (2003). Influence of ammonium to total nitrogen supply ratio on growth, yield and fruit quality of tomato grown in a closed hydroponic system. Euro. J. Hortic. Sci., 68: 204-211.

AOAC (1990). Official methods of analysis, 15th Edn. Association of Official Analytical Chemists, Washington D.C.

Artes F, Conesa MA, Hernandez S, Gil MI (1999). Keeping quality of fresh-cut tomato. Postharvest Biol. Technol., 17: 153-162.

Auerswald H, Schwarz D, Kornelson C, Krumbein A, Brückner B (1999). Sensory analysis, sugar and acid content of tomato at different EC values of the nutrient solution. Sci. Hortic., 82, 227-242.

Bechmann J, Earles R (2000). Post harvest handling of fruits and vegetables. ATTRA Horticultural Technical Note 2000, p. 19.

Barrett DM, Weakley C, Diaz JV, Watnik M (2007). Qualitative and nutritional differences in processing tomatoes grown under commercial organic and conventional production systems. J. Food Sci., 72: 441-451.

Boros-Matovina V, Blaake TJ (2001). Seed treatment with the antioxidant Ambiol enhances membrane protection in seedling exposed to drought and low temperature. Trees, 15: 163-167.

Crisosto CH, Mitchell FG, Johnson S (1995). Factors in fresh market stone fruit quality. Postharvest News Inf., 6: 17-21.

Dorais M, Papadopoulos AP, Gosselin A (2001). Greenhouse tomato fruit quality. In: Janick, J. (ed.), Hort. Rev., 5: 239-319.

El-Saeid HM, Imam RM, El-Halim SMA (1996). The effect of different night temperatures on mophorlogical aspect, yield parameters and endogenous hormones of sweet pepper. Egyptian J. Hortic., 23: 145-165.

FAO (2005). Fertilizer use by crop in Ghana First version, published by FAO, Rome, p. 11.

FAO (2006). Import Surge Brief - Commodities in Developing Countries: No. 5 Ghana: rice, poultry and tomato paste, published by FAO, Rome, p. 2.

Hao X, Papadopoulos AP (2004). Effects of calcium and magnesium on plant growth, biomass partitioning, and fruit yield of winter greenhouse tomato. HortScience, 39: 512-515

Kader AA (1992). Fruit maturity, ripening, and quality relationships. Perishables Handling Newsletter, 80: 2.

Kays SJ (1999). Postharvest factors affecting appearance. Postharvest Biol. Technol., 15: 233-247.

Kays SJ (1991). Postharvest physiology of perishable plant products. Van Nostrand, p. 532.

Lesage P, Destain M (1996). Measurement of Tomato Firmness by using a Non-Destructive Mechanical Sensor. Postharvest Biol. Technol., 8: 45-55.

Lichter A, Dvir O, Fallik E, Cohen S, Golan R, Shemer Z, Sagi M (2002). Cracking of cherry tomatoes in solution. Postharvest Biol. Technol., 26: 305-312.

Maalekuu K, Elkind Y, Tuvia-Alkalai S, Shalom Y, Fallik E (2005). Characterization of physiological and Biochemical factors associated with post harvest water loss in ripe pepper fruits during storage. J. Am. Soc. Hort. Sci., 130 (5): 735-741.

Mondal MF (2000). Production and storage of fruits (in Bangla). Published by Afia Mondal. BAU Campus, Mymensingh-2202; p. 312.

Norman JC (1992). Tropical vegetable production. Macmillan press, pp. 52-67.

Radajewska B, Borowiak ID (2002). Refractometric and sensory evaluation of strawberry fruits and their shelf life during storage. Acta Hortic., 567:759-762.

Sharma RM, Yamdagni R, Gaur H, Shukla RK (1996). Role of calcium in horticulture - A review. Haryana J. Hortic. Sci., 25(4): 205.

Siddiqi MY, Malhotra B, Min X, Glass ADM (2002). Effects of ammonium and inorganic carbon enrichment on growth and yield of a hydroponic tomato crop. J. Plant Nutr. Soil Sci., 165: 191-197.

Steven AS, Celso LM (2005). Tomato, strawberry fruits and their shelf life during storage. Acta Hortic., 567: 759-762. Http://usna.usda.gov/h666/138tomato.

Watkins CB, Pritts MP (2001). The influence of cultivars on postharvest performance of fruits and vegetables. Proceedings of the Fourth International Conference on Postharvest Science. Acta Hortic., 1(553): 59-63.

20

Micro level investigation of marketing and post harvest losses of tomato in Coimbatore district of Tamilnadu

K. Kalidas[1] and K. Akila [2]

[1]Vanavarayar Institute of Agriculture, Pollachi -642103, Tamil Nadu State, India.
[2]Bank of Baroda, Senjerimalai, Coimbatore – 642002, Tamil Nadu State, India.

Most vegetables are perishable in nature, and in that post harvest losses and distribution channel plays a vital role in price fixation of vegetables, especially in tomato which is sensitive to much environment-genetic interaction disorders which may be manifested during post harvest ripening or post harvest inspection. A substantial quantity of production is subjected to post-harvest losses at various stages of its marketing. The quantum of loss is governed by factors like perishable nature, method of harvesting and packaging, transportation, etc. Tomato being a third most cultivated crop, the post-harvest losses is significant In terms of quantity and economic value. This study undertaken in Coimbatore on tomato has suggested marketing loss in the estimation of marketing margins, price spread and efficiency and has used a modified formula for it. It has been observed that a majority of tomato producers sell their produce to the wholesalers facilitated by commission agents at different stages. The aggregate post-harvest losses from farm gate to consumers in tomato ranges from 13 to 26%. It has indicated the necessity of reducing the market intermediaries, for minimizing post-harvest losses and providing remunerative price to the producers. The results have emphasized that efforts should be made to adopt improved packaging techniques, cushioning material at the farm level. The producer's share in consumer's price as estimated by old method has been found higher and the inclusion of marketing loss in the estimation of marketing margins, price spread and efficiency has indicated that the old estimation method unduly over-states the farmers' net price and profit margins to the market middlemen. It is appropriate to use modified method for the estimation of marketing margins and price spread.

Key words: Post harvest loss, price spread, consumer price, marketing margin, marketing loss, marketing efficiency.

INTRODUCTION

Tomato is a major vegetable crop that has achieved tremendous popularity over the last century. Tomato is grown in an area of 8.65 lakh hectares with an average production of 168.26 lakh tones in India (NHB, 2012) and in Tamilnadu majority of the Tomato growers are from Coimbatore, Salem, Dindugal district (Tamil Nadu Agricultural Department). Tomato, aside from being tasty, promotes healthy nutritional balance as it is a good source of vitamins A and C. Tomato is also an excellent source of Lycopene (a very powerful antioxidant) that helps to prevent development of many forms of cancer. Tomatoes are sensitive to much production and environment-genetic interaction disorders which may be manifested during post harvest ripening or post harvest inspection (Kumar, 2010). A substantial quantity of production is subjected to post-harvest losses at various stages of its marketing (Kishore et al., 2006). The quantum of loss is governed by factors like perishable nature, method of harvesting and packaging, transportation, etc. Tomato being a third most cultivated crop (NHB, 2012), the post-harvest losses is significant in terms of quantity and economic value. Keeping this in view, the post-harvest losses of tomato have been estimated in both physical and value terms at different stages during transportation and marketing by using plastic crates as packaging materials. Further, the impact of post-harvest losses on producer's share, marketing margins, price spread and marketing efficiency in different markets has been studied. The specific objectives of this study were: (1) To identify major channels in marketing of tomato in Coimbatore District; (2) To identify the variety preferred in the study area; (3) To analyze the nature of seasonal price fluctuation in tomato; (4) To estimate the price spread of tomato marketing; (5) To study the post harvest losses at various stages of handling in tomato, and (6) To explore the scope for introducing the post harvest technology practices, largely to reduce wastages.

METHODOLOGY

Sampling procedure

Tamilnadu was purposively selected, as it is one of the major producers of tomato in India. In Tamilnadu, the Coimbatore district was selected because of its maximum contribution to the total state production. In Coimbatore District, Pollachi, Kinathukadavu, Madukkarai, Anaimalai blocks were selected based on the area of production and marketing. Then five villages from each block were selected randomly. From each block, thirty tomato growers were randomly selected. Thus, a total of 120 tomato growers were selected randomly, in the entire population 30 commission agents, 30 wholesalers and 30 retailers were covered in the study area. The market functionaries are selected from the vegetable markets such as Pollachi, Kindathukadavu and Anaimalai. The data related to production and marketing practices, post-harvest losses, price received and returns from produce, during the year of 2011 to 2012 were collected through personal interview with the help of survey schedule.

Analytical techniques

Simple averages and percentages were used to calculate the post-harvest losses at different stages of tomato marketing, marketing margins, costs and losses. In this study, post-harvest losses were measured at different stages. The modified formulae used for estimating the post-harvest losses during tomato marketing are given below.

Producer's net price

The net price realized by the tomato grower was estimated as the difference in gross price received by him and the sum of marketing costs incurred and economic value of fruits loss during harvesting, grading, transit and marketing (George, 1972). Thus, producer's net price may be explained mathematically as:

$$NPG = GPG - \{CG + (LG \times GPG)\}$$

Where, NPG is the net price received by the tomato growers (Rs/tonnes); GPG is the gross price received by tomato growers or wholesale price to Traders (Rs/tonnes); CG is the cost incurred by the producers during marketing (Rs/tonnes), and LG is the physical loss in fruits from farm to market (per tonnes).

Marketing margins

The margins of market middlemen include profit, which accrue for trading facility provided and market establishment after adjusting the marketing loss during handling and transit (Gajanana, 2002). The expression for estimating the margins for middlemen is:

Middlemen = Gross price – Price paid – Cost of – Loss in value during marketing transit/wholesaling.

Net marketing margin of the wholesaler is given mathematically as:

$$MMW = GPW - GPG - CW - (LW \times GPW)$$

Where, MMW is the net margin of the wholesaler (Rs/kg); GPW is the wholesaler's gross price to retailers or purchase price of retailer (Rs/kg); CW is the cost incurred by the wholesaler during marketing (Rs/kg), and LW is the physical quantitative loss in produce at wholesaler's level (per kg).
As said by Chandra (1994), mathematically, the net marketing margin to the retailer is given as:

$$MMR = GPR - GPW - CR - (LR \times GPR)$$

Where, MMR is the net margin of the retailer (Rs/kg); GPR is the price at the retail market or purchase price of consumers (Rs/kg); CR is the cost incurred by the retailer during marketing (Rs/kg), and LR is the physical loss in produce at the retailer level (per kg).
The total margins for the market middlemen (MM) are calculated as:

$$MM = MMW + MMR$$

Similarly, the total marketing cost (MC) incurred by the producer/traders and various middlemen is calculated as:

$$MC = CF + CW + CR$$

The total value loss due to damage during handling of fruits from farm till reaching the ultimate consumers is estimated as:

$$ML = \{LG \times GPG\} + \{LW \times GPW\} + \{LR \times GPR\}$$

Marketing efficiency

The conventional methods, Shepherd's method

Channel 1

Producer → Commission agent → Wholesaler → Retailer → Consumer

Channel 2

Producer → Wholesaler → Retailer → Consumer

Channel 3

Producer → Commission agent → Retailer → Consumer

Channel 4

Producer → Consumer

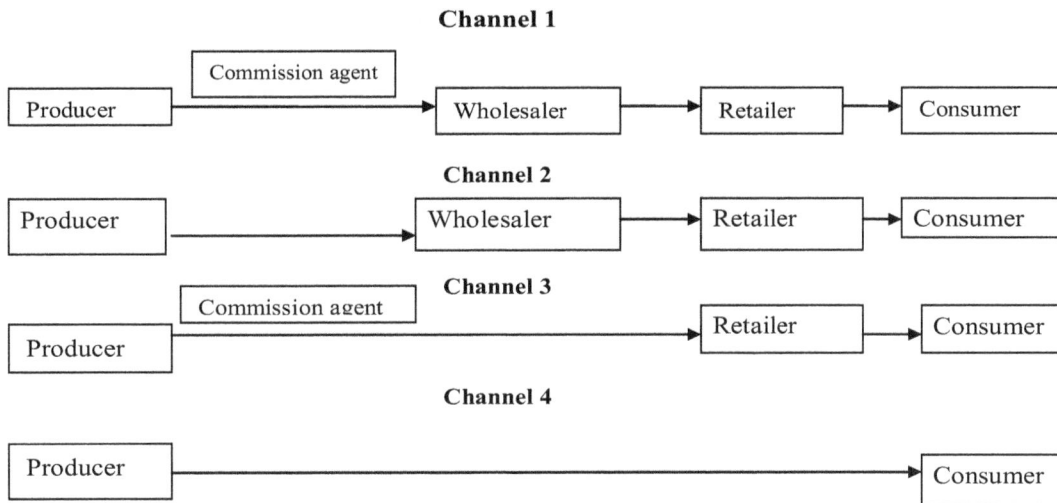

Figure 1. Distribution channels in tomato marketing in Coimbatore.

(Shepherd, 1965) and Acharya's modified formula (Acharya and Agrawal, 2001) do not mention the loss in produce during marketing process as a separate item. However, reduction due to post-harvest losses is one of the efficiency parameters. Therefore, it is pivot to incorporate the loss component explicitly in the existing marketing ratios to get the correct measures of marketing efficiency while comparing the market channels. The post-harvest loss/marketing loss component was incorporated in the formula given by Acharya and Agrawal (2001) and the modified marketing efficiency (ME) was measured as:

RESULTS AND DISCUSSION

Marketing practices and distribution channels

As said by Sreenivasa et al. (2002), the main factor which plays the key role in decision-making of the growers is the price offered by the traders during harvesting season. The producer selling wholesaler is a common marketing practice in the area. The tomatoes are marketed locally in plastic crates, gunny bags, bamboo basket, and wooden box or loose. For distant markets, plastic crates of 15-kg capacity are used by the traders. It was observed that some tomato growers sell the produce directly to the consumers. It was found that tomato producers in Coimbatore follow several marketing channels, as given in Figure 1.

Variety preference

Lakshmi (NP5005) is the most preferred variety in Coimbatore as 60% of the tomato growers go for this

$$ME = \frac{NPG}{MM + MC + ML}$$

Seasonal index

To analyze the seasonal price fluctuation in tomato seasonal index is constructed using average percentage method as: (Season average/over time price average)*100

variety which is followed by US3140 and Red ruby. The reason for preferring NP5005 variety is as follows:

1. The fruits are round, pale green when unripe and capsicum red when fully ripe.
2. The fruits are uniform ripening, very firm, and the ripened fruits store well on the vine.
3. It has very good transport and keeping quality, 8-10 days at room temperature amongst the round-fruited varieties.

Seasonal price fluctuation in tomato

Seasonal index (Cundiff and Still, 1968) calculated for five years taking the farmer's market price of tomato shows that the tomato prices were least in the months of March and February shown in Table 1. Tomato market in Tamilnadu is highly affected by the arrivals from Karnataka and Andhra Pradesh in these months as it is a peak season in Karnataka. The prices were high in the month of May, June, July, November and December. Prices were high in these months as sowing takes up in Coimbatore during June and July month and the tomato

Table 1. Seasonal price fluctuation in tomato in Coimbatore District of Tamilnadu.

Month	Seasonal index
January	96
February	66
March	62
April	75
May	122
June	130
July	105
August	75
September	97
October	94
November	152
December	126

availability in the month of November and December is scarce as it is rainy season. In November prices where 52% higher than base prices in which producer can receive best price.

Post-harvest loss in tomato

At farm level

The post-harvest loss due to harvesting injuries due to pest and disease infection, physiological damage, mechanical damage of tomato fruits were worked out to be 6 to 7% (Table 2). Also the loss during the transit is calculated around 5 to 6%. All the thrown away or discarded fruits at the farm were treated as post-harvest loss. These fruits were neither marketed nor consumed in any form the grower has to bear this post-harvest loss, irrespective of the marketing channel. Since sorting, grading and packaging is the first function to be performed in the marketing process, any loss during this process is considered as post-harvest loss. It is more appropriate in the perishable commodities like tomato, as the entire production is marketable surplus.

At wholesale level

Tomato fruits were packed in different packaging materials such as plastic crates and wooden boxes. The plastic crates and wooden boxes having capacity of 15 kg were used for transportation of fruits to medium- and long-distant markets. Tomato fruits are transported from the study area to distant markets such as Kerala, Chennai and Bangalore, by trucks. The loss in tomato fruits during transportation and wholesalers' level was 6% (Table 2), largely due to bad transportation practices, improper packaging materials, lack of infrastructure facilities, and lack of cold storage and environment conditions.

At retail level

The losses at retailer's level are estimated to be 8%. The main cause of loss in market was the damage due to press/ bumped and physical injury, which accounted for 50%. The discarded tomato fruits fetched no economic value to the retailers. These were eaten by stray animals or thrown away by the retailers. The aggregate post-harvest loss from production to consumption level in ranged from 13 to 26%. The results revealed that the efforts should be made to adopt improved packaging techniques, cushioning material and cold storage facilities at the retailers' level.

Costs, margins, losses and strategies for tomato marketing

The results obtained through new and old methods in the three marketing practices and their implications have been given in Table 3.

Marketing costs

The marketing cost of tomato was estimated to be Rs. 5.81 / kg in channel 1, Rs.4.19 / kg in channel 2, Rs.4.05 / kg in channel 3 and Rs.1 /kg in channel 4. The cost of collecting, sorting, grading packaging, plastic crates, commission and marketing fee were the major components of the marketing cost. The marketing cost is higher in marketing channel 1 as it involves too many market intermediaries. It is evident from Table 3 that the wholesaler had incurred less cost on marketing compared to producer and retailer in all the channels. Thus the overall post harvest loss of produce is less in channel 4, when compared to channel 1, 2 and 3.

Table 2. Aggregate post-harvest loss in tomato produced in Coimbatore district of Tamilnadu.

Particulars	Channel 1 Qty (kg/Qtl)	Channel 2 Qty (kg/ Qtl)	Channel 3 Qty (kg/ Qtl)	Channel 4 Qty (kg/ Qtl)
Farm level				
Collection	3.15	3.17	3.38	3.18
Sorting	2.00	2.52	2.75	2.57
Packaging	1.70	1.87	1.43	1.95
Transportation	2.60	2.50	2.86	3.10
Unloading	2.75	2.69	2.00	1.49
Customer handling				1.58
Subtotal	**12.20**	**12.75**	**12.42**	**13.87**
Wholesaler level				
Repacking	2.50	1.30	-	-
Sorting	-	2.00	-	-
Loading	1.00	-	-	-
Unloading	0.50	-	-	-
Transportation	2.10	-	-	-
Subtotal	**6.10**	**3.33**	-	-
Retailer level				
Repacking	2.00	1.50	1.43	-
Sorting	2.00	2.00	2.50	-
Unloading	1.00	1.60	1.00	-
Transportation	1.46	1.90	2.20	-
Customer handling	1.50	1.30	0.50	-
Subtotal	7.96	8.30	7.63	-
Grand Total	26.26	24.38	20.05	13.87

Table 3. Impact of post-harvest loss on producer's share, marketing costs and Margins in Tomato in Coimbatore District of Tamilnadu

Particulars	Channel 1			Channel 2			Channel 3			Channel 4		
	Old	New	Differ. (%)	Old	New	Differ. (%)	Old	New	Differ. (%)	Old	New	Differ. (%)
Producer's net price (Rs/kg)	5.94	4.98	16	6.8	5.84	14	6.9	5.82	16	9	7.77	14
Marketing cost (Rs./kg)												
Producer	2.06	-	-	1.20	-	-	2.10	-	-	1.00	-	-
Wholesaler	1.86	-	-	1.07	-	-	0	-	-	-	-	-
Retailer	1.89	-	-	1.92	-	-	1.95	-	-	-	-	-
Sub total	5.81	-	-	4.19	-	-	4.05	-	-	1.00	-	-
Profit margin (Rs/kg)												
Wholesaler	1.14	0.48	-	0.93	0.63	-	-	-	-	-	-	-
Retailer	1.11	0.13	-	2.08	0.96	-	2.05	1.14	-	-	-	-
Sub total	2.25	0.61	73	3.01	1.59	47	2.05	1.14	44	-	-	-
Marketing loss (Rs/kg)												
Marketing loss (Rs/kg)	2.74	-	-	2.41		-	1.99		-	1.3		-
Consumer price (Rs./kg)	14	14	-	14	14	-	13	13	-	10	10	-
Marketing efficiency (Rs./Kg)	0.73	0.55	-	0.94	0.71	-	1.13	0.81	-	9	3.37	-
Price spread (Rs./kg)	**8.06**	**9.02**	-	**7.2**	**8.16**	-	**6.1**	**7.18**	-	**1**	**2.23**	-

Table 4. Economics of packaging material.

Particulars	Wooden box	Bamboo basket	Plastic crates
Cost (Rs/box)	25	60	120
Capacity (kg)	15	10	20
Cost (Rs/qtl)	175	600	720
Durability (year)	1	2.5	5
Post harvest loss (Rs/qtl)	80	24	4
Labour requirement (Rs/qtl)	360	120	
Subtotal (Rs/qtl)	440	144	4
Replacement cost (Rs/qtl)	140	500	-
Total cost (Rs/qtl)	**755**	**1244**	**724**

Marketing loss

Marketing loss was calculated at different stages of marketing along with the functionaries who had actually incurred the loss with relevant prices. The total marketing loss due to discarded fruits in channel 1 amounted to Rs.2.74 / kg. The retailer had accounted for 38% of the loss (Re 0.98/ kg), which was higher than that of producers' (36%) and wholesalers' (37.69%) share. The pattern of sharing of marketing loss in channel 2 was similar to that in the market channel 1, with retailer accounting for Rs 1.12/ kg (46%) of market loss. The marketing loss in channel 3 is Rs.1.99/kg in which the producer has major share of 54% (Rs.1.08/kg). Marketing loss in channel 4 is minimum with Rs1.3/kg which constitute 13% of consumer's price.

Profit margins

The producer's net price as calculated was highest when they sold tomato in the farmers market. The tomato producers could reap a substantially higher net price of Rs.9 / kg in the channel 4 as compared to Rs.5.94/ kg in the channel 1 and Rs 5.94/ kg which is least comparatively. When marketing loss was taken into account for the estimation of profit margins of different marketing intermediaries, which was more relevant, it was found that the old estimation method had overestimated the profit margins. The impact of inclusion of marketing loss in estimation of wholesalers' and retailers' margins considerably reduced middlemen margin by 72%. Hence, it was concluded that by excluding one of the prime components in the marketing process, viz. post harvest loss, the profit margins of different market intermediaries were unduly over-estimated.

Price spread

The price spread in tomato was found to range from Rs

8.06 per kg in channel 1 to Rs 1.00 per kg in the channel 4 in the conventional method. The main component of price spread was marketing cost, which accounted for 72% in channel 1. The impact of post harvest loss increases the price spread in channel 1, by 12%, channel 2 by 13%, channel 3 by 17% and channel 4 by 23%.

Marketing efficiency

The marketing efficiency was found higher in farmers market compare to other channels, primarily because of lower marketing costs and higher price realized by the tomato producers in both the methods of estimation. However, by inclusion of marketing loss in the equation, the marketing efficiency declined. It revealed the fact that post harvest loss was also one of the pivot factors in deciding the marketing efficiency and the relationship was found inverse, that is, 'the higher the post-harvest loss, the lower will be the efficiency'. The marketing efficiency index was low in channel 1 because of higher marketing costs and profit margins to the middlemen. Better efficiency could be achieved by reducing the cost of marketing particularly the commission charges, and marketing losses. By providing viable alternate markets, the farmer's net share could be increased.

Packaging material

The additional cost incurred by farmer for purchasing plastic crates from wooden box is Rs.545 and for replacing bamboo basket it is Rs.180/qtl respectively (Table 4). Additional cost incurred in wooden box, is RS.440, bamboo basket is Rs.144, and plastic crate is Rs.4. The plastic crates incur minimum cost and by using plastic crates instead of bamboo basket the farmer can save Rs.520/quintal. The replacement cost incurred for replacing wooden box and bamboo basket are Rs.140 and Rs.500, respectively for one year, which is not incurred in plastic crates. Wooden box is found to be economical than bamboo basket but still it is not used by farmers widely due to the handling inflexibility. Thus plastic crates are found to be economical and the farmers should be encouraged to use plastic crates.

Conclusion

This study undertaken in Coimbatore on tomato has suggested including the marketing loss in the estimation of marketing margins, price spread and efficiency. Conventional wholesale marketing was prevalent in tomato. The post harvest loss was found to be high in marketing channel which involves more intermediaries. Post harvest loses of tomato in each marketing channel was due to lack of storage facilities and improper handling. The overall post harvest losses were estimated to 26% of tomato. Since the tomato is a highly perishable crop and improper handling of produce the post harvest loss is high. Marketing efficiency is high, when the farmer sells his product directly to the consumer which benefits both the producer and the consumer. Necessary steps should be taken by the government to sell the farmers produce directly to the consumer which proportionately raises the farm income level. It is concluded that the marketing loss is inversely proportional to the marketing efficiency. Plastic crates are found to be best packaging material as it incurs minimum loss.

Conflict of Interests

The author(s) have not declared any conflict of interests.

REFERENCES

Acharya SS, Agrawal NL (2001). Agricultural Marketing in India, Oxford & IBH Publishing Company, New Delhi, pp. 98-138.
Chandra P (1994). Marketing efficiency in Indian agriculture", (New Delhi: Allied Publishers Private Ltd.,) pp. 130-150.
Cundiff EN, Still R (1968). Basic marketing concepts, environment and decision, (New Delhi: Prentice Hall of India Private Ltd), pp. 21-28.
George PS (1972). Role of price spreads determining agricultural price policy. Agric. Situat. India. 27(9):617-619.
Gajanana TM (2002). Marketing practices and post-harvest loss assessment of banana var. Poovan in Tamil Nadu. Agric. Econ. Res. Rev. 15(1):56-65.
Kishore Kumar D, Basavaraja H, Mahajanshetti SB (2006). An economic analysis of post-harvest losses in vegetables in Karnataka. Indian J. Agric. Econ. 61(1):134-146.
NHB (2012). Horticultural Information Service", National Horticultural Board, Ministry of Agriculture, New Delhi.
Sreenivasa MD, Gajanana TM, Sudha M, Subramanyam KV (2002). Post-harvest loss estimation in mango at different stages of marketing — A methodological perspective, Agric. Econ. Res. Rev. 15(2):188-200.

21

Effect of low temperature storage on fruit physiology and carbohydrate accumulation in tomato ripening-inhibited mutants

Kietsuda Luengwilai and Diane M. Beckles

Department of Plant Sciences, University of California, Mail Stop 3, One Shields Avenue, Davis CA 95616, USA.

Chilling-sensitive fruits often produce a burst of ethylene when reconditioned at ambient temperature after cold storage. This has led some researchers to propose that chilling injury (CI) may be induced by post-chilling ethylene production. To test this hypothesis, we examined two tomato (*Solanum lycopersicon* L.) mutants, *non-ripening* (*nor*) and *ripening-inhibitor* (*rin*), that do not produce climacteric ethylene, after they were subjected to cold-storage and reconditioning. The response of the mutants differed, and was not as extreme as the parent line cv. Ailsa Craig, but both showed symptoms of chilling stress. Therefore while ethylene production may influence chilling injury, it is not essential for initiating this process in tomato cv. Ailsa Craig.

Key words: Chilling injury, tomato fruit, ripening mutants, rin, nor.

INTRODUCTION

Chilling injury in tomato (*Solanum lycopersicon* L.) is a complex syndrome that is detrimental to fruit quality (Serrano et al., 1996; Sevillano et al., 2009). When tomato is stored at 2-12°C, and is then allowed to ripen at ambient temperature (20°C), a battery of physiological and biochemical responses can be activated that damages the fruit. These responses include a failure to ripen, water-soaking, poor appearance and susceptibility to disease (Morris, 1982). This poses a problem when storing tomato postharvest: low-temperatures are needed to delay senescence but this simultaneously increases the risk of chilling injury (CI).

An interrelationship between ethylene production and CI has been proposed (Wang, 1989). Many aspects of ripening in climacteric fruits like tomato are largely regulated by ethylene. This growth regulator can also hasten senescence, one of the main symptoms of CI (Saltveit, 2003), indicating that extensive cross-talk occurs between the two pathways. Genetic and biochemical evidence also support a connection between the two processes.

In some species, there is a spike in ethylene synthesis when fruits that was previously held in the cold is ripened at warmer temperatures (Sevillano et al., 2009) and avocadoes, pineapples and 'Fortuna' mandarin treated with the ethylene inhibitor 1-MCP each showed enhanced tolerance to CI (reviewed in Pech, 2008). Furthermore, CI is attenuated in transgenic melons with a lesion that severely reduces ethylene biosynthesis in the fruit (Ben Amor et al., 1999). These observations collectively point to a role for ethylene influencing CI in several crops, but this may not be true for all CI-sensitive climacteric fruit. Melon has clearly defined ethylene-independent ripening pathways but this has not been demonstrated unequivocally in tomato, so CI mechanisms may not operate similarly in these crops. This is true even within closely-related species. Ethylene treatment of 'Fortuna' mandarin created CI resistance, but in another citrus, the cultivar 'Shamouti,' ethylene hastened CI symptoms (Porat et al., 1999). The overall picture that emerges is that ethylene production does influence CI in some crops but empirical testing is needed.

The aim of this work is to investigate if climacteric ethylene is an important factor in initiating or accelerating CI in tomato. This was addressed by determining if CI is evident in tomato genotypes that have an impaired ethylene-dependent ripening pathway. *Ripening inhibitor (rin)* and *non-ripening (nor)* are tomato mutants that produce little or no climacteric ethylene (Tigchelaar, 1977). *Rin* lacks a MADS-box transcription factor (MADS-RIN) while *nor* lacks a transcription factor that may regulate MADS-RIN (Giovannoni, 2007a). Both transcription factors act upstream of the ethylene signaling pathway, so that *rin* and *nor* retain their sensitivity to ethylene (Moore et al., 2002). These mutants show physiological and biochemical changes during maturation that may be associated with ethylene-independent ripening pathways (Jeffery et al., 1984; Lelievre et al., 1997).

We hypothesized that if a burst of ethylene synthesis in post-chilled fruit stimulates the onset of CI, then these ripening impaired mutants may be resistant to CI at an equivalent chronological stage of the parental control. If however, CI is not primarily influenced by ethylene, injury will be evident due to an affect of cold on ethylene autonomous-ripening pathways in the mutants. Such pathways in tomato would include starch degradation, sugar, citric and malic acid production, loss of chlorophyll, and some aspects of fruit softening that is, those initiated before ethylene production in tomato (Jeffrey et al., 1984). Studying CI-sensitivity in *rin* and *nor* may therefore deliver new insight into chilling injury in tomato.

MATERIALS AND METHODS

Plant material

Tomato (*Solanum lycopersicon* L.) cv. Ailsa Craig and two near isogenic lines - *rin* (*ripening inhibitor*) and *nor* (*non-ripening*) mutants were obtained from the Tomato Genetic Resource Centre, University of California- Davis. Plants were grown as described previously (Luengwilai and Beckles, 2009b). Fruit from cv. Ailsa Craig (referred to as the 'control genotype') was harvested at 42 days post anthesis (DPA), which approximates to USDA Mature Green 1. Fruits from *nor* and *rin* were harvested when they reached the same chronological age as the parent line that is, at 42 DPA.

Storage conditions

The fruits were stored at different temperature-time regimes in large controlled-temperature rooms. The length of the experiment was 36 days except where noted. One set of fruit (approximately 6-12) of each mutant genotype was stored at 5°C for 28 days and these conditions are referred to as 'cold storage.' After cold storage the fruits were then held at 20°C for 8 days or the 'reconditioning' phase. Another set of fruits (approximately 6-12) was simultaneously stored at 20°C for 36 (28 + 8) days. This treatment is referred to as the 'temperature control conditions.'

Respiration, ethylene production, firmness, total soluble solids (TSS), color and weight loss measurements

A minimum of 3 fruits per replicate was used for measuring respiration and ethylene production, as well as firmness, total soluble solids and color as described by (Luengwilai et al., 2007). Starch was measured as described by (Luengwilai and Beckles, 2009b). Percentage weight loss was calculated as follows:

$$\frac{(W_i - W_f)}{W_i} \times 100$$

Where W_i = Initial weight (prior to storage), W_f = Final weight (as measured on the stated date)

Chilling injury index

Severe symptoms of chilling injury are manifested by the appearance of surface pitting. These symptoms were evaluated in the fruits at days 1 and 8 of the reconditioning period. Symptom severity was scored as 0 = no pitting; 1 = less than 5% of fruit surface pitting; 2 = pitting covering 5-25% of fruit surface; 3 = pitting covering 25-50% of fruit surface and 4 = more than 50% of fruit surface covered with pitting. The extent of chilling injury damage was expressed as a chilling injury index, which was calculated using the formula:

$$\text{Chilling injury index (score 0-4)} = \frac{\sum (\text{CI level}) \times (\text{Number of fruit at the CI level})}{\text{Total number of fruit in the treatment}}$$

Statistical testing

Values are considered different if *P*-values were less than 0.05 by Student's *t*-test.

RESULTS AND DISCUSSION

We began our investigation by first observing the pattern of ethylene emissions and CO_2 production in the tomato genotypes. Increased production of these gases are characteristic of ripening in climacteric fruit, but *rin* and *nor* do not show this trend (Giovannoni, 2007b). Relevant to our question is the observation that CI in fruits is characterized by increased respiration and ethylene production in the initial phases of reconditioning (Jackman et al., 1989). The evolution of CO_2 and ethylene were measured immediately and up to 8 days after the fruits were restored at 20° C after being held in the cold for 4 weeks. Cold-stored Ailsa Craig fruit showed a climacteric peak of CO_2 and the rate of ethylene production increased four-fold (from 2 to 8 µl C_2H_4/kg/h) during post-chilled ripening (Supplementary Figure 1).

In this study *rin* and *nor* fruits did not show any increase in ethylene production regardless of storage condition prior to reconditioning (Figure 1A). In both mutants however, chilling caused an increase in respiration immediately upon transferring the fruit to room temperature (Figure 1B). From this data we can make two conclusions. First, that the mutants showed some signs of stress due to cold-storage, as evidenced by

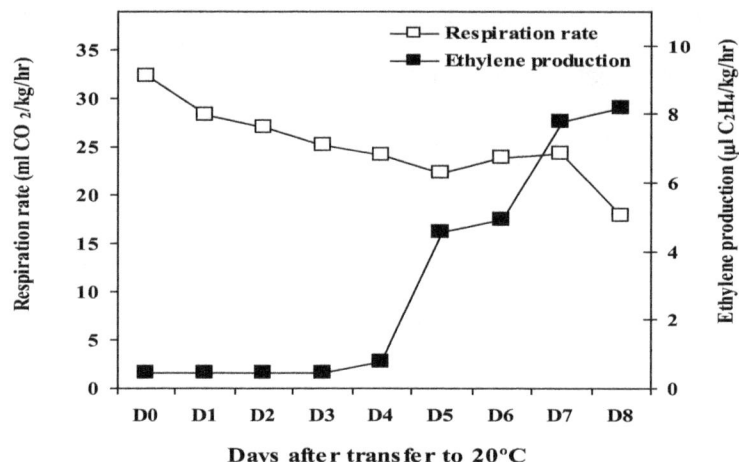

Supplementary Figure 1. Ethylene production and respiration in Ailsa Craig fruit. The data showed the characteristic climacteric burst of CO_2 or ethylene during the reconditioning period. Values were mean ± SE of 6 fruit for all genotypes. Error bars are not shown for clarity.

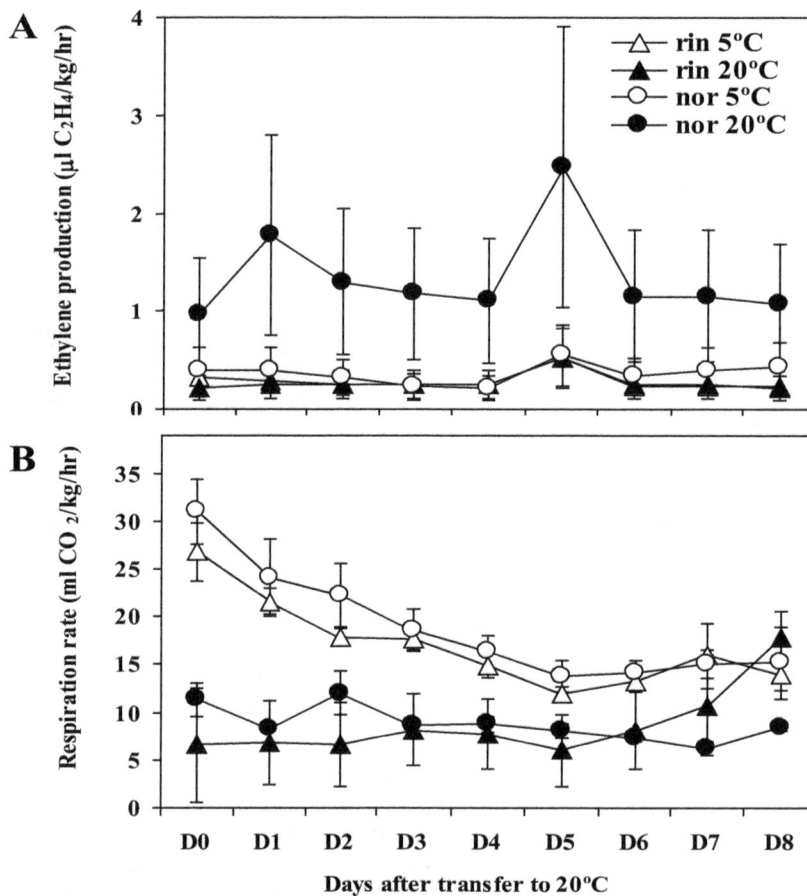

Figure 1(A). Ethylene production in *rin* and *nor* tomato fruit stored at 20°C for 8 days after 28 days of cold storage. Prior to holding at 20°C fruit were kept either at (i) 5°C for 28 days (described as *rin* 5°C or *nor* 5°C in the legend; light symbols) or (ii) were kept in control conditions, that is 20°C for 28 days described as *rin* 20°C and *nor* 20°C in the legend; dark symbols. (B). Respiration rates of *rin* and *nor* tomato fruit held at 20°C for 8 days after cold storage. Values were mean ± SE of 6 fruit for all genotypes.

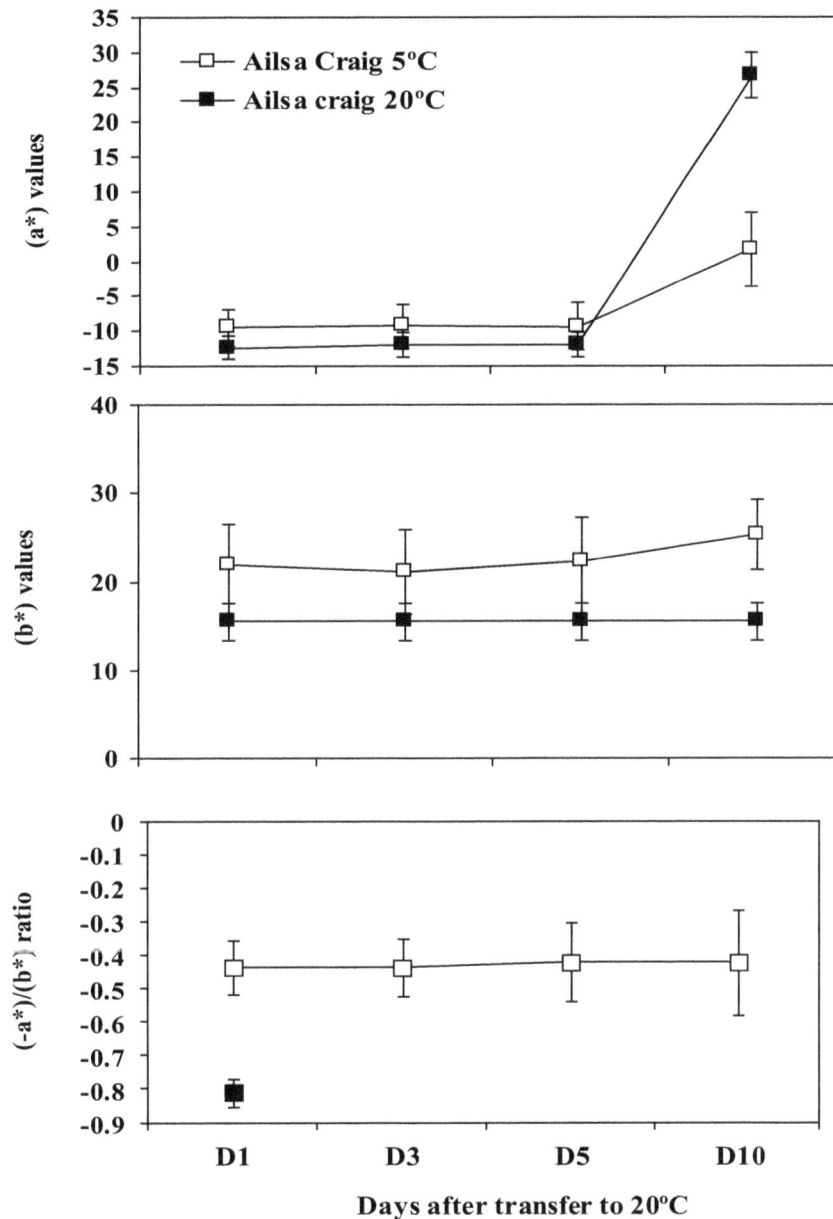

Supplementary Figure 2. Color development of Ailsa Craig fruit indicated by a*, b* and -a*/b* values. Fruit was stored at 5°C or at 20°C. For fruit stored at 20°C, a* values were not negative after day one, hence only one point is shown on the graph. Values are mean ± SE of 6 fruits.

increased CO_2 evolution. Second, because no change in ethylene production was recorded in the mutants regardless of storage conditions, any general changes in maturation seen may be attributable to ethylene-independent ripening processes (Jeffrey et al., 1984). We therefore paid particular attention to how these parameters responded to cold treatment in the mutants.

We observed the extent of de-greening and accumulation of lycopene and other carotenoids in ripening fruit using the CIE dimension scores, where a*

denotes red-green characteristics and b* denotes yellow-blue. The values and their ratios have been used as important indicators of the levels of various pigments in fruits (Batu, 2004). Chilling delayed color development in Ailsa Craig and caused abnormal and uneven pigmentation, but after 5 days of reconditioning a* values increased 3-fold (Supplementary Figure 2). The mutants did not redden in cold storage or at room temperature and as a result all had negative a* values which remained constant while holding at 20°C (Figure 2A). Evaluation of

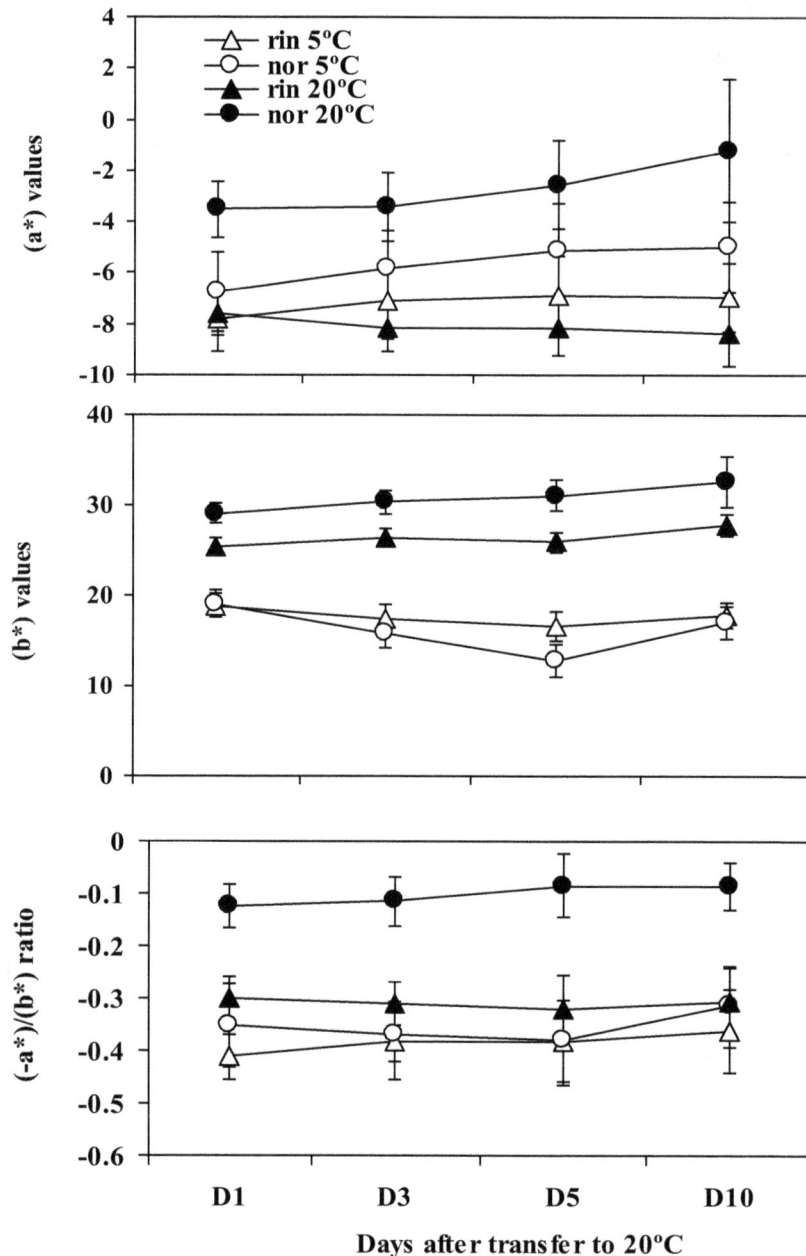

Figure 2. Color development in fruit of *rin*, and *nor* tomato fruit after cold storage and transfer to 20°C for 8 days. (A) a* values represents the greenness (negative) and redness (positive) and is strongly influenced by lycopene content (B) b* values represented the blueness (negative) and yellowness (positive), increased b* values are associated with yellow carotenoids. (C) -a*/b*indicates changes in chlorophyll content. Values were mean ± SE of 12 fruits for all genotypes and storage conditions. Prior to holding at 20°C, fruits were kept either at (i) 5°C for 28 days (described as *rin* 5°C or *nor* 5°C in the legend; light symbols) or (ii) were kept in control conditions that is, 20°C for 28 days (described as *rin* 20°C or *nor* 20°C in the legend; dark symbols).

b* and -a*/b* values suggests that cold treatment simultaneously reduced both values in *nor* but only b* values in *rin* (Figure 2B and 2C). It is possible to use this data to infer an effect of cold on ripening.

This is predicated on the following assumptions:

(i) that -a*/b* scores positively correlate with chlorophyll content (Steet and Tong, 1996; Koca et al., 2007),

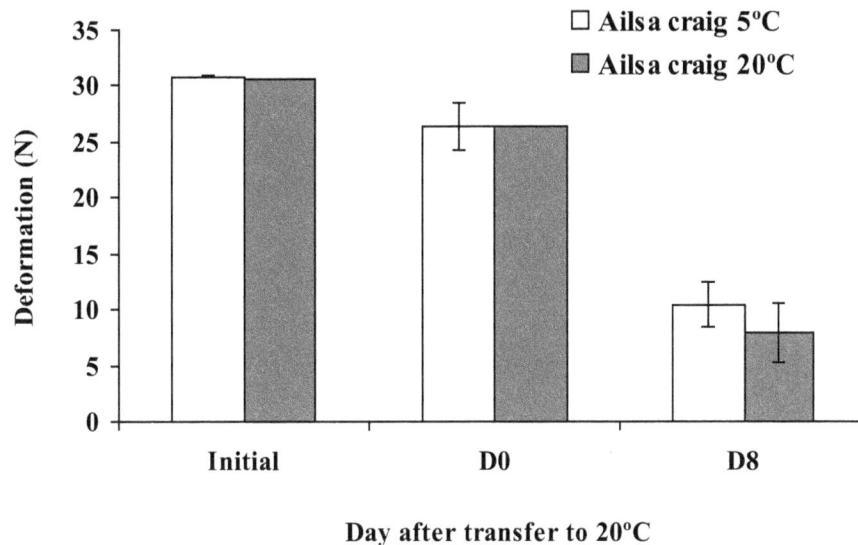

Supplementary Figure 3. Deformation of Ailsa Craig fruit. Fruit were stored at 5°C or 20°C and then reconditioned for 8 days at 20°C. Values are mean ± SE of 6 fruits.

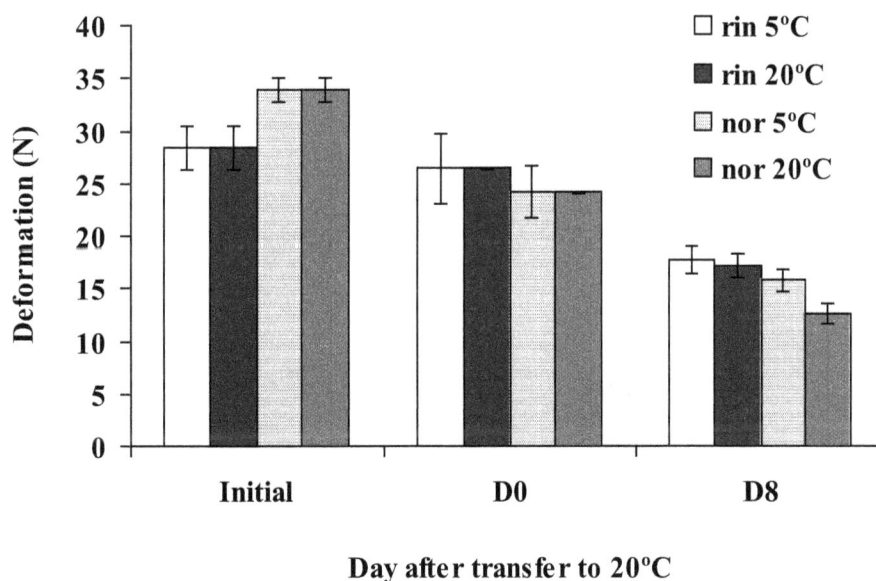

Figure 3. Fruit deformation of *rin*, and *nor* tomato fruit after 5°C storage and transferred to 20°C for additional 8 days. Values were mean ± SE of 12 fruits for both genotypes and storage conditions.

(ii) that b* values reflect yellow-pigmented carotenoids (Artes et al., 1999; Batu, 2004),
(iii) some carotenoids synthesis and loss of chlorophyll are initiated in response to ripening signals that occur before ethylene is produced (Jeffery et al., 1984).

When considered together, this data suggests that the pigment metabolic pathways that are ethylene-independent were affected by chilling in both mutants.

Another signal of ripening includes fruit softening. Softening is stimulated by both the ethylene and non-ethylene ripening pathways in melon (Pech et al., 2008) and this may also be true in tomato. As anticipated, the Ailsa Craig fruits were easier to deform than the mutants but cold-storage did not alter the response (Supplemental Figure 3). Fruits from *nor* held at 5°C were firmer during the subsequent room temperature incubation, but there was no effect on *rin* where the controlled and treated fruit behaved identically (Figure 3).

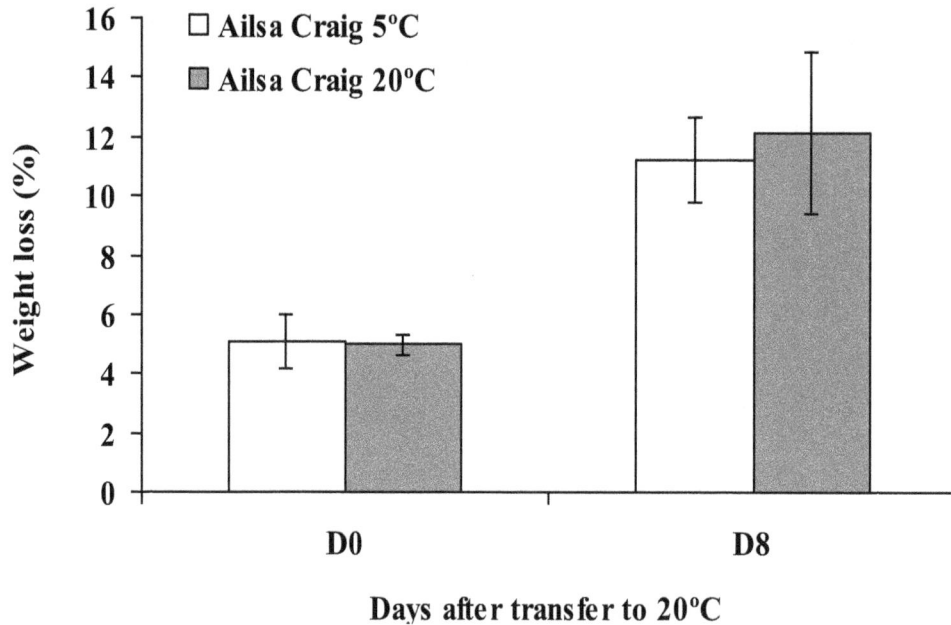

Supplementary Figure 4 .Weight loss of Ailsa Craig fruit. Fruit were stored in cold or at 20°C as indicated on the graph. Values are mean ± SE of 6 fruits.

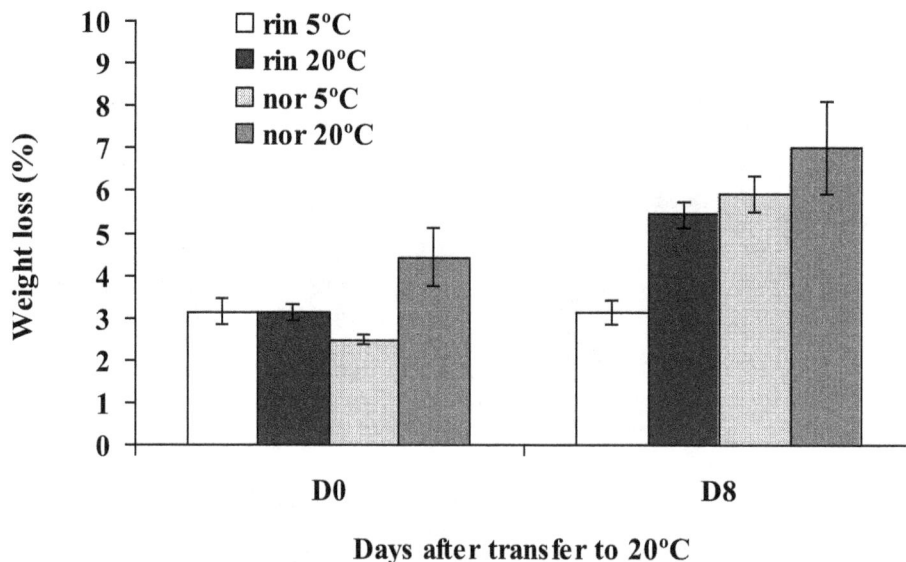

Figure 4. Percentage of fruit weight loss *rin*, and *nor* tomato fruit after 5°C storage and transferred to 20°C for additional 8 days. Values were mean ± SE of 12 fruits.

The results of the weight loss experiment were more complex. A typical manifestation of CI is that fruit will lose mass because of lowered temperature and this will continue when they are allowed to ripen at 20°C (Artes and Escriche, 1994). In our experiment, all fruits were weighed immediately after harvesting. Fruits used for cold treatment were re-weighed 28 days after incubation at 5°C. They were transferred to 20°C, held for 8 days,

and were then weighed again. The temperature controls (20°C) was also weighed immediately after harvest and following 36 days of storage at 20°C. During cold-storage only *nor*, lost fruit mass over that initially measured (Supplemental Figure 4). After reconditioning, fruit mass decreased in *rin*, but was attenuated in *nor* (Figure 4).

Starch degradation is another biochemical process linked to tomato fruit ripening and may contribute to red-

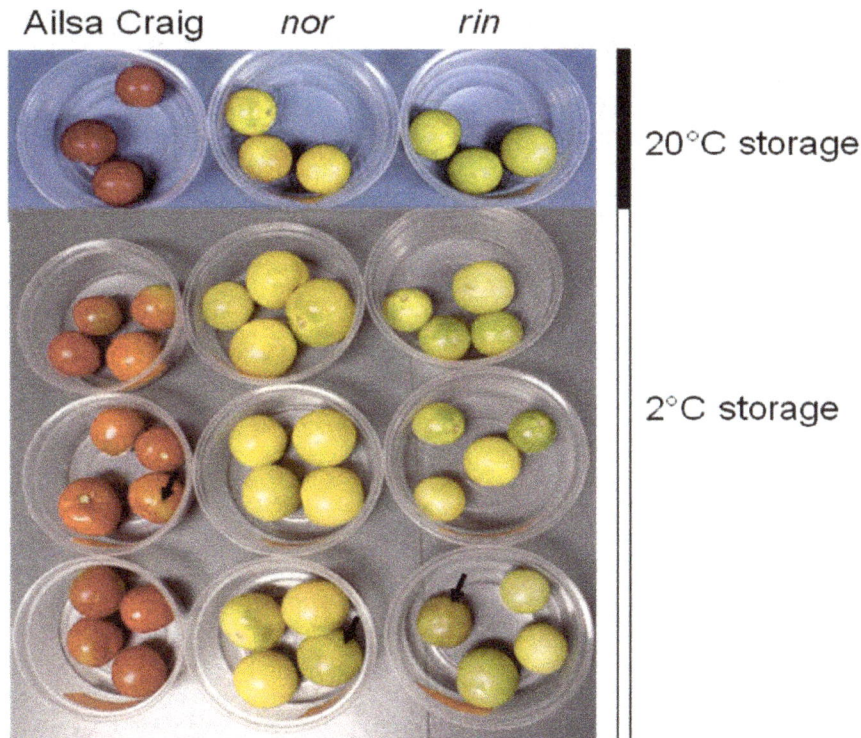

Supplementary Figure 5. Ailsa Craig, *rin*, and *nor* fruit after storage at 2°C or 20°C for 28 days and then, after transfer to 20°C for an additional 28 days. The CI index was 0.75 and 0.3 for Ailsa Craig and *rin* respectively. *Nor* fruit did not show any sign of surface pitting injury. Arrows indicate uneven ripening in Ailsa Craig and evidence of water-soaking in *rin* and *nor*.

fruit total soluble solids (TSS) (Luengwilai and Beckles, 2009a). TSS indicates the level of acids and sugars in the fruit; the biochemical pathways that produce these compounds are stimulated by climacteric ethylene but their initiation precedes this event in fruit development (Jeffery et al., 1984). Our measured values of TSS (5.5 ± 0.5%) were identical in all storage time-temperature regimes measured between the mutants, which were 23% lower than Ailsa Craig (7.2% and 5.5% respectively). Starch accumulation was also identical between genotypes in freshly-harvested mature green fruit of *nor, rin* and Ailsa Craig (11 ± 2, 13 ± 3 and 20 ± 5 mg/gFWT respectively). However after cold storage and reconditioning for 7-21 days starch levels were reduced almost 100-fold, with values of 0.20 ± 0.100 mg/gFW in *nor,* 0.40 ± 0.10 mg/gFW in *rin* and 0.46 ± 0.001 mg/g FW in Ailsa Craig. Starch was degraded post-cold-storage presumably as a substrate for respiration, which peaked immediately after the fruit were incubated at 5°C in all the genotypes (Figure 1B).

In this experiment fruit, starch is degraded in the absence of climacteric ethylene, at similar rates to that in the control cultivar. It is possible that in the mutants, respiration had a big influence on starch degradation, or that there was enough ethylene produced in the mutants

for near complete starch breakdown. These events may not be mutually exclusive. The fruit was evaluated for physiological symptoms of CI including fruit decay and surface pitting. There was no evidence of these abnormalities in any of the genotypes (data not shown).

Although this seemed unusual, a lack of severe damage to tomatoes held at 5°C was also seen in another study (Chomchalow et al., 2002). This type of data exemplifies the difficulty with studying the CI phenomenon because it is highly context-dependent. (Saltveit and Morris 1990). We therefore incubated 12 fruits for each genotype at 2°C for 21 days followed by storage at 20°C for another 21 days to encourage a response. All genotypes showed CI symptoms - failure to fully ripen and water-soaking. Surface pitting was visible in Ailsa Craig and *rin* and the calculated CI index was 0.7 and 0.3 respectively (Supplementary Figure 5).

Conclusion

Our hypothesis was that if post-chilling climacteric ethylene was important in initiating CI, that the tomato ripening mutants *rin* and *nor* would show enhanced resistance to CI. We tested this by focusing on potential changes to putative ethylene-independent pathways that is, ripening changes that are initiated before the onset of the climacteric, in response to chilling. The CI response was mild in control and test genotypes at 5°C. In spite of this, our data suggests that this condition altered characteristics associated with non-ethylene ripening in *nor* and *rin*. In all samples, respiration, changes in fruit mass and non-lycopene fruit pigmentation were characteristic of fruits experiencing cold-stress.

There were differences in the pattern and the severity of physiological and biochemical traits assayed between mutants in response to 5°C, however when the fruit was exposed to 2°C, visible evidence of CI phenotypes were found in both mutants. While we cannot rule out the possibility that some basal level of ethylene, that is from wounding or that produced in vegetative tissues was sufficient in the mutants to cause some ripening changes during cold-storage, we were able to effectively show that there was no climacteric burst of ethylene in *rin* and *nor* and infer that this was not responsible for the results observed. We therefore conclude that the discrete pathways that are regulated by the transcription factors encoded by *rin* and *nor* may engender differences in the response by the mutants to cold, but that CI occurs in the absence of climacteric ethylene in the Ailsa Craig cultivar.

ACKNOWLEDGEMENTS

We thank Dr. Elizabeth Mitcham for the generous use of controlled temperature rooms and a Gas Chromatography machine for ethylene and respiration measurement. We also thank Dr. Marita Cantwell for sharing unpublished data on chilling injury in tomato. KL's graduate fellowship is provided by the Anandamahidol Foundation. This work was funded by National Science Foundation Grant MCB-0620001 to DMB and Hatch Projects CA-D*-PLS-7198-H and CA-D*-PLS-7821-H

REFERENCES

Artes F, Conesa MA, Hernandez S, Gil MI (1999). Keeping quality of fresh-cut tomato. Postharvest Biol. Technol. 17:153-162.

Artes F, Escriche AJ (1994). Intermittent Warming Reduces Chilling Injury and Decay of Tomato Fruit. J. Food Sci. 59:1053-1056.

Batu A (2004). Determination of acceptable firmness and colour values of tomatoes. J. Food Eng. 61:471-475.

Ben Amor M, Flores FB, Latche A, Bouzayen M, Pech JC, Romojaro F (1999). Inhibition of ethylene biosynthesis by antisense ACC Oxidase RNA prevents chilling injury in Charentais cantaloupe melons. Plant Cell Environ. 22:1579-1586.

Chomchalow S, Assi NE, Sargent S, Brecht J (2002). Fruit maturity and timing of ethylene treatment affect storage performance of green tomatoes at chilling and nonchilling temperatures. Hort. Tech. 12:104-114.

Giovannoni J (2007a). Genomics Approaches to Understanding Ripening Control and Fruit Quality in Tomato. In Proceeding IS on Biotechnology of Temperate Fruit Crops and Tropical Species, R. Litz and R. Scorza, eds (Acta Horticulturae).

Giovannoni JJ (2007b). Fruit ripening mutants yield insights into ripening control. Curr. Opin. Plant Biol. 10:283-289.

Jackman RL, Yada RY, Marangoni A, Parkin KL, Stanley DW (1989). Chilling Injury - a Review of Quality Aspects. J. Food Qual. 11:253-278.

Jeffery D, Smith C, Goodenough P, Prosser I, Grierson D (1984). Ethylene-Independent and Ethylene-Dependent Biochemical-Changes in Ripening Tomatoes. Plant Physiol. 74:32-38.

Koca N, Karadeniz F, Burdurlu HS (2007). Effect of pH on chlorophyll degradation and colour loss in blanched green peas. Food Chem. 100:609-615.

Lelievre JM, Tichit L, Dao P, Fillion L, Nam YW, Pech JC, Latche A (1997). Effects of chilling on the expression of ethylene biosynthetic genes in Passe-Crassane pear (*Pyrus communis* L) fruits. Plant Mol. Biol. 33:847-855.

Luengwilai K, Beckles DM (2009a). Starch Granules in Tomato Fruit Show a Complex Pattern of Degradation. J. Agric. Food Chem. 57:8480-8487.

Luengwilai K, Beckles DM (2009b). Structural Investigations and Morphology of Tomato Fruit Starch. J. Agric. Food Chem. 57:282-291.

Luengwilai K, Sukjamsai K, Kader AA (2007). Responses of 'Clemenules Clementine' and 'W. Murcott' mandarins to low oxygen atmospheres. Postharvest Biol. Technol. 44:48-54.

Moore S, Vrebalov J, Payton P, Giovannoni J (2002). Use of genomics tools to isolate key ripening genes and analyse fruit maturation in tomato. J. Exp. Bot. 53:2023-2030.

Morris LL (1982). Chilling Injury of Horticultural Crops - an Overview. Hortscience 17:161-162.

Pech JC, Bouzayen M, Latche A (2008). Climacteric fruit ripening: Ethylene-dependent and independent regulation of ripening pathways in melon fruit. Plant Sci. 175:114-120.

Porat R, Weiss B, Cohen L, Daus A, Goren R, Droby S (1999). Effects of ethylene and 1-methylcyclopropene on the postharvest qualities of 'Shamouti' oranges. Postharvest Biol. Technol. 15:155-163.

Saltveit M (2003). Temperature extremes. In: Postharvest Physiology and Pathology of Vegetables. JA Bartz, JK Brecht ed (New York: Marcel Dekker). pp. 457-483.

Saltveit ME, Morris LL (1990). Overview on chilling injury of horticultural crops. In: Wang CY (ed) Chilling Injury of Horticultural Crops. CRC Press Inc., Boca Raton, FL, pp. 3–15.

Serrano M, MartinezMadrid MC, Martinez G, Riquelme F, Pretel MT, Romojaro F. (1996). Review: Role of polyamines in chilling injury of fruit and vegetables. Food Sci. and Technol. International 2:195-199.

Sevillano L, Sanchez-Ballesta MT, Romojaro F, Flores FB (2009). Physiological, hormonal and molecular mechanisms regulating chilling injury in horticultural species. Postharvest technologies applied to reduce its impact. J. Sci. Food Agric. 89:555-573.

Steet JA, Tong CH (1996). Degradation kinetics of green color and chlorophylls in peas by colorimetry and HPLC. J. Food Sci. 61:924-927.

Tigchelaar EC (1977). Ripening Behavior of Single and Double Mutant Tomato Hybrids of Rin and Nor. Hortscience 12:418-419.

Wang CY (1989). Relation of chilling stress to ethylene production In Low temperature stress physiology in crops, P. Li, ed (Boca Raton, Florida: CRC Press). pp. 177-189.

22

Plant secondary metabolites as source of postharvest disease management: An overview

AMRITESH C. SHUKLA

Department of Horticulture, Aromatic and Medicinal Plants, Mizoram University, Aizawl-796 004, India.
E-mail: amriteshmzu@gmail.com. Tel: 0091-389-2330273. Fax: 0091-389-2330787.

Postharvest losses of stored products are enormous due to fungal deteriorations. Although, there are a number of synthetic fungicides available in the market for checking these deteriorations, they have several side effects such as high toxicity, long degradation periods, their residues in food chain, chronic poisoning through the continuous intake of small quantities, development of new races of pathogens, non-biodegradable nature and exhaustible source. Therefore, we must search the new sources of pesticides, which in addition to their efficiency, must be safe and selective to target specific pathogen. Plants are very rich sources of bioactive chemicals such as phenolics, polyphenols, quinones, flavones, flavonoids, flavonols, tannins, cumarins, terpenoids, lectins and polypeptides. Some plants yield fraction of essential oils, which have inhibitory effects on microorganisms. They are highly enriched with terpenoids. They are volatile, biodegradable, eco-friendly and are easily available In local environment. Several studies have been conducted on the use of such botanicals for controlling postharvest diseases, and hence, the present communication reviews the work done on investigating the fungitoxic potential of essential oils and extracts from higher plants in fungal deterioration of stored products.

Key words: Myotoxins, fungitoxic, biodeterioration.

INTRODUCTION

It is estimated that between 60 to 80% of all grains produced in the tropics is stored by farmers themselves. For small farmers, the main purpose of storing grains is to ensure household food supplies. It also provides a form of saving to cover future cash needs through sale or for barter or gift exchange. Small quantities of grains are also stored for seed purpose. Farmers who produce surplus may also store grains for sale later to take advantage of seasonal price rise.

Traditional storage systems must provide maximum protection against deterioration of commodity by inclement weather and pests. Farm storage systems have been developed to satisfy these requirements. Most of them are well adapted to their environment and losses are generally low, often below 5% of grain weight over a storage season (Tyler and Boxall, 1984). However, for resource, poor farmers even losses of this magnitude have important implications for food security. In addition to storage of food grains at farmer's level, thegovernments of different countries also procure and store them for reasons of food security for its growing population. There was spiral increase in populations of several countries including India. In case of India, it has crossed one billion marks in 2000A.D. and it will give India the dubious distinction of being the most populated country in the world by 2025A.D. Therefore, the challenge of feeding ever growing population shall be a very difficult task. We cannot meet such a challenge with the increase in food production alone but their protection from deterioration caused by fungi and other pests during storage have to be given due emphasis (Shukla, 1997).

In developing countries, the greatest losses during

storage to cereals and other durable commodities are caused by fungal pests. Deteriorations due to fungi are due to unhygienic conditions of storage and this in turn is associated with initial high moisture content of the stored products or absorption of moisture during storage due to defects in the storage system (Stinson et al., 1980).

DETERIORATION IN STORED PRODUCTS BY PESTS

There are several storage fungal pests that cause deteriorations in stored commodities. The most common among them are - *Aspergillus flavus, A. niger, A. clavatus, A. terreus, A. versicolor, A. candidus, Alternaria alternata, Curvularia lunata, Cladosporium cladosporoides, C. herbarum, Epicoccum nigrum, Emericella nidulans, Emericella rugulosa, Fusarium acuminatum, F. moniliforme, Mucor hiemalis, Penicillium citrinum, P. chrysogenum, P. expansum, P. funiculosum, P. italicum, Rhizopus arrhizus, Rhizopus nigricans, Syncephalastrum racemosum* (Shukla, 1997; Shukla et al., 2000; Pandey, 2008; Shukla, 2010). Several of these fungal pests such as species of *Aspergillus, Penicillium, Fusarium, Alternaria* and *Cladosporium* etc. not only bring about deterioration in the quality and quantity of agricultural produce in storage and transit but they also create health hazards in animals and human beings by producing toxic metabolites in the form of mycotoxins in the stored commodities (Samson et al., 1995; Shaaya et al., 1997; Pandey, 2008).

These organisms are capable of growing under diverse conditions of moisture, pH and temperature. If the mould growth occurs, there is always the concomitant possibility of mycotoxin production (Zohri and Abdel-Gawad, 1993). Mycotoxins are dangerous chemicals that cause several complications in the body. They are carcinogenic, hepatoxic, nephrotoxic and teratogenic (Samson et al., 1995; Pandey, 2008). Extreme toxicity of mycotoxins lies in the fact that they are extremely stable and dangerous in minute quantities. Further, once formed, they cannot be removed from the commodity concerned by processing or removal of visible mould growth. They are heat stable, so they cannot be destroyed by cooking. Since mycotoxins are extremely toxic, regulatory and industry guidelines limit are set at very low levels. In developing countries, often the good quality products are exported while substandard produce unacceptable to foreign buyers (because they exceed regulatory limits for mycotoxin content) is sold to the domestic market (Dawson, 1991). Therefore, the mycotoxin contamination of food and feeds is not a particular problem to the developed world, although heavy economic costs are incurred in ensuring low concentrations of mycotoxins (Mannon and Johanson, 1985). In poorer developing countries, such contaminations have more serious consequences, affecting agricultural economies, reducing

annual production and good quality exports and seriously affecting the health of the population. Therefore, the control measures for checking deterioration and mycotoxin production should be such that which occur naturally in the local environment; less toxic to environment, animal and human being and cost effective.

MANAGEMENT OF DETERIORATION CAUSED BY PESTS AND MYCOTOXIN PRODUCTION

To control fungal deterioration of agricultural produce, many organic and inorganic fungicides have been developed and used. The use of many of these has, however, been restricted due to their undesirable side effects such as a high and acute toxicity, the long degradation periods, their concentration in food chain, the suspected dangers of chronic poisoning through the continuous intake of small quantities (Samson et al., 1995; Kumar et al., 2007). Besides, due to development of new races of pathogens, many of these fungicides are gradually becoming out of date (Dikshit, 1980).

As such the development of new effective and harmless fungicides is needed on an increasing scale. According to Brandes (1967) much of our efforts are being wasted in routine testing of the standard fungicides, when there is a pressing need to investigate new sources of effective fungicides (Brandies, 1967).

Furthermore, the sources of these synthetic fungicides are largely petrochemicals which are exhaustible. Therefore, haunt for inexhaustible sources of such chemotherapeutants is highly desirable. Green plants appear to be the reservoir of effective chemotherapeutants and can provide reversible source of useful pesticides (Swaminathan, 1978). Tropical floras, in contrast to their temperate zone counterparts, have developed a more efficient and varied defense mechanism because of the far severe conditions for survival.

They, thus provide a rich and intriguing source for isolating natural secondary plant metabolites, which exhibit interesting antimicrobial properties. Although only some 15,000 secondary plant metabolites have been chemically identified, their total number may exceed 4,000,000 (Saxena, 1993).

They are vast cornucopia of defense chemicals. Recent reports on the possibility of use of higher plants and their constituents have indicated their usefulness in providing fungicides, which are largely non-phytotoxic, more systemic and easily biodegradable (Fawcett and Spencer, 1969; Beye, 1978). They are sustainable and can be continuously propagated year- after-year and do not have any negative impact on the environment as long as care is taken to avoid the propagation of plants from foreign ecosystems which might, therefore, become established as weeds.

Further, where plants are used as storage protectants, they are almost always applied to control insect pests. This is reflected in volumes of research directed to identifying insecticidal or insect repellent plants and plant extracts. Nevertheless, some work has been undertaken to determine whether plants can control storage fungi. Most workers have investigated the properties of spices as inhibiting agents of mycelial growth of *Aspergillus* species and of its toxin production.

Syzygium aromaticum (cloves) have been found to be particularly effective, often completely inhibiting both fungal growth and toxin production (Hitokoto et al., 1980; Mabrouk and El-Shayeb, 1980). Many commercially available spices and herbs, turmeric (*Curcuma* spp), basil (*Ocimum basilicum* L.), marjoram (*Marjorana hortensis* Moench.), anise (*Pimpenella anisum* L.), cumin (*Cuminum cyminum* L.) and coriander (*Coriandrum sativum* L.) are able to completely inhibit toxin production, but only partially inhibit fungal mycelial growth (Hitokoto et al., 1980). Aqueous extracts of weeds and medicinal plants have also been shown to inhibit toxin production by *Aspergillus flavus*.

These include *Ricinus communis, Arnebia nobilis* and *Nicotiana plumbaginifolia* (Bigrami et al., 1980). Other fungi, such as, *Fusarium solani, F. phaseoli* and *Verticillium albo-atrum,* have been shown to be susceptible to tannins extracted from bark of various trees, including chestnut and wattle (Lewis and Papavizas, 1967). In addition to above, several other fungi have been shown to be susceptible to essential oils extracted from higher plants (angiosperms and gymnosperms) (Table 1).

DISCUSSION

Several hundred-research papers are published each year on the antimicrobial activity or other functional activity of botanicals from higher plants, and a complete review of all of them is beyond the scope of this article. The most excellent ones are given above. A brief discussion on efficacy and application of antimicrobial botanicals from higher plants is given below. The antimicrobial botanicals which have the potential to be used as storage protectants can be divided into several useful categories, including phenolics, polyphenols, quinones, flavones, flavonoids, flavonols, tannins, cumarins, terpenoids, lectins and polypeptides (Cowan, 1999). Many herbs, such as thyme, contain multiple active compounds which represent different chemical families. The essential oil fraction of botanicals is often the inhibitoriest chemical fraction to growth and survival of microorganisms. Essential oils are highly enriched with terpenoids. Examples of herbs and spices containing terpenoids which have been shown to have antimicrobial activity include allspice, basil, bay, burdock, cinnamon,

paprika, chilli pepper, clove, eucalyptus, dill, gotu kola, grape fruit seed extract, horseradish, lemon verbena, oregano, paod' arco, papaya, peppermint, rosemary, savory, sweet flag, tansy, tarragon, thyme, turmeric, valerian and willow (Duke,1985; Cate, 2000). The other major chemical group found in plants which has been frequently reported to have antimicrobial and antifungal activity is the sulfoxide/ isothiocyanate family, which includes onion, garlic, mustard and members of the *Brassica* family. Approximately 30% of essential oils which have been examined are inhibitory to bacteria, and more than 60% of essential oil derivatives have been shown to be inhibitory to fungi (Cowan, 1999; Chaurasi and Vyas, 1977; Shaaya et al., 1997; Shukla et al., 2000; Kumar et al., 2007; Pandey, 2008).

The mechanism of action for the antimicrobial activity of botanical storage protectants is not fully understood. However, terpenoids and phenolics are thought to exert inhibitory action against microorganisms by membrane disruption (Cichewick and Thorpe, 1996; Lambert et al., 2001; Schultes, 1978). Simple phenols and flavonoids appear to inhibit growth by binding to biochemicals essential for metabolism (Peres et al., 1997). Both coumarins and alkaloids are thought to inhibit growth of microorganisms at the genetic level (Hoult and Paya, 1996; Rahman and Chaudhary, 1995; Shukla, 2010).

Although numerous studies have been done *in vivo* to evaluate the antimicrobial activity of botanicals, only a few studies have been done with stored products for preventing or controlling mould growth. Inhibition of fungal growth on coriander and fennel seeds dressed with 0.5% concentration of *Cedrus* oil has been reported (Dikshit, 1980). Seeds of coriander showed good result when fumigated with essential oil from *Citrus media* and *Ocimum canum* at their MIC (Dubey et al., 1993). *In vivo*, application *Cymbopogon citratus* oil showed that growth of *Aspergillus flavus* was greatly checked (Mishra and Dubey, 1994). Seeds of wheat and groundnut fumigated with oil from *Eucalyptus citriodora* showed excellent result (Shukla, 1997). Oil and leaf powder of *Cymbopogon citratus* significantly reduced deterioration and aflatoxin production in shelled melon seeds inoculated with toxigenic *Aspergillus flavus, A. niger, A. tamarii* and *Penicillium citrinum*. Use of *Trachyspermum ammi* oil inhibits growth of dominant storage fungi such as spp. of *Aspergillus, Penicillium, Alternaria* etc. *in vivo* condition (Shahi et al., 2002; Shukla, 2010).

These studies show that some botanicals have the potential to be effective storage protectant although product development to optimize functionality and flavour will be challenging. More studies are needed on applications of botanicals from higher plants in storage protection to fully understand how best to optimize their use.

Use of many plants in storage protection is commonly Generally Recognized As Safe (GRAS) but some plants

Table 1. Effect of essential oils and extracts on stored pests and diseases.

Plant species and plant part used for extraction	Chief findings	Reference (s)
Oil from roots and flowers of *Raphanus sativus*	Effective against *Fusarium avenaceum*, *Phoma* spp. and *Alternaria brassicae*	Nehrash (1961)
Oil from *Juniperus communis*	Effective against *Aspergillus niger*	Slavenas and Razinskaite (1962)
Oil from *Mentha piperata* and *M. officinalis*	Both oils exhibited antimicrobial activity	Kovacs (1964)
Oil from *Mentha arvensis* var. *piperascens*	Oil of the plant from Formosa showed the highest antibacterial and antifungal activity	Sanyal and Verma (1969)
Some extracts and volatile oils.	Volatile oils showed much stronger fungicidal and fungstatic effect than the extracts	Cresan and Hodisan (1975)
Oils from *Cymbopogon citratus*, *Mentha arvensis* and Sweet basil	*Mentha arvensis* was effective against *Penicillium italicum* causing fruit rot of *Citrus reticulata*	Arora and Pandey (1976)
Oil from rhizome of *Curcuma angustifolia*	Effective against some saprophytes, plant pathogens and dermatophytes	Banerjee and Nigam (1977)
Oils from seeds of *Carum bulbocastanum*, *C. carvi*, *Trachyspermum ammi*, *T. roxburghinum*, *Cuminum cyminum*, *Nigella sativa*, leaves of *Psidium guajava* and galls of *Thuja orientalis*	All the oils except *Carum bulbocastanum* and *Psidium guajava* were found active against nine fungi and six bacteria	Nigam and Rao (1977)
Oils from *Cymbopogon citratus*, *C. martini*, *C. winterianus Ocimum basilicum*, *O. citriodorum*, *O. gratissimum* and *Mentha citrata*	Showed antifungal activity against *Penicillium notatum* and some derimatophytes	Sawhney et al. (1977)
Oil from *Nepeta hindostana*	Effective against *Aspergillus* and *Penicillium* spp	Sharma and Gautam (1977)
Oils from seeds of *Ammomum subulatum* and *Azadirachta indica*, from flower buds of *Sygygium aromaticum* and bulb of *Allium sativium*.	*Azadirachta indica* and *Allium sativum* possessed good antifungal activity against eight species of fungi	Thind and Dahiya (1977)
Oil from seeds of *Lantana camara*	Effective against *Curvularia lunata*, *Fusarium oxysporum* and some other fungi	Avadhoot and Verma (1978)
Oils from *Piper nigrum*, *Avapana triplinerve*, and *Mentha arvensis*	Antifungal activity of the oil was investigated against *Curvularia lunata*, *Rhizopus* spp., *Aspergillus* spp., and *Penicillium* spp. Oil from *M. arvensis* inhibited the growth of all fungi. Oils from *P. nigrum* and *A. triplinerve* were inactive against *A. fumigatus* and *P. decumbens*	Chaurasia and Kher (1978)
Oil from *Cedrus deodara* roots	Showed antifungal response against the fungi tested	Dikshit et al. (1978)
Oils from *Cinnamomum tamala* leaves, *Boswellia serrata* and *Nardostachys jatamansi*.	Showed antifungal activity against several fungi	Girgune et al. (1978)
Oils from *Aster thomsoni*, *A. peduncularis*, *Cymbopogon jwarancusa*, *Selinum tenuifolium*	Showed antifungal activity against five fungi with varying sensitivity. Oil of *C. jwarancusa* exhibited the best response	Mathela and Sinha (1978)
Oils from *Ageratum conyzoides*, *Feronia elephantum* and *Blumea membranosa*	Oil of *B. membranosa* exhibited the strongest toxicity as compared to other two oils against storage fungi	Sharma et al. (1978)
Oil from *Cymbopogon martini* var. *sofia*	Effective against *Helminthosporium oryzae* at 0.7% and also toxic to 15 other fungi tested	Singh et al. (1978)
Oil of *Mentha arvensis* var. *piperascens*	Strong antifungal activity against 17 out of 23 fungi tested; and was more active than some fungicides tested	Singh et al. (1978)
Oils from leaves of *Caesalpinia sappan*	Strong efficacy against *Aspergillus nidulans*	Yadava et al. (1978)
Oil from seeds of *Nigella sativa*	Showed antifungal activity against *Aspergillus* spp. and *Curvularia lunata*	Agrawal et al. (1979)
Oil from leaves of *Adinocalymma allicea*	Effective against *Helminthosporium oryzae* at 500 ppm, killed 12 fungi out of 21 tested and proved to be non-phytotoxic to host; and much more active than some	Chaturvedi (1979)

Table 1. Contd.

	commercial fungicides tested	
Oil from *Blumea membranacea*	Fungitoxic against *Cladosporium cladosporoides, Aspergillus sydowi* and *A. luchuensis* while in effective against *Fusarium oxysporum*	Geda and Bokadia (1979)
Oils from leaves of *Abutilon indicum, Bothriochloa pertusa, Murraya exotica* and *Dalbergia sisso.*	Only the oils of *A. indicum* and *Bothriochloa pertusa* showed fungitoxicity.	Jain et al. (1979)
Oils from tuber of *Cyperus scariosus* and leaves of *Ocimum basilicum*	Oil of *C. scariosus* was more active than that of *O. basilicum* against certain bacteria and various fungi	Lahariya and Rao (1979)
Oils from *Anethum graveolens, Apium graveolens, Carum carvi, Coriandrum, sativum, Cuminum cyminum, Foeniculum vulgare, Oenanthe stolonifera, Trachyspermum ammi, Parthenium hysterophorus, Eupatrium ayapana, Clerodendron interme, Lantana camara, Psoralea corylifolia, Zingiber officinale* and *Cymbopogon martini.*	Oils of *T. ammi, O. stolonifera, Anethum graveolens, Apium graveolens, P. hysterophorus* and *P. corylifolia* showed significant antifungal activity against all the fungi tested	Sharma and Singh (1979b)
Oil from seeds of *Oenanthe javanica*	Effective against *Aspergillus fumigatus, A. nidulans, Trichothecium roseum Microsporum gypseum, M. cocci.*	Sharma and Singh (1979a)
Oils from *Cinnamomum camphora, Eucalyptus camaldullensis, Ocimum kilimandscharicum* and *Valeriana wallichii*	Showed antifungal activity against certain plant & human pathogen	Suri and Thind (1979)
Oil from leaves of *Eucalyptus citriodora*	Effective against *A. niger* and *Clathridium corticola* at 1:1000 dilutions	Suri et al. (1979)
Oil from the leaves of *Cestrum diurnum*	Fungicidal activity against *Rhizoctonia solani* at MIC of 0.7%. At this concentration it exhibited the mycelial growth of all the 39 fungi tested indicating thereby wide range of activity	Renu et al (1980)
Oil from *Cymbopogon martini, C. oliveri var. rosasofia, Trachyspermum ammi* (dethymelated oil) and *Ocimum kilimandschericum* (Campherized oil)	All the oils showed wide range of activity (except Campherized oil) and were more active than some synthetic fungicides	Singh et al. (1980)
The volatile fractions of leaves of 131 species of higher plants were screened. The oil of *Peporomia pellucida* was found to exhibit the strongest fungitoxicity against *Helminthosporium oryzae.*	The MIC of the *P. pellucida* oil against *Helminthosporium oryzae* was 2000 ppm at which it showed broad fungitoxic spectrum, quick in killing activity, heavy inoculum density themostable, nonphytotoxic, non systemic, and self life up to 150 days. It also prevents the appearance of disease during preliminary *in vivo* testing	Singh (1980)
Different parts of 15 angiospermic plants were screened. The volatile oil extracted from the rhizomes of *Alpinia galanga* showed the highest fungitoxicity.	Oil showed highest fungitoxicity against *Helminthosporium oryzae*. The MIC was 0.4% of the medium. The oil was as fungitoxic as quintozene and Zeneb and gave more inhibition of *H. oryzae* than dinocap and Copper oxychloride. The oil was also fungitoxic against *Alternaria alternata, Aspergillus flavus, A. fumigatus, A. niger* and *Pestilotia* spp.	Tripathi (1980)
Essential oils and extracts from seeds of *Putranjiva roxburghi.*	Oil was effective against broad spectrum of storage fungal pests. It was thermostable and remained toxic for at least 150 days	Saxena (1980)
Oil from leaves of *Ocimum canum*	The oil at 3000 ppm exhibited broad range of activity inhibiting all the 31 fungi tested	Bhargava et al. (1981)
Oil from leaves of *Ocimum canum*	Showed fungitoxicity against *Aspergillus flavus, A. vesicolor* and number of other fungi	Dubey et al. (1981)
Oil from fruits of *Cinamomum cecidodaphne*	Showed fungitoxicity against all the storage fungi tested.	Chandra et al. (1982)
Essential oils from epicarp of *Citrus medica*	Showed fungitoxicity against *A. flavus, A. vesicolor* and several other storage fungi. The oil was thermostable and broad spectrum	Dubey et al. (1982)
Oil from epicarp of *Citrus medica* and leaves of	Showed toxicity against *A. flavus* and *A. vesicolor* and	Dubey et al.

Table 1. Contd.

Ocimum canum	many other storage fung	(1983)
Oil from leaves of *Schinus molle*	Showed toxicity against *A. flavus, Alternaria alternata, Penicillium italicum*. Oil was thermostable and toxicity lasts for at least 12 months, the maximum time taken into consideration	Dikshit et al (1986)
Oil from Pericarp of *Prunus persica*	Showed toxicity against all the storage fungal pests tested	Mishra and Dubey (1990)
Oil from epicarp of *Citrus sinensis*	Showed fungitoxicity against some important storage fungi tested	Singh et al. (1993)
Oil from leaves of *Cymbopogon citratus*	Showed toxicity against *A. flavus, A. niger* and many other storage fungi	Misra and Dubey (1994)
Essential oils from *Eucalyptus citriodora, E. dalarympleana, E. labeopinea, E. pauciflora*	Oil at 1000 ppm showed complete inhibition of *Penicillium italicum*. The oil of E. *dalarympleana* and *E. laveopinea* showed fungistatic activity against the test fungus at 3000 ppm; but the oil of *E. laveopinea* showed partial inhibition at 3000 ppm. The oil of *E. citriodora* at 1000 ppm exhibited fungicidal nature and withstood heavy inoculum	Shahi et al. (1997)
Essential oils from leaves of *Melaleuca alternifolia* and *Monarda citriodora* var. *citriodora*.	Showed fungitoxicity against several storage fungi tested	Bishop and Thornton (1997)
Essential oil from leaves of *Callistimon lanceolatus*	Showed fungitoxicity against A. *flavus, A. niger* and many other storage fungi	Mishra et al. (1997)
The oil from leaves of *Cymbopogon flexuosus*	Effective against *Aspergillus flavus, Penicillium italicum* and *Alternaria alternata*. The oil showed broad spectrum, inhibited heavy doses of inocula, thermostable and toxicity persisted for at least 12 months	Shukla et al. (2000)
Oil from leaves of *Ocimum sanctum* and *O. gratissimum*	*Ocimum sanctum* showed absolute toxicity against *A. flavus* but was moderately active against *A. niger*. However, *O. gratissimum* was found to exhibit absolute toxicity against both the tested fungi	Sharma (2001)
The oil from epicarp of *Citrus sinensis*	Oil exhibited strong fungitoxicity at 0.5% concentration against *A. flavus, P. italicum* and *Alternaria alternata* as a contact toxicant and inhibited heavy doses of inocula with quick killing action. The pesticidal action of the oil was thermostable up to 80 °C and lasted even up to 24 months with broad spectrum	Shukla et al. (2000)
Oil from the flower buds of *Eugenia caryophylata* (clove).	Clove oleoresin at 0.2 to 0.8% (v/v) was tested against *Candida albicans, Pencillium citrinum, Aspergillus niger* and *Trichophyton mentagrophytes* and was highly effective against *T. Mentagrophytes* and *Candida albicans*, however, *P. citrinum* and *A. niger* were relatively more resistant. Clove oleoresin was first dispersing in sugar solution and then used for antifungal testing	Nunez et al. (2001)
Oil from leaves of *Cymbopogon flexuosus*.	MIC was 0.2 µl/ml against *Alternaria alternate* 0.4 µl/ml against *A. flavus, A. fumigatus, A. parasiticus, Cladosporium cladosporoides, P. italicum, P. digitatum;* and 0.5 µl/ml against *Borytis cyneria* and *Helminthosporium oryzae*. The efficacy persists broad spectrum, thermostable, self life up to 48 months. The oil was used for in *vivo* controlling post harvest spoilage of *Malus pumilo*	Shahi et al. (2002b)
Essential oil extracted from leaves of *Eucalyptus pauciflora*	MIC was 0.3, 0.4, 0.5 and 0.6% against *Alternaria, Aspergillus, Penicillium,* and *Rhizopus* respectively	Shahi et al. (2002a)
Oil from aerial parts of *Ammoides pusilla*.	Oil showed antimicrobial activity against eight strains of bacteria, several fungi and yeast such as *Aspergillus*	Laouer et al. (2003)

Table 1. Contd.

	niger and *Candida albicans*. GC and GC/MS of oil showed 46 constituents among which thymol (44.5%), Y-terpinene (32.9%) and p- cymene (13.5%) were the chief	
Oil extracted from dried, crushed flowering plants of *Thymus serpyllum*	Oil showed antifungal properties against *A. flavus, A. awamori, A. niger, A. foetidus* and *A. oryzae*. It also inhibited all the three stages of asexual reproduction, that is, spore germination, mycelial growth and spore formation	Rahman (2003)
Essential oil and phenolic extracts of *Dinnetia tripetala* (pepperfruit)	Oil and phenolic extracts inhibited growth of several food borne microorganisms including *Penicillium* spp. and *Aspergillus* spp. etc.	Ejechi and Akpomedaye (2005)
Oil from *Lippia alba, Lippia microphylla, Citrus lemon, Cymbopogon citratus* and the phytochemicals citral, eugenol and mircene	Oil as well as phytochemicals showed significant antimould activity. Among the products that evidenced the antimould activity citral and eugenol showed the lowest minimum inhibitory concentration which was 1% and 4% respectively. The mould strains assayed are *Fusarium* spp. *Rhizopus* spp. *Aspergillus flavus, A. niger* and *Penicillium* spp.	de Souza et al. (2005)
Oil from seeds of *Cuminum cyminum* (cumin)	Oil contained more than 60 compound principal among them were cumin aldehyde (36 %), β-pinene (19.3%), P-cymine (18.4%) and Y-terpinene (15.3%). Antimicrobial testing showed high activity against *A. niger*, the Gram+ bacteria *Bacillus subtils and Staphylococcus epidermidis* as well as the yeast *Saccharomyces cereviceae* and *C. albicans*	Jirovetz et al. (2005)
Oil of cinnamon bark (*Cinnamomum zeylanicum*)	The oil contained 61% cinnamaldehyde, 29% cinnamic acid, and two minor unidentified compounds. The oil's efficacy at 300 and 100 µl/l completely inhibits the growth of *A. flavus* and *A. ruber* respectively	Jham et al. (2005)
Oil of *Foeniculum vulgare* sp. *piperitum*	The GC-MS of the oils showed estragole (53.08, 56.11 and 61.08%), fenchone (13.53, 19.18 and 23.46), and α-phellandrene (5.77%, 3.30%, and 0.72%), respectively. Strong antifungal property against *Alternaria alternata, Fusarium oxysporum*, and *Rhizoctonia solani* at 40 ppm.	Ozcan et al. (2006)
Five essential oils viz., thyme, sage, nutmeg, eucalyptus and cassia	The cassia oil inhibited completely the growth of *Alternaria alternate* at 300 to 500 ppm, while, the thyme oil exhibited a lower degree of inhibition 62.0% at 500 ppm, only	Feng and Zheng (2007)
Essential oil from the leaves of *Chenopodium ambrosioides* Linn	The oil completely inhibited the mycelial growth of *Aspergillus flavus* Link. at 100 µ/ml. Further, the oil exhibited broad fungitoxic spectrum against *Aspergillus niger, A. fumigatus, Botryodiplodia theobromae, Fusarium oxysporum, Sclerotium rolfsii, Macrophomina phaseolina, Cladosporium cladosporioides, Helminthosporium oryzae* and *Pythium debaryanum* at 100 µg/ml	Kumar et al. (2007)
The oil of *Putranjiva roxburghii* exhibited the greatest toxicity	The oil was found to be fungicidal and thermostable against *A. flavus* and *A.niger*, at its minimum inhibitory concentration (MIC) of 400 ppm	Tripathi and Kumar (2007)
The essential oils of oregano (Origanum vulgare), thyme (*Thymus vulgaris*) and clove (*Syzygium aromaticum*)	Oregano essential oil showed the highest inhibition of mold growth, followed by clove and thyme. *A. flavus* was more sensitive to thyme essential oil than *A. niger*. Clove essential oil was a stronger inhibitor against *A. niger* than against *A. flavus*.	Viuda et al. (2007)
The essential oil of *Citrus medica* L.	The oil exhibited a wide spectrum of fungitoxicity, inhibiting all 14 fungus species of *Arachis hypogea*	Pandey (2008)

Table 1. Contd.

| The essential oil of *Cymbopogon flexuosus, Trachyspermum ammi* and their active constituents | Oil of *C. flexuosus* and its major constituents Citral 38% and Geraniol 24.56% as well as oil of *T. ammi* and its constituents Thymol 80.7%, ρ-cymene 11.4% and α-pinene 7.9% were found effective against *A. flavus* and *Penecillium italicum* | Shukla (2010) |

contain noxious compounds, which may render them unsafe for both animals and humans to consume. Toxicology axiom." The dose makes the poison" also apply in some cases. It means that a substance that is safely consumed in the diet at low levels may be unsafe if consumed at a higher level in the diet. Therefore, the data demonstrating that the botanical is safe when consumed at the higher level is needed. And it is virtually difficult to find typical toxicological data such as Acceptable Daily Intake and No Effect Level. Further, here may be unusual sensitivities of some parts of the population to specific herbal compounds or strong aromatic ingredients. Therefore, while using botanicals in storage protection, these points should be duly emphasized to avoid negative nutritional or health consequences.

Conclusion

A small number of antimicrobial agents have been used for many years with little expansion, and there is a real need to expand the list of storage protectants which can be used to ensure safety and quality of stored products. These systems may have synergistic or additive uses with one another or may also be used with conventional antimicrobial compounds. The future of naturally occuring antimicrobial system seems be sure, as new storage protection systems are being rapidly developed and used in a variety of storage products.

REFERENCES

Agrawal R, Kharya MD, Srivastava R (1979). Antimicrobial and antihelmentic activities of the essential oil of *Nigella sativa* Linn. Ind. J. Exp. Biol. 17:1264-1265.
Arora R, Pandey GN (1976). The application of essential oils and their isolates for blue mold decay control in *Citrus reticulata*, J. Food. Sci. Technol. 14:14-16.
Avadhoot Y, Verma KC (1978). Antimicrobial activity of essential oil of seeds of *Lantana camara* var. *aculeata* Linn. Indian Drugs Pharm. 13:41-42.
Banerjee A, Nigam SS (1977). Antifungal activity of the essential oil of *Curcuma augustifolia*. Ind. J. Pharm. 39:143-145.
Beye, F. (1978). Insecticides from Vegetable Kingdom. Plant Res. Develop. 7:13-31.
Bhargava KS, Dixit SN, Dubey N.K, Tripathi RD (1981). Fungitoxic properties of Ocimum canum, J. Ind. Bot. Soc. 60:24-27.
Bigrami SK, Misra RS. Sinha,KK, Singh P (1980). Effect of some wild

and medicinal plant extracts on aflatoxin production and growth of *Aspergillus flavus* in liquid culture. Indian Bot.Soc. 59:123-126.
Bishop CD, Thornton IB (1997). Evaluation of antifungal activities of essential oils of *Monardo citriodora* var. *citriodora* and *Melaleuca alternifolia* on post-harvest pathogens. J. Essent. Oil Res. USA 9(1):77-82.
Brandies GA (1967). Commercial development of fungicides (Discussion): 246-247. In Holten *et. al.* (eds.). Plant Pathology Problems and progress, 1908-1958. *Indian University Press,* Allahabad, India.
Cate M (2000). Antimicrobial and Toxicological characteristics of commercial herbal extracts and the antimicrobial efficacy of herbs in marinated chicken. *M.S. thesis* directed by F. A. Draughon, May, Univ. of Tennessee, Knoxville.
Chandra H, Asthana A, Tripathi RD, Dixit SN (1982). Fungitoxicity of a volatile oil from the fruits of Cinnamomum cecidodaphne Meissn. Phytopathologia Mediterranea 21(1):35-36.
Chaturvedi R (1979). Evaluation of higher plants for their fungitoxicity against *Helminthosporium oryzae*. Ph.D.Thesis, University of Gorakhpur, Gorakhpur, India.
Chaurasi SC, Vyas KK (1977). *In vitro* effect of some volatile oil against *Phytophthora parasitica* var. *piperina*. J. Res. Indian Med. Yoga Homeoath. 1977:24-26.
Chaurasia SC, Kher A (1978). Activity of essential oils of three medicinal plants against pathogenic and nonpathogenic fungi. East Pharma. 21:183-184.
Cichewick RH,Thorpe PA (1996). The antimicrobial properties of chile pepper (*Capsicum* species) and their use in Mayan medicine, J. Ethanopharmacol. 52:61-70.
Cowan MM (1999). Plant products as antimicrobial agents. Clin. Microbiol. Rev. 12:564-582.
Cresan A, Hodisan V (1975). Investigations on fungistatic and fungicidal action of some extracts and volatile oils from medicinal plants. In Contributti Botanice. pp. 171-179.
Dawson RJ (1991). A global view of the mycotoxin problem. In: Champ B. R., Higley E., Hocking, A. D., Pitt, J. I., (eds.). Fungi and mycotoxin in stored products. Canberra: ACIAR Proc. 36:22-28.
de Souza EL, de Oliveira Lima E, de Luna KR, de Sousa F, de Sousa CP (2005). Inhibitory action of some essential oils and phytochemicals on the growth of various mould isolated from foods. *Brazilian* Arch. Biol. Technol. 48(2):245-250.
Dikshit A, Singh AK, Tripathi RD, Dixit SN (1978). The volatile fungitoxic activity of *Cedrus deodara. Proceedings of Symposium On Plant Disease Problems* Abdel-. Jaipur. p. 47.
Dikshit A (1980). Fungitoxic evaluation of some plants, *Ph D thesis*. University of Gorakhpur, Gorakhpur, India.
Dikshit A, Naqvi, AA, Hussain A (1986). *Schinus molle*: A new source of fungitoxicant. Appl. Environ. Microbiol. 15:1085-1088.
Dubey NK, Bhargava KS, Dixit SN (1983). Protection of some stored food commodities from fungi by essential oils of *Ocimun canum* and *Citrus medica*. Int. J. Trop. Plant Dis.1:177-179.
Dubey, NK, Kishore N, Tripathi NN, Tripathi RD, Dixit SN (1981). Fungitoxicity of the essential oils of *Ocimum canum* against *Aspergillus flavus* and A. vesicolor. Indian Perfumer 25(2):1-5.
Dubey NK, Kishore N, Tripathi N N., Tripathi RD, Dixit SN (1982). Fungitoxicity of the essential oils of *Citrus medica* against storage fungi. Ann. Appl. Biol. (suppl.) 100:58-59.
Duke PJ (1985). Hand book of Medicinal Herbs." *CRC Press, Inc.,* Boca

Raton, Florida..

Ejechi BO, Akpomedaye DE (2005). Activity of essential oil and phenolic acid extracts of Pepperfruits (*Dennetia tripetala* G. Baker;*Anonaceae*) against some food-borne microorganisms. Afr. J. Biotechnol. 4(3):258-261.

Fawcett CH, Spencer DM (1969). Natural antifungal compounds: 637-669. In Torgeson, D. C. (eds.). Fungicides an Advance Treatise Vol. II. *Academic Press*, New York and London.

Feng W, Zheng X (2007). Essential oils to control *Alternaria alternata in vitro* and *in vivo*. Food Control 18 (9):1126-1130.

Geda A, Bokadia MM (1979). Antifungal activity of the essential oil of *Blumea membranacea*. Ind. Drugs Pharma. 14:21-22.

Girgune IB, Jain NK, Garg BD (1978). Antifungal activity of some essential oils. Indian Drugs 16:24-26.

Hitokoto H, Morozumi S, Wauke T, Sakai S, Kurata H (1980). Inhibitory effects of spices on growth and toxin production of toxigenic fungi. Appl.. Environ. Microbiol. 39:818-822.

Hoult JRS, Paya M (1996). Pharmacological and biochemical action of simple coumarins: Natural products with therapeutic potential. Gen. Pharmacol. 27:713-722.

Jain PK, Charia AK, Sharma SK, Bokadia MM (1979). Study of some essential oils for their antifungal activities. Indian Drugs 16:122-123.

Jham GN, Dhingra OD, Jardim CM, Valente VMM (2005). Identification of the major fungitoxic component of cinnamon bark oil. Fitopatol. Bras 30:404-408.

Jirovetz L, Buchbauer G., Stoyanova, AS, Georgiev EV, Damianova ST (2005). Composition, quality control and antimicrobial activity of the essential of cumin (*Cuminum cyminum*) seeds from Bulgaria that had been stored for up to 26 years International J. Food Sci. Technol. 40(3):305-310.

Kovacs G (1964). Studies on antibiotic substances from higher plants with special reference to their pathological importance. *K. Vet. Hjsk, Arsskr*, pp.47-92 .

Kumar R, Dubey NK, Tiwari OP, Tripathi YB, Sinha KK (2007). Evaluation of some essential oils as botanical fungitoxicants for the protection of stored food commodities from fungal infestation. J. Sci. Food Agric. 87(9):1734-1743.

Lahariya AK, Rao JT (1979). *In vitro* antimicrobial studies of the essential oil of *Cyperus scariosus* and *Ocimum basilicum*. *Indian Drugs*. 16 (7): 150-152.

Lambert RJW, Skandamis PN, Coote, PJ, Nychas GJE (2001). A study of the minimum inhibitory concentration and mode of action of oregano essential oil, thymol and carvacrol, J. Appl. Microbiol. 91:453-562.

Laouer H, Zerroug MM, Sahli F, Chaker AN, Valentini G, Ferretti, G, Grande M, Anaya J (2003). Composition and antimicrobial activity of Ammodes pusilla (Brot.) Breistr. essential oil. J. Essent. Oil Res.. 15(2): 135-138.

Lewis JA, Papavizas GC. (1967). Effects of tannins on spore germination and growth of *Fusarium solani* f. *phaseoli* and Verticillium albo-atrum. Canadian J. Microbiol. 13:1655-1661.

Mabrouk SS, El-Shayeb NMA (1980). Inhibition of aflatoxin formation by some spices. *Lebensm*. Unters. Forsch. 171:344-347.

Mannon J, Johanson E. (1985). Fungi down on the farm. New Scient, 28:12-16.

Mathela CS, Sinha GK (1978). Antibacterial and antifungal study of some indigenous essential oils. J. Res. Ind. Med. Yoga Homeopath. 13(3):122-124.

Mishra AK, Dubey NK (1994). Evaluation of some essential oils for their toxicity against fungi causing deterioration of stored food commodities. Appl. Environ. Microbiol. U.S.A. 60:1101-1105.

Mishra AK, Dubey NK (1990). Fungitoxic properties of *Prunus persica* oil. Hindustan Antibiot. Bull. 32(3-4):91-93.

Misra D, Mishra, M, Tiwari SN (1997). Toxic Effect of volatiles from *Callistemon lanceolatus* on six fungal pathogens of rice. Indian Phytopathol. 50(1):103-105.

Nehrash AK (1961). The antimicrobial activity of extracts and essential oils from cultivated and wild radish. J. Microbial. Kiev. 23:32-37.

Nigam SS, Rao TSS (1977). Antimicrobial efficacy of some Indian essential oils. In Kapoor, L. Ramkrishanan D 1977 (eds.). *Advances*

in essential oil industry. Today and Tomorrow's Printers and Publishers , New Delhi, India. pp. 177-180.

Nunez L, Aquino MD, Chirife J (2001). Antifungal properties of Clove oil (*Eugenia caryophylata*) in sugar solution. *Braz. J.* Microbiol. 32(2):123-126.

Ozcan MM, Chalchat JC, Arslan D, Ates A, Unver A (2006). Comparative essential oil composition and antifungal effect of bitter fennel (*Foeniculum vulgare* ssp. *piperitum*) fruit oils obtained during different vegetation. J. Med. Food 9(4):552–561.

Pandey RK (2008). Physiological and pathological studies of certain fungi. *D.Phil Thesis*, University of Allahabad, Allahabad.

Peres MT, Monache, FD, Cruz A B, Pizzolatti MG, Yunes RA (1997). Chemical composition and antimicrobial activity of *Croton urucurana* Baillon (*Euphorbiaceae*). J. Ethnopharmacol. 56:223-226.

Rahman A, Chaudhary MI (1995). Diterpinoid and steroidal alkaloids. Nat. Prod. Rep. 12:361-379.

Rahman MU (2003). Mycotoxic effect of *Thymus serpyllum* oil on the asexual reproduction of *Aspergillus* species. J. Essent. Oil Res. 15(3):168-171. (Do not correspond with that cited in the article)

Renu,Tripathi RD,Dixit SN (1980). Fungitoxic properties of *Cestrum diurnum*. Naturwiss. 67(3): 150-151.

Samson RA, Hoekstra ES, Frisvad JC, Filterborg O (1995). Introduction to food borne fungi (*eds*). Pub. Ponsen and Looyen, Wageningen, The Netherlands.

Sanyal A, Verma KC (1969). *In vitro* antibacterial and antifungal activity of *Mentha arvensis* var. *piperescenes* oil obtained from different sources. Indian J. Microbiol. 9:23-24.

Saxena AR (1980). Evaluation of higher plants for their fungitoxic properties. *Ph. D. Thesis*, University of Gorakhpur, Gorakhpur, India.

Saxena RC (1993) Neem as a source of natural insecticides: An Update- Proc. Symp. Botanical Pesticides at Central Tobacco Research Institute, Rajamundry.

Schultes RE (1978). The kingdom of plants. In "Medicines from the Earth," ed. W.A.R.Thompson. p. 208. McGraw-Hill Book Co., New York.

Shahi SK, Shukla AC, Dikshit A (2002b). *Eucalyptus pauciflora*: a potential source of ecofriendly pesticide, PROTECTON. *Proc. National Seminar on Plant Biotechnology for Sustainable Hill Agriculture*. (eds.). Kumar, N., Negi, P.S. and Singh, N.K. pp.149-152.

Shahi SK, Shukla AC, Patra M, Dikshit A (2002a). Use of essential oil as botanical-pesticide against post harvest spoilage in fruits, *Malus pumilo.Bio Control*, Kluwer Academic Publishers. 48:223-232.

Shahi S, Shukla, AC, Dikshit S, Dikshit A (1997). Modified spore germination inhibition technique for evaluation of candidate fungitoxicant (*Eucalyptus* spp) H. W. Dehne *et. al.* eds. Proc. 4th Int. *Symp. Diagnosis and Identification of Plant Pathogens*, Kluwer Academic Publishers, Netherlands. pp. 253-263.

Shaaya E, Kostjukvshi M, Eilberg J, Sukprakaran, C (1997). Plant oils as fumigants and contact insecticides for the control of stored product insects. Journal of Stored-product Research, UK, V: 7-15.Sharma, A, Gautam MP (1977). Investigation on the antifungal activity of volatile oil derived from *Nepeta hindostana* (Roth) Hains. Indian Drugs Pharm. Ind.12:33-34.

Sharma GP, Jain NK, Garg BD (1978). Antifungal activity of some essential oils. Indian Drugs 16:21-23.

Sharma N (2001). Preservation of dried fruits and nuts from biodeterioration by natural plant volatiles. *Donahaye, E. J. et. al. eds. (2001) Proc. Int. Conf. CAF in stored Products,* Executive Printing Services, clovis California, USA, pp. 195-208.

Sharma SK, Singh VP (1979b). Antifungal study of some essential oils of Ganan the Javanica . Ind. Drugs 16(2):289-291.

Sharma SK, Singh VP (1979a). Antifungal activcity of some essential oils. *Ind. Drugs and Pharma. Ltd.* 13(1):3-6.

Shukla AC (1997). Fungitoxic studies of some aromatic plants against storage fungi, *Ph D Theis,* University of Allahabad, Allahabad, India.

Shukla AC, Shahi SK, Dikshit A (2000). Epicarp of *Citrus sinensis*: A Potential source of natural pesticide, Indian Phytopathol. 53(3):318-322.

Shukla AC (2010). Bioactivities of the major active constituents isolated from the essential oil of *Cymbopogon flexuosus* (Steud.) Wats and *Trachyspermum ammi* (L) Sprague as a herbal grain protectant(s). *D.Sc. Thesis,* Allahabad University.

Singh AK (1980). Antifungal activity of volatile fractions of some higher plants. *Ph. D. Thesis,* University of Gorakhpur, Gorakhpur, India.

SinghAK, Dikshit A, Tripathi SC (1978). Antifungal properties of ginger grass oil. *Proc. Nat. Acad . Sci .* India , 48th Annual session. p.148.

Singh AK, Dikshit A, Sharma ML, Dixit SN (1980). Fungitoxic activity of some essential oils. Econ. Bot. 34:186-190.

Singh AK, Tripathi SC, Dixit SN (1978). Fungitoxicity of volatile fraction of some Angiospermic plants. Proc. Symp. Environmental Science and Human Welfare, Ujjain. p. 13.

Singh G, Upadhyaya RK, Narayanan CS, Padmkumar, KP, Rao GP (1993). Chemical and fungitoxic investigations on the essential oil of *Citrus sinensis* (L) Pers, *Zeitschriftfuer Pflanzenkrankheiten-und-Pflanzenschutz,* Germany. 100 (1):69-74.

Slavenas J, Razinskaite D (1962). Some studies of phytocidal substances of Juniper oil from the common *Juniper* Leit. *TRS.* Moks. Acad. Darbal. Ser. C. 2:63-64.

Stinson EE, Billis DD, Osman S. F, Sciliano J, Ceponis MJ, Heisler EG (1980). Mycotoxin production by Alternaria species grown on apples, tomatoes and blueberries. Agric. Food Chem. 28:960-963.

Suri RK, Thind TS (1979). *In vitro* antifungal efficacy of some essential oils. *East* Pharm. 22:109-110.

Suri RK, Nigam SS, Thind TS. (1979). *In vitro* antimicrobial efficacy of essential oil of *Eucalyptus* citriodora. Ind. Drugs Pharm. India 14 (3):35-37.

Sawhney SS, Suri RK, Thind TS (1977). Antimicrobial efficacy of some essential oils *in vitro.* Ind. Drugs 15:30-32.

Swaminathan MS (1978). Inaugural Address First Bot. Conference, Meerut, India. pp.1-31.

Thind TS, Dahiya MS (1977). Inhibitory effect of essential oils of some medicinal plants against soil inhabiting dermatophytes. *Indian Drugs.*14: 17-20.

Tripathi NN (1980). Fungitoxicity in some higher plants. *Ph. D. Thesis,* University of Gorakhpur, Gorakhpur, India.

Tripathi NN, Kumar N (2007). *Putranjiva roxburghii* oil—A potential herbal preservative for peanuts during storage. J. Stored Prod. Res. 43(4):435-442.

Tyler PS, Boxall RA (1984). Post-harvest loss reduction programmes: a decade of activities: - what consequences? *Tropical* Stored Products Inf. 23:13-28.

Viuda MM, Navajas YR, Lopez JF, Alvarez JAP (2007). Antifungal activities of thyme, clove and oregano essential oils, J. Food Safe.. 27(1):91–101.

Yadava RN, Saxena, VK Nigam, SS (1978). Antibacterial activity of the essential oil of *Caesalpinia sappan* Linn. Indian Perfumer 22(2):73-75.

Zohri AA, Abdel-Gawad KM, (1993). Survey of mycoflora and mycotoxins of some dry fruit in Egypt. Basic Microbiol. 33:279-288.

23

Use of information and communication technologies by rural farmers in Oluyole local government area of Oyo State, Nigeria

J. M. Usman[1]*, J. A. Adeboye[1], K. A. Oluyole[2] and S. Ajijola[3]

[1]Federal College of Forestry, PMB 5087, Jericho, Ibadan, Nigeria.
[2]Cocoa Research Institute of Nigeria, PMB 5244, Idi-Ayunre, Ibadan, Nigeria.
[3]Institute of Agricultural Research and Training, PMB 5029, Moor Plantation, Ibadan, Nigeria.

This study investigated the role, impact, level of use and potentials of integrating Information and Communication Technologies (ITCs) into agricultural development process in Oluyole local government area of Nigeria. The study observed that as much as
developing agriculture, much more were required in the form of ICTs to adequately extend innovations to effectively employ resources, take advantage of new methodologies and markets to better the lots of living standard of the farmers. Based on the findings, it was recommended that ICTs should be incorporated into all endeavours related to agricultural development. Awareness should be generated among young and middle-aged farmers about availability of ICT services in order to increase farmers' participation in ICT initiatives. Also, since small or and marginal farmers were using ICTs services, more emphasis should be given to providing information strictly relevant to their farming systems. Strong interfaces should be developed at village level so that the problem of computer illiteracy among farmers may be resolved.

Key words: Nigeria, Oyo state, information, communication, technologies, rural farmers.

INTRODUCTION

Information and Communication Technology (ICT) is the scientific, technological and engineering discipline of management technologies used in the handling of information, processing and application related to computers. ITC is also concerned with interactions between man and machines; and associated socio-economic and cultural matters (Osuagwu, 2001). Information technology could be regarded as the coming together of computing and telecommunications for the purpose of handling information. The bottom-line is that information technology is an application that is computer-based for the purpose of sharing ideas, data, and other relevant information and the improvement of the status quo for development. However, in the recent past, there has been revolution

with regards to information and communication technologies (ICTs). The world is going through an information technology revolution that has drastically changed many facets of human life, from politics, education, and entertainment to industry (Ajayi, 2002). Agriculture and its associated natural resources management are not likely to be exceptions. Omotayo (2005) observed that agricultural extension depends largely on information exchange between and among farmers and a broad range of other actors. However, Oluwadare and Okunlola (2006) pointed out that Nigeria's economy is rural-based, with over 70% of the population deriving their means of livelihood from agriculture either directly or indirectly and further stated that these rural areas are still starved of most modern facilities such as potable water, electricity, good roads, decent housing, educational facilities, modern health facilities, storage facilities and most especially communication facilities.

Frontline extension workers who are the direct link

---
*Corresponding author. E-mail: usmanj05@yahoo.com.

Table 1. Socio economic characteristics of the respondents (farmers).

Variable	Frequency	Percentage
Age distribution (in years)		
21-30	20	28.6
31-40	21	30.0
41-50	24	34.3
51-60	5	7.1
Total	70	100
Sex distribution		
Male	69	98.6
Female	1	1.6
Total	70	100
Highest academic qualification		
First school leaving certificate	30	429
National Diploma (OND)	9	12.9
National Certificate of Education (NCE)	5	7.1
Higher National Diploma (HND)	15	21.4
First Degree	2	2.9
Adult Education	9	12.9
Total	70	100
Marital status		
Single	19	27.1
Married	46	65.7
Divorced	3	4.3
Widowed	2	2.9
Total	70	100
Religion		
Christianity	38	54.3
Ilam	32	45.7
Total	70	100

between farmers and other actors in the agricultural knowledge and information system could be well positioned to make use of ICT to access expert knowledge or other types of information that could facilitate the accomplishment of their routine activities. In this respect the objectives of the study were to:

(i) Determine the socio-economic characteristics of the farmers in Oluyole Local Government Area.
(ii) Examine selected ICTs devices and determine the most frequently used by the farmers.

METHODOLOGY

The study area for this survey was Oluyole Local Government Area of Oyo State, Nigeria. Primary data on personal characteristics of farmers' awareness, access and utilization, problems militating against the use of ICT and types of ICT needed were collected by means of 90 questionnaires administered and filled directly or indirectly (by oral interview) at seminars, meetings and visits to farmers in Oluyole local government area of Oyo State, Nigeria.

Oyo State is an inland state in south western Nigeria, with its capital at Ibadan. It is bounded in the north by Kwara state, in the east by Osun State, in the south by Ogun state and in the west partly by Ogun state and partly by The Republic of Benin. Visits were paid to the farmers in question during the meetings, seminars, and in their respective villages. Determining variables of socio-economic importance used were age, sex, educational/academic qualification, marital status and religion of farmers. Respondents' usage (daily, weekly, monthly or occasionally) of ITC (radio, internet, mobile phone, television or combinations) by farmers was respectively noted. Of the ninety (90) copies of the questionnaire administered, seventy (70) copies were retrieved for collation of data for this study. Descriptive and inferential statistics including frequencies, percentages and means were used to analyze the data collected.

RESULTS AND DISCUSSION

The socio-economic characteristics of the respondents considered (age, sex, educational/academic qualification, marital status and religion) crucial in influencing the adoption of new technologies behaviour of the farmers is shown in Table 1. The results showed that majority of the

Table 2. Respondents frequency usage of ICT.

Variable	Frequency	Percentage
Daily	16	22.9
Weekly	19	27.1
Monthly	1	1.4
Occasionally	30	48.6
Total	70	100

Table 3. Respondents usage of ICTs.

Variable	Frequency	Percentage
Radio	48	68.6
Internet	5	7.1
Mobile phones	9	12.9
TV	3	4.3
Multiple response	5	7.1
Total	70	100

farmers (34%) were aged between 41 and 50 years followed by farmers within the age range of 31 to 40 years (30%). This age range could be regarded as middle age, thus, they were expected to be innovative and economically active. Those of the age range 21 to 30 years constituted 20% while 5% formed those of the age range of 51 to 60 years. This implies that on the average, 64.3% of the respondents were between 31 and 50 years of age. On gender distribution, the study revealed that 98.6% were males while 1.4% was women. This means that majority of the people that engage in agriculture and use ICT for agricultural developments were males. The low presence of women in Agriculture at Oluyole was similar to the findings of Odewale (1995) who noted that only about a quarter of farmers sampled were female. This may be due to the drudgery nature of agricultural activities. The study showed that 42.9% of the farmers had first school leaving certificate. Those with Higher National Diploma (HND) and National Diploma (ND) constituted 21.4% each, while 12.9% had NCE certificate while those with B.Sc. were 2.9%. These results showed some level of literacy among the farmers. Okigbo (1998) reported that there was a strong positive correlation between the level of farmers' education and their ability to meaningfully utilize ICT for agricultural development. On marital status, the study revealed that 65.7% of the respondents were married, 27% single, 4.3% divorced while 2.9% were widowed. This showed that most of the respondents were family men and women who require family income to cater for their families.

With regards to religion, distribution of the respondents revealed that practicing Christians constituted 54.3% while 45.7% were Muslims. Table 2 showed that 48.6% of the respondents use ICT in getting agricultural information occasionally, followed by weekly users (27.1%),

daily users (22.9%) and 1.4% were monthly users. Radio had remained the most important medium for communicating with the rural populations of developing countries (Helen and Amin, 2002). Agriculture oriented programmes were often held weekly in most parts of Nigeria, such as, on both radio and television stations of Broadcasting Corporation of Oyo State (BCOS) like Agbeloba and FADAMA Korede. The daily and occasional users' use radio in getting political news, general news, sports etc. From Table 3, it can be seen that the highest percentage of respondents (68.6%) used radio as a source of information, followed by mobile phone users (12.9%) and Internet users were 7.1%. Farmers that used more than one device and television were 7.1 and 4.3% respectively. The high percentage of radio users showed that radio was relevant to any strategy that involves rural development in Nigeria. Radio remained one of the most important medium for communicating with the rural populations of developing countries. This remained particularly true in Africa where, according to the BBC World Service, there were estimated 65 million radio receivers in 1996. Also, Niang (2001) revealed that by the end of the 1990s there were approximately 12 newspapers, 52 televisions and 198 radios for every 1000 Africans. The limited access of African farmers to newspapers, televisions and the internet merely reinforced the importance of radio in Africa.

From Table 4, it revealed that the farmers perceived market information including daily updates on the prices of agricultural commodities in the markets of the area as one of the most relevant ICTs services. This was seen as "most essential" by 44.3% of the respondents, 51.4% saw it as "essential" while 4.3% viewed it as "less essential". This enabled them take decision for sale of their products at those markets where their produce would command the best prices and reduced shortage.

Conclusion

This study attempted to contextually highlight the role, impact, level of use and potentials of integrating ICTs into agricultural development process in Oluyole local government of Nigeria. The study observed that in as much as finance and other infrastructure were important in developing agriculture, much more is required in the form of ICTs to adequately translate innovations, effectively employ resources and take advantage of new methodologies and markets to better the lots of living standard of the farmers. While the study called on the Nigerian government to be more committed to funding research and development in the agricultural sector, it however, cautioned that such innovations should be adequately planned for continuity and it must be locally oriented in order to touch nook and cranny taking into cognizance their socio-economic potentials and capabilities. Based on the findings, it is recommended

Table 4. Information needs of farmers.

Variable	Most essential frequency (%)		Essential frequency (%)		Less essential frequency (%)	
Marketing information	31	44.3	36	51.4	3	4.3
Accounting and payment	12	17.1	8	11.4	50	71.4
Input prices and availability	38	54.3	32	45.7	0	0.0
Soil testing and soil sampling information	0	0.0	2	2.9	68	97.1
Dairying and marketing of milk and milk products	13	18.6	5	7.1	52	74.3
Early warning and management of diseases and pest	35	50.0	30	42.9	5	7.1
Farm business and management information	7	10.0	9	12.9	54	77.1
Crop insurance information	2	2.9	5	7.1	63	90.0
General agricultural news	61	87.1	9	12.9	0	0.0
Post-harvest technology	21	30.0	38	54.3	11	15.7
Latest packages of practices	13	18.6	3	4.3	54	77.1
Weather forecasting	17	24.3	52	74.3	1	1.4
Question-and-answer services	8	11.4	3	4.3	59	84.3
Information about rural development programmes and subsidies	67	95.7	3	4.3	0	0.0
Facilitating access to land records/online registration	7	10.0	9	12.9	54	77.1

that efforts should be made to incorporate ICTs in all endeavours related to agricultural development. Awareness should be generated among young and middle-aged farmers about availability of ICT services in order to increase farmers' participation in ICT initiatives. Older farmers should be brought into the chain of ICT networks at a later stage. Also, since small and marginal farmers are using ICTs services, more emphasis should be given to providing information relevant to their farming systems.

Strong interfaces should be developed at village level so that the problem of computer illiteracy among farmers may be resolved. There is the need for ICT policy and regulatory framework that would expend ICT infrastructure, faster ICT-Led innovation, increase access to capital and increase the flow of information and knowledge.

REFERENCES

Ajayi GO (2002). Information and Communication Technologies in Africa- African Response to the Information and Communication Technologies Revolution: Case Study of ICT Development in Nigeria. Paper delivered at International Center for Theoretical Physics (ICTP) Trieste, Italy, 11th-16th Feb, 2002.
Helen HO, Amin K (2002). Listening to stakeholders. Agricultural Research and Rural Radio Linkages. International Service for National Agricultural Research (ISNAR). Briefing paper No. 48. April, 2002. The Hague, Netherlands.
Niang T (2001). Radio in Action: A CTA Experience Presentation to the First International Workshop on farm Radio Broadcasting. February 19-22, 2001, FAO, Rome, Italy.
Odewale OA (2005). Effect of Special Programme for Food Security (SPFS) on farmers' production in Akufo farm settlement of Oyo State. An unpublished HND project work Federal College of Forestry, Ibadan. pp. 33-34.
Okigbo BN (1998). Rural information in agriculture credit administration reading in agricultural finance. Longman Nigeria PLC. Edited by Aja Okorie and Martins O. Ijere. p. 170.
Oluwadare SA, Okunlola JO (2006). Transforming Nigerian Agriculture through Information and Communication Technologies: Challenges and Opportunities. Proceedings of 20th Annual National Conferences of Farm Management Association of Nigeria, Forestry Research Institute of Nigeria, Federal College of Forestry, Jos, Nigeria. 18th-21st Sep., 2006. pp. 246-250.
Omotayo OM (2005). ICT and Agricultural Extension. Emerging issues in transferring agricultural technology in developing countries. In: Adedoyin SF (ed), Agricultural Extension in Nigeria. Ilorin: Agricultural Extension Society of Nigeria.
Osuagwu DE (2001). 'New Technologies and Services in Internet Business'. J. Prof. Administrator, July - Sept, 29-41.

24

Post-harvest losses and welfare of tomato farmers in Ogbomosho, Osun state, Nigeria

Abimbola O. Adepoju

Department of Agricultural Economics, University of Ibadan, Oyo State, Nigeria.

Crop losses, especially along the post-harvest food supply chain, have been identified as one of the major causes of food shortage problems in most developing countries and in Nigeria in particular. Vegetable farmers such as those that grow tomatoes often record great amount of produce loss which translates to a waste of resources, a reduction in their income and ultimately their welfare. This study examined the effects of post-harvest losses on the welfare of 107 tomato farmers in Ogbomosho selected through a multi-stage sampling procedure. The analytical tools used in the study include descriptive statistics, gross margin analysis, Ordinary Least Square (OLS) and regression model. Results revealed that majority of the tomato farmers were male, married and had no formal education. The average gross margin values of ₦3, 229.45 and ₦72, 905.80 were obtained with and without post-harvest losses for the tomato farmers respectively. This implied a 95.5% post-harvest loss incurred by the farmers. Household size and the total value of post-harvest losses were found to significantly affect the per-capita income and hence welfare of the tomato farmers negatively. The study recommends that farmers engaged in tomato production be adequately trained on post-harvest crop handling techniques. In addition, priority should be given to investment in post-harvest processing technologies and establishment of processing industries especially in the production areas.

Key words: Post-harvest loss, tomato, welfare, farmers, Nigeria.

INTRODUCTION

Global efforts in the fight against hunger to raise farmers' income and improve food security especially in the world's poorest countries should give priority to the issue of crop losses (FAO, 2010). This is due to the adverse effects of crop losses on food quality, environment and generally on economic development. Crop losses indicate a waste of productive agricultural resources such as land, water, labour, managerial skills and other inputs that could have been channelled into more viable ends.

Roughly, about one-third of food produced for human consumption is lost or wasted globally, which amounts to about 1.3 billion tons per year (FAO, 2011). In addition, 30 to 40% of the food crops produced in the world is never consumed as a result of damage, rotting as well as pest and diseases which affect crops after harvest (Meena et al., 2009). As a consequence, post-harvest losses have thus been identified as one of the determinants of food problem in most developing countries.

Tomato, a relatively short duration crop, is one of the most important vegetables worldwide (Shankara et al., 2005; Seisuke and Neelima, 2008). It is high yielding and economically attractive, hence, the area under cultivation is increasing daily. For instance, world tomato production in 2001 was about 105 million tons of fresh fruit from an estimated 3.9 million hectares (FAO, 2005). In Nigeria, tomato accounts for about 18% of the average daily consumption of vegetables (Babalola et al., 2010). This makes it a very important food crop to an average Nigerian. In 2008, 1,701,000 tons of tomato was produced. As a result, Nigeria became the 13th largest producer of tomato in the world (FAO, 2010). It is grown in the South-western part of the country in small holdings under rain-fed conditions and in the North under irrigation systems (Ayandiji et al., 2011) and consumed both in fresh and in paste form. However, because of its highly perishable nature, many problems are encountered in its production. These problems include diseases, nematodes, insect pests, high flower drop, all these resulting in low yield and poor quality fruits. These, coupled with poor post-harvest handling as a result of lack of storage facilities, good road network, good marketing channel amongst others, brings to the fore the need for efficient harvesting, handling, transportation and marketing techniques to curb postharvest losses in tomato production.

Poor post-harvest handling of perishable farm produce by the farmers can be traced to the negligence on the part of some farmers when rotten fruits are mixed with healthy ones. This tends to have a multiplying effect of rot on the healthy fruits. Also, fruits and vegetables vegetables which include tomatoes produced in rural areas are hardly taken to the market either due to lack of access to nearby markets or inadequate market information by these farmers. Since the farmers have little or no capacity to process their produce and coupled with the fact that there are no modern storage facilities, their products are prone to damages and post-harvest losses (Kader, 2005). Even when the farmers decide to take their produce to the market, they are often constrained by problems of transportation such as poor road network and inefficient mode of conveying their produce to the market. In Nigeria, this includes the use of dilapidated trucks. All these problems together reduce the quality of the farmers' products and force them to sell the rotten tomatoes (popularly known as *esa* among the Yoruba) at ridiculously low prices. This in-turn reduces their income and ultimately their welfare as they are not able to afford other basic necessities of life.

However, one of the country's agricultural policy thrust specifies that farmers be encouraged to use simple but effective on-farm, off-farm storage facilities and agro-processing technology in order to add value to farm produce and increase their shelf life. In line with this, the Nigerian Stored Products Research Institute (NSPRI) together with Food and Agriculture Organization (FAO)

developed techniques for the storage of fruit and vegetables especially tomatoes. Many of the techniques would require high-energy sources like refrigeration which are not available and affordable to the local farmers. These techniques could help increase the shelf-life of the crops and make them stay longer before they are sold.

However, the non-availability of these facilities to local farmers implies that farmers will always have to sell at reduced prices as they cannot keep the highly perishable products for an extended period of time. This has grave implications on the income of farmers and could consequently result into a rapid decline in their welfare. Also, it is distressing to note that while many resources are being devoted to planting crops, irrigation, fertilizer application and crop protection measures for increased productivity, little is being done to minimize post-harvest losses. A reduction in post-harvest food loss could guarantee increase in food availability thereby reducing the need for food importation and consequently impact positively on the welfare of farmers (Adesina, 2012). This is pertinent if the country is to meet its goal of food self-sufficiency by 2015. Most of the available studies on post-harvest losses in Nigeria were carried out using the urban markets as case studies (Adeoye et al., 2009). The near neglect of the farmers who constitute the upstream sector of agricultural production and who are most affected by post-harvest losses in terms of decreased market efficiency, severe reduction in income and consequently loss of welfare, therefore justifies the need for this study. This study apart from examining the value of loss and returns accruing to the tomato farmers, also seeks to know the extent to which post-harvest losses affect the welfare of the farmers in question.

MATERIALS AND METHODS

This study was carried out in Ogbomosho area of Oyo state. The choice of this area was premised on the fact that tomato is one of the major crops produced in the area. Ogbomosho, an ancient Yoruba city founded in the mid-17th century is located in the South Western part of Oyo State, Nigeria. The area is situated on Latitude 8.133°N and on Longitude 4.250°E of the equator. The two major seasons are dry and rainy seasons with an average temperature of 27°C. Located at an elevation of 342 m above sea level, the predominant vegetation zone in Ogbomosho is derived savannah with mean annual rainfall of 1.247 mm. Employing 3.2% growth rate from 2006 census figures, the 2011 estimated population for the Local Government is 1,200,000 persons. The area is densely populated with 536 people per km[2]. Their major occupation is farming while notable crops grown include cassava, tomatoes and maize.

Primary data used for the study were obtained from 115 randomly selected tomato producers. Sample information were collected on variables such as socio-economic characteristics of respondents including: gender, household size, access to credit, educational level, years of experience, mode of transportation of farm produce, scale of operation, labour used, and working hours per day. Finally, information on prices of output as well as fixed and variable costs incurred in the production of tomatoes were obtained from the respondents. However, owing to incomplete questionnaire information, data from 107 respondents were used for analysis in this study. Data were analyzed using descriptive statistics, gross margin and the ordinary least squares regression analysis. Descriptive statistics such as frequency distribution, percentages and mean were used in analyzing the socio-economic characteristics of respondents and the constraints faced by the farmers, while gross margin analysis was used to estimate the profit made by the tomato farmers in Ogbomosho. The gross profit of a business is estimated as the difference between the total sales and the variable cost incurred.

$$GM = TR - TVC$$

Where: GM = Gross Margin, TR = Total Revenue, TVC = Total Variable Cost, TR = Value of output (amount realized from the sale of tomatoes). This was obtained by multiplying the quantity of tomatoes sold by the unit selling price. TVC = Cost of all inputs (Pre – harvest and harvest labour wage, transportation costs and other input costs).

The gross profit margin was obtained by dividing the gross profit by sales. The gross profit margin is given as:

$$\text{Gross profit margin} = \frac{\text{Sales} - \text{Variable cost}}{\text{Sales}}$$

Gross profit margin is particularly useful in evaluating the profitability of tomato production amongst the farmers in Ogbomosho. This is because many of the farmers' practice subsistence farming which often times involve a small amount of fixed capital.

Ordinary least square model

The ordinary least square model was used to examine the effects of post-harvest losses on the welfare of tomato farmers in Ogbomosho, Osun state, Nigeria. The model is stated as follows:

$$y_i = \alpha + \beta x_i + \varepsilon_i.$$

Where: y_i= Dependent variable, α = Constant term, β = Regression Coefficients, ε_i= Error term.

The model is explicitly stated as:

$$Y = \alpha + \beta_1 X_1 + \beta_2 X_2 + \beta_3 X_3 + \beta_4 X_4 + \beta_5 X_5 \text{------} + \beta_{11} X_{11} + \varepsilon_i$$

Where:
Y= Per Capita income (in Naira), X_1=Market participation rate, X_2=Marital Status (Married=1, otherwise=0), X_3= Years of formal education, X_4=Gender (Male=1, otherwise=0), X_5= Age (Years), X_6= Size of Farmland (in Hectares), X_7= Farming Experience (in years), X_8= Time spent on the farm (in Hours), X_9= Primary Occupation (Farming=1, otherwise=0), X_{10}= Household size, X_{11}= Value of Post-harvest losses (in Naira), ε = error term.

The market participation rate was defined as the ratio of the value of the total quantity of tomatoes sold to the price value of the total harvested tomato for the planting season. That is,

$$MPR = \frac{\text{Value of tomatoes sold}}{\text{Value of total tomatoes harvested}}$$

RESULTS AND DISCUSSION

Table 1 presents the socio-economic characteristics of the respondents. More than three-quarters of the respondents were males while females accounted for the remaining 24.3%. This implies that tomato farming is dominated by male farmers and could be attributed to the cultural setting of the area in which land is mainly allocated to males while females are deprived of direct land ownership. More than half (52.3%) of the respondents were below 41 years of age with the mean age of all the interviewed farmers being 31 years, implying that a good number of the farmers in the area are in their economic active age. Also, more than half of the interviewed farmers were married and had household sizes of between 7 and 12 members. The mean average household size stood at 8 while about 52.3% of the respondents had between 1 and 10 years of farming experience. With respect to the educational status of the respondents, more than three-fifths (67.3%) of the farmers had no formal education while only 2.8% had tertiary education. This result agrees with the findings of Ayandiji (2011) who discovered that only 2.2% of citrus farmers had tertiary education but contrasts with it in the percentage of farmers who had formal education (66.7%). With only about 3.7% of the farmers cultivating less than 1 ha and about three-fifths (60.6%) between 1 and 10 ha, results showed that the production of tomatoes in this study area is rather on a large scale. Unfortunately, owing to the lack of storage and processing facilities, coupled with the inefficient mode of packaging and transportation, post harvest losses in this area is inevitable. Further, the main mode of transportation of produce to the point of sale is the use of 'pick-up' vans employed by more than four-fifths of the farmers while other forms of transportation which include the use of car and motorcycle were used mainly by the small scale producers. This corroborates the findings of Muhammad et al. (2012) in which the main mode of transportation employed by fruits and vegetable farmers in Garun Mallam Local Government Area of Kano State, Nigeria was found to be the use of open pick-ups.

The major post-harvest constraints reported by respondents in the study area as presented in Table 2 include the lack of storage facilities, long distance to market, poor transportation network, pests and diseases and lack of access to credit facilities while the least reported constraint was theft of produce. This is consistent with the findings of Seid et al. (2013) and Basappa et al. (2007) who found inadequate storage facilities, inadequate transport facilities, pests and diseases to be significant factors contributing to post-harvest losses of maize and commercial horticultural crops respectively. Lack of storage facilities was reported as a major constraint by all the farmers. As a result, farmers' are usually forced to take their produce to the market directly from the farm. This results most times into

Table 1. Socio-economic characteristics of respondents.

Variable		Frequency	Percentage
Sex	Male	81	75.7
	Female	26	24.3
Age	<21	14	13.1
	21-40	42	39.2
	41-50	45	42.1
	>50	6	5.6
Marital status	Single	25	13.1
	Married	58	39.2
	Separated/Divorced	13	42.1
	Widowed	11	5.6
Household size	1-6	35	32.7
	7-12	58	54.2
	>12	14	13.1
Farming experience	1-10	56	52.3
	11-20	50	46.8
	>20	1	0.9
Educational status	No Formal Education (NFE)	72	67.3
	Primary	23	21.5
	Secondary	9	8.4
	Tertiary	3	2.8
Farm size	<1ha	4	3.7
	1-5	35	32.7
	6-10	30	28.0
	>10	38	32.5
Mode of transportation	Motorcycle	15	14.0
	Van pick-up	89	83.2
	Car	3	2.8
	Total	107	100.0

Source: Field survey, 2012.

a glut in the market and consequently increased incidence of post-harvest losses. The long distance to market coupled with the poor transport network were also major constraints reported by all and about 99.1% of the farmers respectively. This could be as a result of the fact that the longer the distance from the farm to the market, the higher the transportation costs that will be incurred. Some farmers, because of their inability to pay the transportation costs, harvest their produce and sell as much as they can at the farm gate. Whatever is left is utilized by the household while the remaining is left to rot. For instance, the distance from *iluju* and *tewure* (which were part of the communities sampled), to Central Ogbomosho where the middle-men from Ilorin buy in bulk, is about 50 km. Apart from the high transportation costs, the perishable nature and high moisture content of tomatoes account for a sizeable amount of post-harvest losses during transportation. This could be attributed mainly to the inefficient packaging of the produce during transportation as the produce are usually packed in baskets which are heaped on top of one another. Also, pests and disease infestation as well as lack of access to credit were major constraints faced by almost all (99.1%) and about 95.3% of the farmers respectively. Lack of credit access was mainly due to the inability of the farmers to meet the requirements of credit institutions for

Table 2. Post-harvest constraints encountered by the farmers.

Constraints	Frequency	Percentage
Lack of storage facilities	107	100
Long distance to market	107	100
Poor transport network	106	99.1
Pests/diseases	106	99.1
Lack of access to credit	102	95.3
Insufficient Working capital	66	61.7
Low Government support	58	54.2
Theft	36	33.6

Source: Field survey, 2012.

Table 3. Gross margin analysis result.

Analysis	Total variable cost (₦)	Total revenue (₦)	Gross margin (₦)	Average gross margin (₦)
Without Loss	1,655,029.00	9,455,950.00	7,800,921.00	72,905.80
With Loss	1,655,029.00	2,000,580.00	345,551.00	3,229.45

Source: Field survey, 2012.

obtaining loans. Other post-harvest constraints reported by the farmers include insufficient working capital and low government support.

Table 3 present the results of the gross margin analysis for evaluating the profitability of tomato production in Ogbomosho. The unit selling price was used to value the post-harvest losses incurred. The Gross margin after post-harvest loss of ₦345, 551.00 was much lower than the gross margin without loss of ₦7, 800,921.80. This implied a 95.5% post-harvest loss incurred by the farmers and showed the great extent to which post-harvest losses reduced the income of the farmers in Ogbomosho and consequently their welfare. This is in agreement with the findings of Babalola et al. (2010) and Ayandiji et al. (2011). The percentage gross profit margin without and with loss stood at about 82.5% and 17.3% respectively. The low percentage gross profit margin is an indication that the farmers retained a low percentage of each naira of sales, with little left over for other expenses and as net profit. This in-turn has negative welfare implications for the farmers.

One of the essential determinants of the welfare of the household, which requires that the quantity of tomatoes sold to tomatoes harvested be high, is the Market Participation Rate (MPR). As presented in Table 4, the positive relationship between MPR and per-capita income implies that farmers with high market participation rate fare better than those with lower participation rate. Similarly, the positive association between farm size and per-capita income implies that the larger the size of the farmland, the higher the output, per capita income and consequently, improved welfare of the farmers. This

corroborates the findings of Mujib et al. (2007) that the use of more land to cultivate tomato will result into more income and hence, increased household welfare. The amount of time spent daily on the farm also had a positive effect on per-capita income. Households in which their household heads were mainly engaged in tomato production had a higher level of welfare compared with households in which tomato production was not their primary occupation.

On the other hand, the value of post-harvest losses had a negative impact on the per- capita income and consequently welfare of the farmers'. The large expanse of land on which tomato was cultivated in this area which although increased productivity, also increased the chances of losses due to poor handling and packaging techniques as well as lack of proper processing and storage facilities. In addition, the negative relationship between household size and the per-capita income of the household implies that the smaller the household size, the higher the per-capita income of the farmers and ultimately a higher level of welfare.

CONCLUSION AND RECOMMENDATIONS

It is discouraging and counter-productive for farmers after channeling so much of their limited resources to production, to lose the harvested produce before it gets to the market or consumers due to factors beyond their control. This connotes a waste of productive resources as well as a significant reduction in expected income and consequently welfare of the farmers. The problem of

Table 4. Ordinary least squares regression result of the effects of post-harvest losses on the welfare of tomato farmers.

Variable	Coefficient	t-value
Market participation rate	62.185	1.73***
Marital Status	7.543	1.19
Formal Education	-5.294	-1.00
Gender	14.021	2.06**
Age	3.0123	0.64
Size of Farmland	1.634	2.39**
Experience	- 0.690	-0.87
Time spent on farm	3.101	-2.16**
Primary Occupation	25.162	4.19*
Household size	-5.618	-7.31*
Value of Post-Harvest loss	-0.002	-3.45*
Constant	164.270	4.78*

Source: Regression result, 2012. Level of significance at 1% is *, 5% is ** and 10% is ***, Number of observations = 107, $F_{(11,95)} = 11.15$, Prob> F = 0.000, R^2(Adjusted) = 0.5129.

post-harvest losses, which has long not been recognized as one of the major factors responsible for food insecurity in Nigeria, should be of utmost priority in any effort at achieving food self-sufficiency. The constraints encountered by the farmers also need to be effectively addressed. This could be through;

(i) The adequate training of farmers on post-harvest crop handling techniques as well as the provision of good storage facilities that could help prevent crop losses especially at the farm level.
(ii) The improvement of linkage roads and the resuscitation of the moribund Nigerian rail transportation system to help curb losses during transit to the market.
(iii) Investment in postharvest processing technologies and establishment of processing industries.

With respect to the welfare of the farmers, a reduction in post-harvest losses will lead to increased market participation, per-capita income and consequently improved welfare of the farmers. Farmers should also be enlightened on the benefit of small household size especially on its positive effect on per-capita income and household welfare. From the foregoing, it is obvious that the reduction in post-harvest losses is the key to improved welfare of the farmers, increased food availability and ultimately national food security.

Conflict of Interests

The author(s) have not declared any conflict of interests.

REFERENCES

Adeoye IB, Odeleye OMO, Babalola SO, Afolayan SO (2009). Economic analysis of tomato losses in Ibadan metropolis, Oyo State, Nig. Afr. J. Basic Appl. Sci. 1(5-6):87-92.

Adesina A (2012). Agricultural Transformation Agenda, ATA: Growing agriculture through value chain, November, 2012.
Ayandiji A (2011). Effects of post harvest losses on income generated in citrus spp in Ife adp zone of Osun state, Nigeria. Afr. J. Food Sci. Technol. 2(3):052-058.
Ayandiji A, Adeniyi OR, Omidiji, D (2011). Determinant post harvest losses among tomato farmers in Imeko-Afon Local Government Area of Ogun State, Nigeria. Global. J. Sci. Front. Res. 11(5):23-27.
Babalola DA, Makinde YO, Omonona BT, Oyekanmi, MO (2010). Determinants of post harvest losses in tomato production: A case study of Imeko – Afon local government area of Ogun state. J. Life. Phys. Sci. Acta SATECH 3(2):14-18.
Basappa G, Deshmanya JB, Patil BL (2007). Post- Harvest Losses of Maize Crop in Karnataka - An Economic Analysis. Karnataka. J. Agric. Sci. 20(1):69-71.
FAO (2005). The State of Food and Agriculture. Food and Agriculture Organization of the United Nations, Rome, 2005. FAO agriculture series.
FAO (2010). Global hunger declining but still unacceptably high: International Hunger Targets Difficult to Reach. Economic and Social Development Department September 2010. Rome: FAO, Rome.
FAO (2011) .Global food losses and food waste. Food and Agriculture Organization of the United Nations, Rome, 2011
Kader AA (2005). Increasing food availability by reducing postharvest losses of fresh produce." Proceedings of the 5th International Postharvest Symposium. Eds. F. Mencarelli and P. Tonutti Acta Hort. P. 682.USA.
Meena MS, Ashwani Kumar, Singh KM, Meena HR (2009). Farmers' Attitude Towards Post-Harvest Issues of Horticultural Crops. Indian Res. J. Ext. Edu. 9(3), September, 2009.
Muhammad RH, Hionu GC, Olayemi FF (2012). Assessment of the post harvest knowledge of fruits and vegetable farmers in Garun Mallam L.G.A of Kano, Nigeria. Int. J. Develop. Sustain. 1(2):510-515.
Mujib ur R, Naushad K, Inayatullah J (2007). Post-harvest losses in tomato crop: A Case of Peshawar Valley. Sarhad J. Agric. 23(4):1279-1284.
Seid H, Hassen B, Yitbarek WH (2013). Postharvest Loss Assessment of Commercial Horticultural Crops in South Wollo, Ethiopia" Challenges and Opportunities". Food Sci. Quality Manage. 17:34-39.
Seisuke K, Neelima S (2008). Tomato (Solanum lycopersicum): A model fruit bearing crop.
Shankara N, Joep Van Lidt de Jeude, Marja de Goffau, Martin H, Barbara Van Dam (2005). Cultivation of tomato production, processing and marketing. Agrodok 17, Agromisa Foundation and CTA, Wageningen.

25

Effect of low temperature storage on conservation varieties of Chrysanthemum cutting

Marcos Ribeiro da Silva Vieira[1] Adriano do Nascimento Simões[1], Glauber Henrique Sousa Nunes[2] and Pahlevi Augusto de Souza[3]

[1]Universidade Federal Rural de Pernambuco, Unidade Acadêmica de Serra Talhada, CEP: 59909-460, Serra Talhada, PE, Brasil.
[2]Departamento de Ciências Vegetais/UFERSA, Caixa Postal 137, CEP: 59625-900, Mossoró, RN, Brasil.
[3]Instituto Federal de Educação, Ciência e Tecnologia do Ceará, CEP: 62930-000, Limoeiro do Norte, CE, Brasil.

The objective of this research was to evaluate postharvest quality of 'Lona' and 'Garfield' varieties chrysanthemums, stored at different temperatures. The experiment was carried out in a plastic greenhouse at Pouso Alegre, Minas Gerais State, Brazil (22° 13'48" S, 45° 56'11" W and 832 m in height). The inflorescences were kept at 1.5, 2.5 and 5.0°C. The evaluated parameters were senescent flowers and necrosed ligules. The evaluations were performed in the open storage room at 4, 8, and 12 days, at room temperature. It was observed that chrysanthemum 'Lona' flower senescence was accelerated at 2.5 and 5.0°C; while for 'Garfield', the senescence was larger at 1.5°C. For 'Lona' and 'Garfield' chrysanthemums, the temperature of 1.5°C favored the development of necrosis.

Key words: Dendranthema grandiflora Tzvelev, varieties, pompom, conservation.

INTRODUCTION

The cultivation of flowers and ornamental plants in Brazil is an important activity because it generates employment opportunities and improves income levels in several states. The main species include: rose, kalanchoe, violet, begonia, gerbera, ficus, fern and chrysanthemum (Mitsueda et al., 2011). Chrysanthemum is ranked as one of the cut flowers that feature a variety of colors and inflorescences. However, the lack of specific care during harvesting, transport and storage causes a lot of damage which impairs the quality of flowers and increased post-harvest losses. The use of low temperature during storage is important for conservation of the flowers, because in addition to inhibiting bacterial and fungal infections, it reduces degradation of certain enzymes and ethylene production, decreases perspiration, respiration, and delays related to the different processes of growth and senescence (Nowak and Rudnicki, 1990; ASHRAE,

1994). The temperature in the preservation of chrysanthemum varies with the variety and the shelf. Nowak (1991) recommend 1°C as the best temperature for storage, however, Sacalis (1993) and Vieira and Souza (2009) recommend a wider temperature of 0-5°C. Some authors have used temperatures recommended above, for several species (Ichimura et al., 1989; Hastenreiter et al., 2006; Vieira and Lima, 2009; Vieira et al., 2010).

The aim of this study is to evaluate the effect of different storage temperatures on the postharvest quality of cut chrysanthemums.

MATERIALS AND METHODS

The experiment was conducted in a greenhouse in Pouso Alegre /

MG (22° 13'48" S, 45° 56'11" W) and 832 m in height. Two varieties of cut chrysanthemum (*Dendranthema grandiflora* Tzvelev), 'Lona' (inflorescence type pompom, globular, formed by small ligules with purple coloring and reaction time of seven and a half weeks) and 'Garfield' (with identical phenotypic characteristics, but with ligules orange staining) were used. The experimental design was completely randomized with six replications and three stems experimental. The flowers were harvested when they had nearly 50% of ligules expanded, which corresponds to the commercial harvest. Thereupon, the stems were standardized to a length of 75 and 15 cm defoliation of the base of the stem.

The following were placed in plastic containers containing 1 L of water and stored at temperatures of 1.5, 2.5 and 5.0°C and relative humidity (RH) of 90%. After 7 days of storage, the inflorescences were transferred into plastic containers containing 300 ml of water not distilled (renewed every 48 h). The evaluations were carried out in the chambers and after 4, 8 and 12 days of exposure at room temperature with an average of 25.2 °C, where they were assessed with the following parameters:

Senescent flowers: those who had more than 50% of the disk flowers with anthers mature and attenuation of dark purple to light purple to chrysanthemum 'Lona' and attenuation of dark orange to light orange to chrysanthemum ' Garfield '.
Ligules darkened: were considered those that had blackened necrotic spots on the edges or in the center. The calculations were subjected to analysis of variance, and means were compared by Duncan test at 5% probability of error.

RESULTS AND DISCUSSION

In evaluating the data senescent flowers of chrysanthemum 'Lona', it was observed that during storage there was no difference between the temperatures (Table 1). However, for evaluation at 4, 8 and 12 days flowers that remained in temperature of 1.5°C, this process developed more slowly compared with those stored at 2.5 and 5.0°C. The stems to come out of cold storage had a rate of 5 to 7% of senescing flowers, but to those stored for 12 days at room temperature, this index remained below 50% at a temperature of 1.5°C; while those stored at 2.5 and 5.0°C rose to 49 and 56% respectively.

These results are explained by the retardation of physiological processes (Taiz and Zeiger, 2004), as reported by Brackmann et al. (2000) in chrysanthemum 'Red refocus', who noted that the percentage of senescent flowers was lower in stems stored at low temperature.

This fact was also investigated by Vieira and Souza (2009) in chrysanthemum Yoko Ono, which reported that storage above 1.5°C had accelerated senescence process. Vieira and Lima (2009) studied the postharvest chrysanthemum Faroe, and observed an increase in the percentage of senescent flowers during storage at 10°C. According to Ferguson et al. (1990), elevated temperatures may directly or indirectly injure plant protein by inactivation of enzymes, changes in the conformation of peptides or disruption of complexes in the membrane. Chrysanthemum 'Garfield "(Table 2) showed no difference between the temperatures during storage.

Metabolic activity observed in flowers during the period, demonstrated that the sensitivity grows at low temperatures, which requires the use of temperatures less than 5.0°C during storage. These results are comparable with the data reported by Vieira and Souza (2009), who observed greater symptoms of senescence in chrysanthemum Statesman stored at 1.5°C. However, these results are not in accordance with other studies by these authors, which reported a higher percentage of senescent flowers above 1.5°C in chrysanthemum Yoko Ono.

According to Nowak and Rudnicki (1990), the post-harvest treatment is related to the genetic, physiological and anatomical differences in species and varieties, confirming the results observed in this study. When assessing the darkening of ligule (Tables 3 and 4), results showed that it was higher in temperature of 1.5°C for both cultivars of chrysanthemum cutting. The flowers removed from cold storage had on average, 2 to 4% of ligules with darkened spots for chrysanthemum 'Lona' and 'Garfield' respectively, a value that has evolved to 8 and 10% in the first 4 days at room temperature and 18 and 23% at last review. Similar results were observed by Brackmann et al. (2000) for chrysanthemum during storage of 'Red refocus', where the percentage of darkened ligules were observed at -0.5°C compared with the temperature of 2.5°C.

In evaluating the ligules of chrysanthemum Yoko Ono and Statesman, Vieira and Souza (2009) observed the temperature of 1.5°C favored the development of browning of ligules. In other species Joyce and Shorter (2000) found the temperature range of security for the storage of flowers Anigozanthos spp., Cvs. H1 and Bush Dawn is between 2 and 5°C; for when kept at 0°C showed chilling injury whose symptoms were wilting and discoloration of the petals.

There was a reduction in the life of the flowers of potted Campanula medium stored at 2°C in that the storage time increased from 1 to 3 weeks (Bosma and Dole, 2002). In *Curcuma alismatifolia* (curcuma, Tulip and Tulip siam) Bunya-Atichart et al. (2004) observed dryness and change in color of the bracts of pink to dark violet.

According to Kays (1991), the sensitivity of a plant or part thereof to chilling (chilling injury) varies depending on the species, cultivar of the plant and the time of exposure to low temperature.

However the mechanisms of tolerance to chilling injury are complex. It may occur along with other biochemical and physiological mechanisms to maintain normal physiological functions under stressful conditions, or it may be promoted by chilling injury (Pennycooke et al., 2005). Overall, these results suggest that low temperature storage can activate more intensely, degradative enzymes cell wall tissue of ligules. According to Buchanan et al. (2000), these enzymes are responsible for the first signs of senescence by altering metabolism. This shows that temperature is the most

Table 1. Percentage of senescent flowers at three different temperatures and times of evaluation of chrysanthemum (*Dedranthema grandiflora Tzvelev*) 'Lona'. Pouso, MG. 2011.

Seasons	Senescent flowers			
	1.5°C		2.5°C	5.0°C
Output Storage	5.13 ^{dA}		5.42 ^{dA}	7.17 ^{dA}
4 days in room °C	11.46 ^{cC}		17.11 ^{cB}	23.25 ^{cA}
8 days in room °C	23.17 ^{bC}		32.48 ^{bB}	38.39 ^{bA}
12 days in room °C	36.58 ^{aC}		53.05 ^{aB}	59.27 ^{aA}
CV%	4.38%	6.52%	6.26%	

[1] Means not followed by the same lowercase letters on the line and letters in the same column differ by Duncan test (α=0.05).

Table 2. Percentage of senescent flowers at three different temperatures and times of evaluation of chrysanthemum (*Dedranthema grandiflora* Tzvelev) 'Garfield'. Pouso Alegre, MG. 2011.

Seasons	Senescent flowers		
	1.5°C	2.5°C	5.0°C
Output Storage	10.47 ^{dA}	11.79 ^{dA}	4.85 ^{dB}
4 days in room °C	25.83 ^{cA}	28.42 ^{cB}	17.64 ^{cC}
8 days in room °C	44.63 b^{A}	37.71 ^{bB}	33.84 ^{bC}
12 days in room °C	64.73 ^{aA}	53.65 ^{aB}	46.19 ^{aC}
CV%	7.84%	6.71%	6.92%

[1] Means not followed by the same lowercase letters on the line and letters in the same column differ by Duncan test (α=0.05).

Table 3. Percentage of ligules darkened at three temperatures and different times of evaluation of chrysanthemum (*Dedranthema grandiflora* Tzvelev) 'Lona' Pouso Alegre, MG. 2011.

Seasons	Ligules darkened		
	1.5°C	2.5°C	5.0°C
Output Storage	2.29 ^{dA}	1.71 ^{cA}	1.24 ^{cA}
4 days in room °C	8.53 ^{cA}	1.59 ^{cB}	1.63 ^{cB}
8 days in room °C	12.58 ^{bA}	2.83 ^{bB}	5.27 ^{bB}
12 days in room °C	18.05 ^{aA}	10.63 a^{B}	9.89 ^{aB}
CV%	5.52%	4.50%	5.83%

[1] Means not followed by the same lowercase letters on the line and letters in the same column differ by Duncan test (α=0.05).

Table 4. Percentage of ligules darkened at three temperatures and different times of evaluation of chrysanthemum (*Dedranthema grandiflora* Tzvelev) 'Garfield'. Pouso Alegre, MG. 2011.

Seasons	Ligules darkened		
	1.5°C	2.5°C	5.0°C
Output Storage	4.48 ^{dA}	2.83 ^{cA}	2.54 ^{cA}
4 days in room °C	10.41 ^{cA}	4.74 ^{cB}	3.25 ^{cB}
8 days in room °C	16.22 ^{bA}	8.06 ^{bB}	8.23 ^{bB}
12 days in room °C	23.64 ^{aA}	14.52 ^{aB}	12.88 ^{aB}
CV%	6.19%	4.53%	6.84%

[1] Means not followed by the same lowercase letters on the line and letters in the same column differ by Duncan test (α=0.05).

important environmental factor in the conservation of vegetables because it directly affects the natural processes of respiration, perspiration and other biochemical and physiological aspects of growth.

Conclusion

Under the conditions of the test, the temperature of 1.5°C slows senescence for Chrysanthemum 'Lona', but decreases the shelf life for chrysanthemum 'Garfield'. At a temperature of 2.5 to 5.0°C there was found to be decrease in the percentage of darkened ligules for both genotypes.

REFERENCES

ASHRAE (1994). Commodity Storage Requirements. Refrigeration Systems and Applications Handbook. Atlanta: American Society of Heating, Refrigerating and Air-Conditioning Engineers (ASHRAE).

Brackmann A, Bellé AR, Vizzoto M, Lunardi R (2000). Chrysanthemuns Dedranthema grandiflora cv. red refocus storage under different temperatures and preservative solutions. Revista Brasileira de Agrociência. 6(1):65-69.

Bosma T, Dole JM (2002). Postharvest handling of cut Campanula medium flowers. HortScience. 37:954-958.

Bunya-Atichart K, Ketsa S, Doorn WG (2004). Postharvest physiology of Curcuma alismatifolia flowers. Postharvest Biology and Technology. 34:219-226.

Buchanan BB, Gruissem W, Jones RL (2000). Biochemistry and molecular biology of plants. 3rd ed. Rockville: American Society of Plant Physiologists. P. 1367.

Ferguson JM, Tekrony DM, Egli DB (1990). Changes during early soybean seed and axes deterioration: II. Lipids. Crop Sci. 30:179-182.

Hastenreiter FA, Vieira JGZ, Faria RT (2006). Post-harvest longevity of Oncidium varicosum (Orchidaceae) flowers. Semina. 27(1): 27-34.

Ichimura K, Kohata K, Koketsu M, Shimamura M, Ito A (1989). Effects of temperature, 8-hydroxyquinoline sulphate and sucrose on the vase life of cut rose flowers. Postharvest Biology and Technology. 15:33–40.

Joyce DC, Shorter AJ (2000). Effects of cold storage on cut Grevillea 'Sylvia' inflorescences. Postharvest Biology and Technology. 18:49-56.

Kays SJ (1991). Postharvest physiology of perishable plant products. New York: Avi Book. P. 532.

Nowak J, Rudnicki RM (1990). Postharvest handling and storage of cut flowers, florist greens and potted plant. Portland: Timber Press P. 210.

Nowak RM (1991). Walker's Mammals of the World. v. I, 5th ed. The Johns Hopkins Press. London. P. 642.

Mitsueda NC, Costa EV, Oliveira PSD (2011). Environmental aspects in flower and ornamental plant agribusiness. Revista em Agronegócios e Meio Ambiente. 4(1):9-20.

Pennycooke JC, Cox S, Stushnoff C (2005). Relationship of cold acclimation, total phenolic content and antioxidant capacity with chilling tolerance in petunia (Petunia x hybrid). Environmental and Experimental Botany. 53:225-232.

Sacalis NJ (1993). Prolonging freshness: postproduction care & handling. In: BALL, V. (Org.). Cut flowers. 2 rd ed. Illinois: Ball: 47-49.

Taiz L, Zeiger E (2004). Fisiologia vegetal. 3. Ed, P. 720.

Vieira MRS, Souza B (2009). Storage of cut chrysanthemums at different cutting temperatures. Pesquisa Agropecuária Tropical 4(39):356-359.

Vieira MRS, Lima GPP (2009). Shelf life of stems chrysanthemum faroe followed cold storage. Magistra 4(21):360-363.

Vieira MRS, Teixeira da Silva JA, Lima GPP, Vianello F (2010). Changes in polyamine, total protein and total carbohydrate content and peroxidase activity during the lifetime of chrysanthemum 'Faroe'. Floriculture and Ornamental Biotechnology. 4:48-52.

26

Modified atmosphere packaging and active packaging of banana (*Musa* spp.): A review on control of ripening and extension of shelf life

Sen C.*, Mishra H. N. and Srivastav P. P.

Agricultural and Food Engineering Department, Indian Institute of Technology Kharagpur, Kharagpur-721302, West Bengal, India.

Banana is one of the most important fruit crops in the world and is considered by millions of people as their main energy source. Post harvest shelf life extension has been a problem for years due to its climacteric respiration pattern and sensitivity to low temperature. Extensive research has been done in this area for many decades. Among various existing technologies, Modified Atmosphere Packaging (MAP) which extends shelf life of fresh fruits and vegetables by reducing their respiration rate is gaining popularity. Active packaging is a kind of MAP where different scrubbing/releasing materials are used. This review summarises important aspects behind the ripening physiology of banana, controlling factors and recent advance technologies of MAP and active packaging.

Key words: Banana, ripening, biochemical changes, modified atmosphere packaging, active packaging.

INTRODUCTION

Banana

Banana is the fifth most important commodity in world trade after cereals, sugar, coffee and cocoa (Uma, 2008). Banana is popular because of its easy availability, low cost, various usage and high nutritive content. Total banana and plantain production all over the world is calculated as 95.8 million metric tons (FAOSTAT, 2009). India ranks first in banana production (26.4 million MT, 28% of total production) followed by Philippines, China, Ecuador and Brazil (FAOSTAT, 2009). Banana is monocotyledon and belongs to family Musaceae. Two basic types of genome groups are there, *Musa acuminata*

(A genome, 2n = 22) and *Musa balbisiana* (B genome, 2n = 22). Polyploidy and hybridization of A and B genomes have given rise to diploid (AA, AB, BB), triploid (AAA, AAB, ABB, BBB), and tetraploid (AAAA, AAAB, ABBB, AABB) banana (Mahapatra et al., 2010). Various other varieties also exist naturally or developed by hybridization of these genomes which have different nomenclatures (Robinson, 1996). Three common species of Musa (*M. cavandishii*, *M. paradisiaca* and *M. sapientum*) are widely grown in the world. Commercially popular banana is usually triploids (2n = 33). Economically important sweet bananas which are eaten as a dessert fruit are from the AAA genome and the plantains which are usually cooked before consumption because of their higher starch percentage are from the AAB genome (Seymour, 1993). Banana is used as a treatment for gastric ulcer and diarrhoea in everyday life. It is also beneficial for preventing cancer and heart disease because of high vitamin A and B6 contents. High potassium content reduces cardiovascular disease and thus controls blood pressure.

Average nutritional composition of raw banana (AAA

---

*Corresponding author. E-mail: chandanisn@gmail.com.

Abbreviations: ACC, 1-aminocyclopropane-1-carboxylic acid; CI, chilling injury; MA, modified atmosphere; MAP, modified atmosphere packaging; PAL, phenylalanine ammonia lyase; PPO, polyphenol oxidase; SAM, S-adenosyl methionine.

Table 1. Nutritional composition of raw banana.

Nutrient and unit	Value	Nutrient and unit	Value
Proximate (%)		Riboflavin (mg)	0.073
Water	74.91	Niacin (mg)	0.665
Energy (kcal)	89	Pantothenic acid (mg)	0.334
Protein (N x 6.25)	1.09	Vitamin B-6 (mg)	0.367
Total lipid (fat)	0.33	Choline, total (mg)	9.8
Ash	0.82	Betaine (mg)	0.1
Carbohydrate, by difference	22.84	Vitamin A, RAE (mcg_RAE)	3
Fiber, total	2.6	Phytosterols (mg)	16
Sugars, total	12.23	Carotene, beta (mcg)	26
Sucrose	2.39	Vitamin A, IU (IU)	64
Glucose (dextrose)	4.98	Vitamin E (alpha- tocopherol) (mg)	0.10
Fructose	4.85		
Starch	5.38		
		Lipid (%)	
Mineral (mg/100 g)		Fatty acids, total saturated	0.112
Calcium, Ca	5	Fatty acids, total monosaturated	0.032
Iron, Fe	0.26	Fatty acids, total polyunsaturated	0.073
Magnesium, Mg	27		
Phosphorus, P	22		
Potassium, K	358	**Amino acid (%)**	
Sodium, Na	1	Leucine	0.068
Zinc, Zn	0.15	Lysine	0.050
Copper, Cu	0.078	Phenylalanine	0.049
Manganese, Mn	0.270	Valine	0.047
		Arginine	0.049
		Histidine	0.077
Vitamin		Alanine	0.040
Vitamin C, total ascorbic acid (mg)	8.7	Aspertic acid	0.124
Thiamin (mg)	0.031	Glutamic acid	0.152

Source: USDA National Nutrition Database (2010).

genome) is given in Table 1. In spite of being such an important fruit in the world economy, due to its highly perishable nature, local consumption of banana is more popular than export market (Pillay and Tripathi, 2007).

RIPENING PROCESS

Different biochemical and physiological changes take place within a short span of time during ripening, so it is necessary to understand the ripening process of banana in order to extend its shelf life.

Climacteric nature of respiration

Respiration is the oxidative breakdown of complex substrates like carbohydrates to simpler molecules like CO_2 and H_2O. The glycolytic pathway which occurs in the cytoplasm of the cell is the first step of respiration followed by tricarboxylic acid cycle, pentose phosphate pathway and electron transport system. Fruits can be classified into two broad categories - climacteric and non climacteric according to their respiration pattern (Figure 1). In climacteric fruit, respiration rate shows a decreasing trend to the lowest value termed as pre-climacteric minimum followed by a sharp rise in respiration rate to the climacteric peak. This sudden rise is called as respiratory climacteric followed by a decrease in respiration rate in the senescence period. Banana being a climacteric fruit shows a sudden increase of respiration rate and consequently a burst of ethylene production at the onset of climacteric peak. Enzymes play a key role in the climacteric rise in respiration rate during gycolytic pathway (Ball et al., 1991; Seymour, 1993). Role of various enzymes and possible biochemical changes during respiration and carbohydrate metabolism is given in Figure 2.

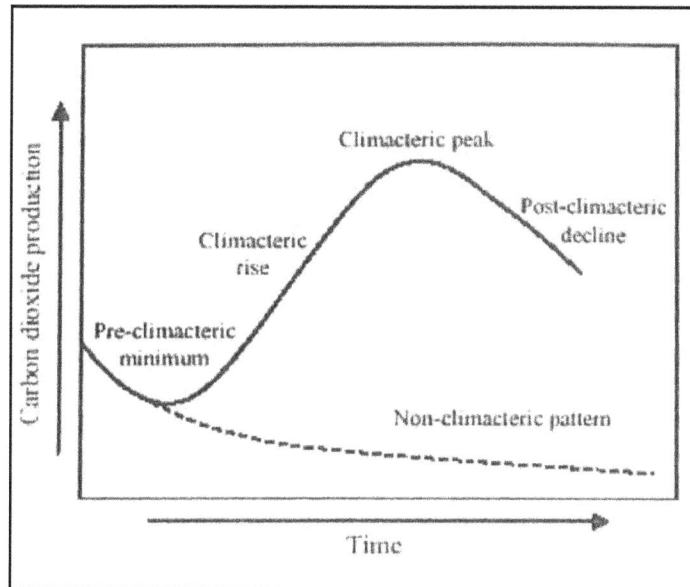

Figure 1. Climacteric and non climacteric ripening (Salveit, 2004).

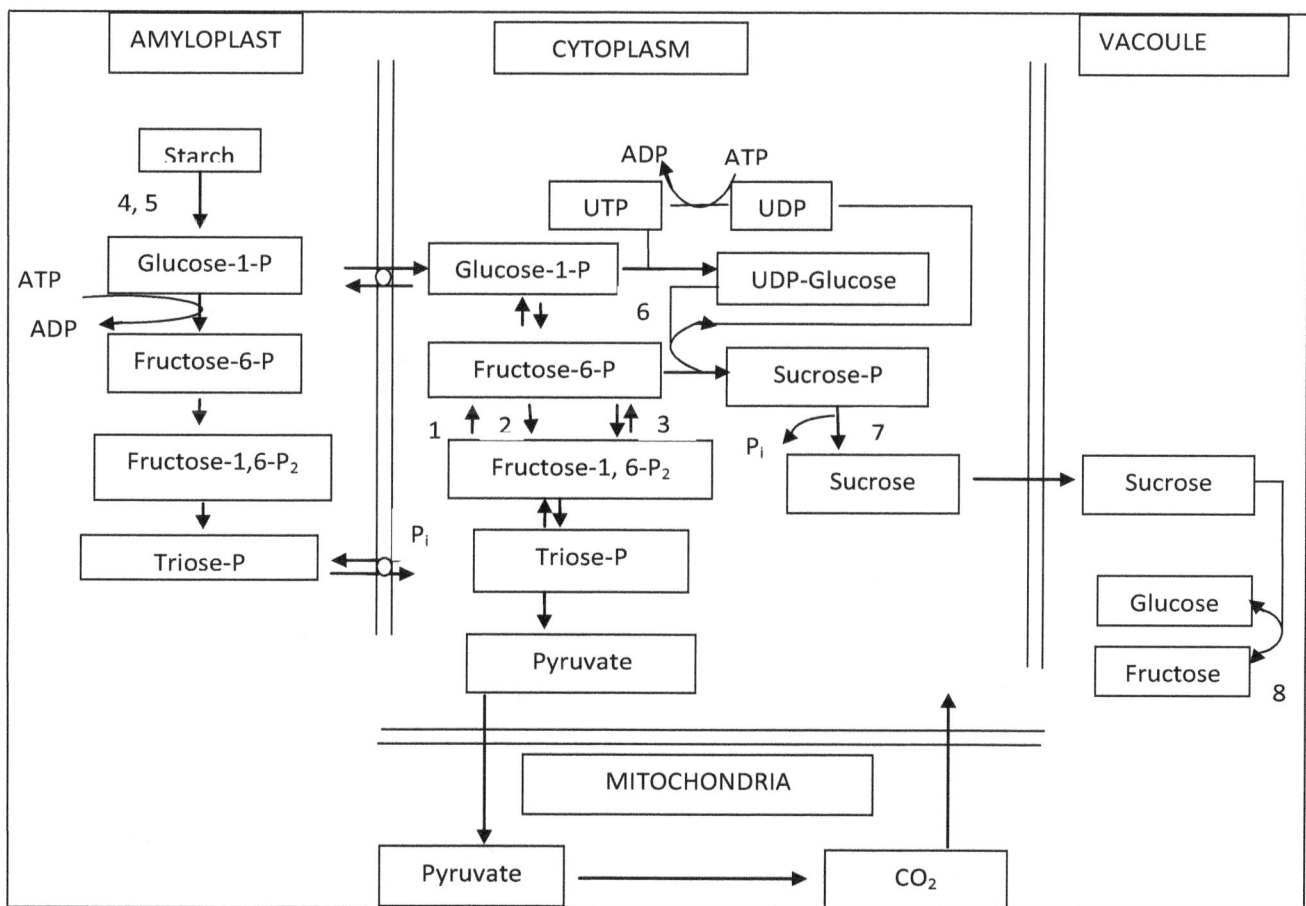

Figure 2. Outline of biochemical pathways and role of enzymes likely to be involved in respiration and carbohydrate metabolism in ripening bananas (Beaudry et al., 1987, 1989; Hubbard et al., 1990; Seymour, 1993). 1) Fructose 1, 6-bisphosphatase, 2) ATP phosphofructokinase, 3) Pyrophosphate-dependent phosphofructokinase, 4) Phosphorylase, 5) α − amylase, β − amylase, α 1,6 − glucosidase, 6) Sucrose phosphate synthase, 7) Sucrose phosphate phosphatise and 8) Invertase.

Table 2. Physical and chemical nature of ethylene.

Property		References
UV light absorbance	161, 166 and 175 nm	
Molecular weight	28.05 g/mol	
Freezing point	-181°C	
Boiling point	-103.7°C	Yañez et al. (2004)
Fusing point	-169°C	
Relative density in air	0.978	
Specific volume	861.5 ml/g at 21°C	
Compounds formed after degradation	Acetylene, butane, ethane, hydrogen and methane	Abeles et al. (1992)
Peak production of ethylene during climacteric phase	3 $\mu l.kg^{-1}.h^{-1}$	McMurchie et al. (1972)

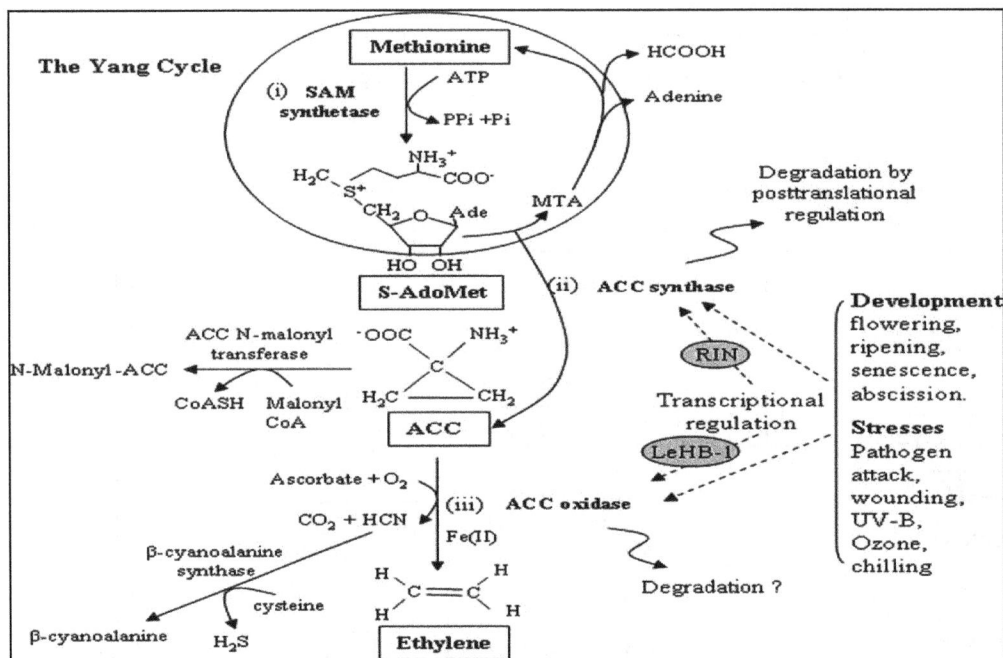

Figure 3. Ethylene biosynthesis pathway (Lin et al., 2009).

The average respiration rates of banana before, during and after climacteric peak are reported as 20 to 70, 120 to 125 and 100 to 125 ml CO_2 kg^{-1} h^{-1} respectively (Palmer, 1971; Yañez et al., 2004).

Effect of ethylene

Climacteric fruits ripen in the presence of ethylene which is a colourless, odourless, tasteless gas that has many effects in plant physiology and is active in such a small amount (part per million)that it is considered as a plant hormone (Theologis, 1993; Brandenburg and Zagory, 2009). Ethylene is transported by diffusion from its production site to active site. Ethylene diffusion gradients are determined by the surface to volume ratio, diffusion resistance and ethylene production rates (Abeles et al., 1992). Physical and chemical nature of ethylene are summarised in Table 2. The pathway of ethylene biosynthesis involves the steps linking S-adenosyl methionine (SAM), its conversion to 1-aminocyclopropane-1-carboxylic acid (ACC) and the subsequent production of ethylene from ACC (Seymour, 1993). Two most important enzymes which are responsible for ethylene production are ACC synthase (EC 4.4.1.14) which converts SAM to ACC and ACC oxidase (EC 1.4.3) which is known as Ethylene Forming Enzyme (EFE), it converts ACC to ethylene (Figure 3).

A thorough understanding of the biochemical basis of ripening is essential to ensure predictable ripening and good quality ripe fruit. Different biochemical changes that take place during ripening are described in Table 3.

MODIFIED ATMOSPHERE PACKAGING (MAP)

Immediately after harvest, the sensorial, nutritional and organoleptic quality of fresh produce start to decline as a result of altered plant metabolism. This quality deterioration is the result of produce transpiration, senescence, ripening-associated processes and the development of postharvest disorders (Gorris and Peppelenbos, 2007). Low temperature and proper hygienic handling of the material are the prime factors that can control these processes. MAP is a preservation technique that further minimizes the physiological and microbial decay of perishable produce. Modified atmosphere (MA) refers to any atmosphere different from the normal air   (20 to 21% O_2, about 0.03% CO_2, about 78 to 79% Nitrogen and trace quantities of other gases) (Yahia, 2009). The modification of the internal package atmosphere may take place at the level of total pressure and/or partial pressures of gas components. MA can be created either by direct gas flushing (active MAP) or by respiration of the enclosed product (passive MAP). In both cases, the permeability of the packaging material is important to maintain atmospheres within desired limits. For respiring products, continuous depletion of O_2 and/or increase of CO_2 and water vapour create MA within the package that is known as passive MAP (Yahia, 2009). In active MAP, gas mixture of desired composition is introduced within the package either after evacuation or by a continuous flow of gas mixture to replace the air (Lee et al., 2010).

Several researchers recommended optimal MA conditions for fresh banana as 13 to 15°C, 90 to 95% relative humidity with 2 to 5% CO_2 and O_2 concentration (Ahmad et al., 2001; Yahia and Singh, 2009; Lee et al., 2010).

Respiration rate modelling in MAP

In MAP, the O_2 partial pressure is typically reduced while that of CO_2 is often increased. The purpose of this is to lower the respiration rate and slow down the metabolic pathways that negatively affect the quality of the stored product. The respiration rate is a good indicator of the physiological stage of the fruit and its storage potential, so, many efforts have been put to model respiration or gas exchange of fruit and vegetables (Nicolaï et al., 2009). Temperature is generally recognized as the most important external factor influencing respiration. Biological reactions generally increase two - or threefold for every 10°C rise in temperature. The respiration rate is usually maximal at moderate temperatures between 20

and 30°C, but decreases considerably to almost zero around 0°C, depending on the genus, species and even cultivar. At high temperatures over 30°C, enzymatic denaturation may occur resulting in reduced respiration rates. The respiration rate is affected by the development stage and the respiration pattern (climacteric/non climacteric) of the fruit as well (Nicolaï et al., 2009). The more recent approach for modelling respiration rate by employing Michaelis–Menten type equation is based on enzyme kinetics (Lee et al., 1991). McLaughlin and O'Beirne (1999) described the respiration rate in three ways as no inhibition, competitive inhibition and uncompetitive inhibition.

Heydari et al. (2010) developed a Microsoft Visual Basic computer package (MM-Calculator) using object linking and embedding (OLE) technique to fit the respiration rate of banana. This is reported to be a comprehensive and graphic program which computes parameters of three types of Michaelis-Menten equation of respiration rate. The related mathematical equations are presented in Equations 1, 2 and 3 respectively:

$$RO_2 = \frac{(VmO_2.[O_2])}{(KmO_2 + [O_2])} \qquad (1)$$

$$RO_2 = \frac{(VmO_2.[O_2])}{KmO_2.(1 + [CO_2]/KmCCO_2) + O_2)} \qquad (2)$$

$$RO_2 = \frac{(VmO_2.[O_2])}{KmO_2 + [O_2].(1 + [CO_2]/KmuCO_2)} \qquad (3)$$

Where RO_2 is respiration rate [ml (O_2) $kg^{-1}h^{-1}$], (O_2) and (CO_2) are O_2 and CO_2 concentrations (%), Vm_{O2} is the maximum O_2 consumption rate, Km_{O2} the Michaelis constant for O_2 consumption (%), Kmc_{CO2} the Michaelis constant for the competitive inhibition of O_2 consumption by CO_2 (%) and Kmu_{CO2} the Michaelis constant for the uncompetitive inhibition of O_2 consumption by CO_2 (%).

Among the three types of enzymatic kinetics, banana respiration supports the uncompetitive model of inhibition (Bhande et al., 2008; Heydari et al., 2010). However, sufficient experimental verification in this area is scanty.

ACTIVE PACKAGING

MAP is one of the most accepted methods for extending the shelf-life of perishable and semi-perishable food products by altering the relative proportions of atmospheric gases that surround the produce. However, high capital cost of gas packaging machinery has limited the use of this technology. This gives rise to the concept of active packaging. Besides being independent of

Table 3. Compositional changes during ripening.

Composition	Summary of biochemical changes during ripening	References
Moisture	Moisture content in pulp increases because of respiratory breakdown of starch to sugar and migration of water from peel to pulp.	Mariott et al. (1981); Mahapatra et al. (2010).
Carbohydrate	Starch decreases (20-30 to 1- 2%) Sugar increases (1-2 to 15-20%). The main starch degrading enzymes are phosphorylases (EC 2.4.1.20), α-amylase (EC 3.2.1.1), β-amylase (EC 3.2.1.2) and α-1, 6-glucosidase (EC 3.2.1.11) (Figure 3).	Seymour et al. (1993); Mahapatra et al. (2010).
Organic acids	Organic acid content increases. Oxalic acid which is responsible for the astringent taste of unripe banana, undergoes significant decarboxylation by the action of oxalate oxidase. During the climacteric, malic acid becomes the major acid (65% of total acidity). pH decreases from 5.4 ±0.4 to 4.3±0.3.	Seymour et al. (1993); Yanez et al. (2004).
Proteins	Increases from 1-2.5% to 3.8-4. 2%. Glutamate oxaloacetate transaminase (EC 2.6.1.1) and glutamate pyruvate decarboxylase (EC 2.6.1.2), involved in aspartate and alanine synthesis respectively are found to be maximum at the initiation of climacteric.	Seymour et al. (1993); Mahapatra et al. (2010)
Lipids	No substantial changes. In the phospholipid fraction, the proportion of linolenic acid increases and proportion of linoleic acid decreases. Increases in the unsaturation of the phospholipid fraction results in increased fluidity in cellular membranes during ripening.	Colinas (1992); Yanez et al. (2004).
Pectins	Insoluble protopectin is converted into soluble pectin. Main pectin degrading enzymes are: pectin methyl esterase (EC 3.1.1.11) poly galacturonase (EC 3.2.1.15), cellulase (EC 3.2.1.4) and hemicellulase.	Abdullah and Pantastico (1990); Kotecha and Desai (1995); Yanez et al. (2004).
Phenolic compounds	Main phenolics compounds are: 3, 4-di hydroxyphenylethyla- mine and 3, 4-dihydroxyphenylalanine. Main browning causing enzymes are polyphenol oxidase (EC 1.10.3.1), and phenyl alanine ammonialyase. Tannins of banana peel cause astringent fruit taste which gradually disappears due to tannin depolymerisation as fruit ripens.	Seymour et al. (1993); Nguyen et al. (2003).
Pigments	Chlorophyll decreases due to increased activity of chlorophyllase (EC 3.1.1.14) at the onset of the climacteric peak. The principal agents responsible for this degradation are pH changes and oxidative systems. After chlorophyll degradation, carotenoid pigments (mainly Xanthophylls and carotene) become visible.	Yanez et al. (2004); Mahapatra et al. (2010).
Volatiles	Volatile production increases until the onset of peel browning. Among almost 200 volatile components in ripe banana, Acetates and butyrates are predominating. 'Banana like' flavour is due to the presence of amyl esters, and distinctive 'fruity' like flavour is due to butyl esters.	Seymour et al. (1993); Yanez et al. (2004).

package permeability and gas packaging equipment, this packaging system is more rapid and accurate than traditional MAP. The concept of active packaging is one of the emerging technologies in food packaging. It has been defined as a system in which the product, the package and the environment interact in a positive way to extend shelf life or to achieve some characteristics that cannot be obtained otherwise (Miltz et al., 1995). The active system can be an integral part of the package or be a separate component placed inside the package (Yahia, 2009). Different substances that can either absorb or release a specific gas, control the internal atmosphere within the package. Active packaging components can work as either absorbing or releasing system.

For banana, or any fresh fruits, absorbents are generally used as active packaging components to remove undesired gases and substances (O_2, CO_2, moisture, ethylene and taints) in order to extend the shelf life. Commonly used active packaging systems for fresh fruits/vegetables are described in Table 4.

EFFECTS OF MAP AND ACTIVE PACKAGING ON QUALITY OF BANANA

It is highly challenging to keep banana in good and fresh condition for long duration. Over the years, several researchers have worked on this problem and both MAP and active packaging have been proved to be beneficial.

Effect on chilling injury (CI)

Storage of mature green 'Cavendish' bananas in low-density polyethylene (0.05 mm thickness) bags for up to 30 days at 8, 11, and 14°C developed an in-package atmosphere of 3 to 11 kPa O_2 and 3 to 5 kPa CO_2 (Hewage et al., 1995). However, these authors reported that these storage conditions did not affect ripening and sensory quality, nor did they alleviate CI symptoms developed at 8 and 11°C. In contradiction to their result, Nguyen et al. (2004) reported that MAP (about 12 kPa O_2 and 4 kPa CO_2) resulted in less visible CI in 'Kluai Khai' bananas at 10°C. Total free phenolics in the peel of MAP bananas decreased slowly and the fruits had low phenylalanine ammonia lyase (PAL) and polyphenol oxidase (PPO) activities. Pulp softness, sweetness and flavour of MAP fruit were good.

Effect on peel and senescent spotting

Banana fruit (Sucrier) packed in polyvinyl chloride film and held at 29 to 30°C prevented the early senescent peel spotting, typical for this cultivar (Choehom et al., 2004). CO_2 and ethylene concentrations within the

packages increased, but the addition of CO_2 scrubbers or ethylene absorbents had no effect on spotting. Experiments with continuous low O_2 concentrations confirmed that the effect of the package was mainly due to low O_2. The positive effect of MAP on peel spotting was accompanied by reducing PAL and increased PPO activity in the peel. Therefore, senescent spotting of banana peel seems to require rather high O_2 levels. Maneenuam et al. (2007) confirmed that high O_2 (90 ± 2 kPa) is associated with the enhanced peel spotting in banana. The in vitro activities of both PAL and PPO measured both in the whole peel and peel spots were lower in high O_2 levels. It has been concluded that peel spotting was not correlated with in vitro PAL and PPO activities, but decrease in the dopamine levels correlated with peel spotting, indicating that it might be used as a substrate for the browning reaction.

Effect of silicone membrane, diffusion channel and N₂O treatments

Stewart et al. (2005) reported that silicone membrane system offers an inexpensive and easy to use alternative to the traditional MAP of bananas. 'Cavendish' bananas were stored for 42 days at 15°C under MA conditions using silicone membrane and diffusion channel systems. Fruit in these atmospheres had fresh appearance, good colour, minimum mould and excellent marketability. According to Poubol and Izumi (2005), N_2O treatments extended the storage life of banana fruit without causing adverse effects on physicochemical quality. The capability of N_2O to slow down fruit ripening is thought to be due to its anti-ethylene activity as suggested by the delay in the climacteric associated rise in ACC oxidase activity.

Effect of ethylene scrubber

Ripening in bananas can be delayed by using an ethylene scrubber. There are several compounds that can be used as inhibitors of ethylene, for example aminoethoxyvinylglycine (AVG), an inhibitor of ethylene synthesis; 1- Methylcyclopropene (1-MCP), an inhibitor of ethylene action and potassium permanganate ($KMnO_4$), an oxidising agent. For banana, 1-MCP and $KMnO_4$ are the most commonly used ethylene scrubbers.

1-Methylcyclopropene (1-MCP)

The application of the anti-ethylene compound 1-MCP in combination with the use of polyethylene bags was reported to extend the postharvest life of banana fruit (Yueming et al., 1999). 1-MCP treatment delayed peel colour change and fruit softening, and extended shelf life

Table 4. Active packaging system for fresh fruits and vegetables.

Active packaging components	Scavengers/absorbers	Working principle	Purpose	References
O_2 absorbers (sachet, labels, films, corks)	Ferro-compound (iron powders), ascorbic acid, metal salt, glucose oxidase, alcohol oxidase.	The most successful oxygen scavengers of sachet form on commercial scale are based on iron. Powdered iron is contained alone or with other catalysts in oxygen permeable film pouch. The basic oxidation reaction for absorbing oxygen is: $4\ Fe + 3O_2 + 6\ H_2O \rightarrow 4\ Fe(OH)_3$ Rough estimation of iron's capacity to absorb oxygen is around 300 ml O_2 per gram of iron.	Reducing respiration rate, mould, yeast and aerobic bacteria growth, prevention oxidation of fats, oil, vitamins, and colours. Prevention damage by worms, insects and insect eggs.	(Mangaraj and Goswami, 2009; Lee et al., 2010).
CO_2 absorbers (sachets)	Calcium hydroxide, Sodium hydroxide, Potassium hydroxide, Calcium oxide and Magnesium oxide, activated charcoal and silica gel	The most versatile commercial CO_2 absorber is Calcium hydroxide, which reacts with carbon dioxide to produce calcium carbonate: $Ca(OH)_2 + CO_2 \rightarrow CaCO_3 + H_2O$ Sodium carbonate can also absorb CO_2 under high humidity condition: $Na_2CO_3 + CO_2 + H_2O \rightarrow 2NaHCO_3$ Active charcoal and zeolite acts as Physical adsorbents of carbon dioxide.	Removing excess CO_2 formed during storage to prevent fruit damage and bursting of package.	(Lee et al., 2001; Shin et al., 2002; Mangaraj and Goswami, 2009).
Ethylene absorbers (sachets/ films)	Aluminum oxide and potassium permanganate (sachets), activated hydrocarbon (squalane, apiezon) + metal catalyst (sachets), Builder- clay powders (films), zeolite films, japanese oya stone (films) and other compound like silicones (phenyl- methyl silicone)	The most popular method is oxidation of ethylene by potassium permanganate ($KMnO_4$) adsorbed on an inert carrier with large surface area such as silica gel, alumina, and activated carbon. $3\ C_2H_4 + 12\ KMnO_4 \rightarrow 12\ MnO_2 + 12\ KOH + 6CO_2$	Prevention fast ripening and softening.	(Vermeiren et al., 2003; Zagory, 1995; Lee et al, 2010; Mangaraj and Goswami, 2009)

Table 1. Cont'd

Humidity absorbers (drip absorbent sheets, films, sachets)	Silica gel (sachets), clays (sachets), sucrose, xylitol, sorbitol, potassium chloride, calcium chloride and sodium chloride.	Desiccants of silica gel, calcium chloride, and calcium oxide are most widely used sachet sealed with the moisture permeable spun bonded plastic film. Silica gel removes moisture by physical adsorption mechanism which can be reversible by temperature change. Calcium oxide reacts to remove water irreversibly as: $$CaO + H_2O \rightarrow Ca(OH)_2$$ Sachets of hygroscopic sugar and in organic salt works by buffering water activity inside the fresh produce package.	Excess moisture control in packed produce. Water activity reduction on food surface to check moulds, yeast and spoilage bacteria	(Shirazi et al., 1992; Lee et al., 2010; Mangaraj and Goswami, 2009)

in association with suppression of respiration and ethylene evolution. Banana fruit ripening was delayed when exposed to 0.01 to 1.0 µL 1-MCP/L for 24 h. Increasing concentrations of 1-MCP were generally more effective for longer periods of time. The greatest longevity of about 58 days was obtained by packing the fruit in sealed polyethylene bags with 1-MCP. Pelayo et al. (2003) conducted detail experiments with 1-MCP on bananas at intermediate stages of ripeness after 36 to 48 h which has been commercially treated with ethylene. Several conditions for the application of 1-MCP including concentrations (100, 300 and 1000 nl/L); temperatures (14 and 20°C) and durations of exposure (6, 12 and 24 h) were studied and the authors concluded that, under the conditions tested in this study, the efficacy of 1-MCP in delaying ripening of partially ripened bananas was too inconsistent for commercial application.

Potassium permanganate (KMnO₄)

A MAP system to extend the shelf life of 'Kolikuttu' bananas at room temperature (approximately 25°C and 85% relative humidity) was developed

by Chamara et al. (2000). The effect of various MAP conditions on sensory properties of fruit was evaluated and efficacy of $KMnO_4$ as an ethylene absorber was also examined. MAP systems were created using low density polyethylene bags with no ethylene absorber or with wrapped or unwrapped bricks impregnated with permanganate. In-package O_2 and CO_2 levels were determined on 10, 14, 17, and 20 days; physicochemical properties and colour of fruits were monitored. MAP with wrapped ethylene absorber produced the optimal results with minimal changes in firmness, total soluble solids, weight, titratable acidity and pH. Chauhan et al. (2006) studied the effects of active packaging using $KMnO_4$ as ethylene scrubber and reported that this method could extend the shelf life of banana to 36 days at 13 ± 1°C. The researchers used impregnated $KMnO_4$ in an inert matrix consisting of white cement and lime stone powder packed in sachet form using high density polyethylene woven fabric.

The study also showed that the synergistic effects of the developed ethylene and CO_2 scrubber could restrict the accumulation of excessive CO_2 within pouches lowering the cytotoxicity and symptoms of anaerobiosis in the

ripened banana.

Conclusions

Low temperature storage of banana brings undesirable changes to the fruit physiology. MAP lowers, but could not remove the chances of physiological disorders, so it is necessary to find out the process by which banana could be stored at ambient temperatures. Active packaging is a simple and cost effective alternative to traditional MAP. In some countries like USA, Japan and Australia, the active packaging concepts are successfully being applied. In other parts, active packaging could not be popular because of the lack of legislations and fear of consumer restrictions. More research work is needed about the dosage, health hazards and environmental impact of the components used. Further advancements are also needed to enlighten the synergistic effects of MAP and active packaging to delay the ripening and to maintain good quality.

REFERENCES

Abdullah H, Pantastico BE (1990). Banana: Fruit

Development, Postharvest Physiology, Handling and Marketing in ASEAN. ASEAN Food Handling Bureau, pp. 33 - 103.

Abeles FB, Morgan PW, Salveit ME Jr (1992). Ethylene in Plant Biology. Academic Press. New York. 2: 26 - 83.

Ahmad S, Thompson A, Asi AA, Khan M, Chatha GA, Shahid MA (2001). Effect of reduced O_2 and increased CO_2 (Controlled Atmosphere Storage) on the ripening and quality of ethylene treated banana fruit. Int. J. Agric. Biol., 3(4): 486 - 490.

Ball KL, Green JH, ap Rees T (1991). Glycolysis at the climacteric of bananas. Eur. J. Biochem., 197: 265 - 269.

Beaudry RM, Paz N, Black CC, Kays SJ (1987). Banana ripening: Implication of changes in internal ethylene and CO_2 concentration and activity of some glycolytic enzymes. Plant physiol., 85(1): 277 - 282.

Beaudry RM, Paz N, Black CC, Kays SJ (1989). Banana ripening: Implication of changes in glycolytic intermediate concentrations, glycolytic and gluconeogenic carbon flux and fructose 2, 6-biphosphate concentration. Plant physiol., 91: 1436-1444.

Bhande SD, Ravindra MR Goswami TK (2008). Respiration rate of banana fruit under aerobic conditions at different storage temperatures. J. Food Eng., 87: 116 - 123.

Brandenburg JS, Zagory D (2009). Modified and Controlled Atmosphere Packaging Technology and Applications. In: Yahia EM (Ed.). Modified and Controlled Atmospheres for the Storage, Transportation, and Packaging of Horticultural Commodities. CRC Press, Taylor & Francis Group, 1: 73 - 92.

Chamara D, Illeperuma K, Galappatty PT (2000). Effect of modified atmosphere and ethylene absorbents on extension of storage life of Kolikuttu banana at ambient temperature. Fruits, 55: 381 - 388.

Chauhan OP, Raju PS, Dasgupta DK, Bawa AS (2006). Modified atmosphere packaging of banana (cv. pachbale) with ethylene, carbon dioxide and moisture scrubbers and effects on its ripening behaviour. Am. J. Food Tech., 1(2): 179 - 186.

Choehom R, Ketsa S , van Doorn WG (2004). Senescent spotting of banana peel is inhibited by modified atmosphere packaging. Postharvest Biol. Technol., 31(2): 167 - 175.

Colinas LMT (1992). Desórdenes fisiológicos de productos hortícolas. In: Yahia EM and Higuera CI (Eds). Fisiología y tecnología postcosecha de productos hortícolas. Centro de Investigaciónen Alimentación y Desarrollo. Limus. Grupo Noriega Editores. México. pp.65 - 71.

FAOSTAT (2009). -Agriculture. http://faostat.fao.org/site/339/default.aspx. As viewed on 3rd February, 2012.

Gorris LGM, Peppelenbos HW (2007). In: Rahman MS (Ed.). Modified Atmosphere Packaging of Produce. Handbook of Food Preservation. CRC Press, Taylor & Francis Group. LLC., 2: 316 - 329.

Hewage SK, Wainwright H, Wijerathnam SW, Swinburne T (1995). The Modified atmosphere storage of bananas as affected by different temperatures. In: Ait-Oubahou A and El-Otmani M (Eds.). Postharvest Physiology and Technologies for Horticultural Commodities: Recent Advances. Institut Agronomique and Vetérinaire Hassan II, Agadir. Morocco. pp. 172 - 176.

Heydari A, Shayesteh K, Eghbalifam N, Bordbar H (2010). Studies on the respiration rate of banana fruit based on enzyme kinetics. Int. J. Agric. Biol., 12: 145 - 149.

Hubbard NL, Pharr DM, Huber SC (1990). Role of sucrose phosphate synthase in sucrose biosynthesis in ripening bananas and its relationship to the respiratory climacteric. Plant physiol., 94: 201 - 208.

Kotecha PM, Desai BB (1995). Banana. In: Salunkhe DK and Kadam SS (Eds). Handbook of fruit science and technology. Production, composition, storage and processing. Marcel Dekker Inc. New York. pp. 67 - 90.

Lee DS, Hagger PE, Lee J, Yam KL (1991). Model for fresh produce respiration in modified atmosphere based on principles of enzyme kinetics. J. Food Sci., 56(6): 1580–1585

Lee DS, Shin DH, Lee DU, Kim JC, Cheigh HS (2001). The use of physical carbon dioxide absorbents to control pressure build up and volume expansion of kimchi packages. J. Food. Eng., 48: 183 - 188.

Lee DS, Yam KL, Piergiovanni L (2010). Active and Intelligent packaging. In: Lee DS, Yam KL and Piergiovanni L (Eds). Food Packaging Science and Technology. CRC Press, Taylor & Francis Group, 1: 445 - 473.

Lin Z, Zhong S, Grierson D (2009). Recent advances in ethylene research. J. Exp. Bot., 60(12): 3311 - 3336.

Mahapatra D, Mishra S, Sutar N (2010). Banana and its by product utilisation: an overview. J. Sci. Ind. Res., 69: 323 - 329.

Maneenuam T, Ketsa S, van Doorn WG (2007). High oxygen levels promote peel spotting in banana fruit. Postharvest Biol. Technol., 43: 128 - 132.

Mangaraj S, Goswami TK (2009). Modified atmosphere packaging - An ideal food preservation technique. J. Food Sci. Technol., 46(5): 399-410.

Mariott J, Robinson M, Karikari SK (1981). Starch and sugar transformation during ripening of plantains and bananas. J. Sci. Food Agric., 32: 1021 -1026.

McLaughlin CP, O'Beirne D (1999). Respiration rate of a dry coleslaw mix as affected by storage temperature and respiratory gas xconcentrations. J. Food Sci., 64: 116 - 119.

McMurchie EJ, McGlasson WB, Eaks IL (1972). Treatment of fruit with propylene gives information about the biogenesis of ethylene. Nature, 237: 235 - 237.

Miltz J, Passy N, Mannheim CH (1995). Trends and applications of active packaging systems. In: Ackerman P, Jagerstad M and Ohlsson M (Eds). Food and packaging Materials – Chemical Interaction. The Royal Soc. Chemistry, pp. 201 - 210.

Nguyen TBT, Ketsa S, van Doorn WG (2004). Relationship between browning and the activities of polyphenol oxidase and phenylalanine ammonia lyase in banana peel during low temperature storage. Postharvest Biol. Technol., 30: 187 - 193.

Nicolaï BM, Hertog MLATM, Ho QT, Verlinden BE, Verboven P (2009). Gas Exchange Modeling. In Yahia EM (Ed). Modified and Controlled Atmospheres for the Storage, Transportation, and Packaging of Horticultural Commodities. CRC Press, Taylor & Francis Group. 1: 93-110.

Palmer JK (1971). The banana. In Hulme AC (Ed). The biochemistry of fruits and their products. Academic Press, A.R.C. Food Research Institute. Norwich. England. 2: 65 - 105.

Pelayo C, Vilas-Boas EVD, Benichou M, Kader AA (2003). Variability in responses of partially ripe bananas to 1-methylcyclopropene. Postharvest Biol. Technol., 28: 75 - 85.

Pillay M, Tripathi L (2007). Banana. In: Kole C (Ed). Genome Mapping and Molecular Breeding in Plants. Fruits and Nuts. Springer- Verlag Berlin Heidelberg. 4: 281-301.

Poubol J, Izumi H (2005). Physiology and microbiological quality of fresh-cut mango cubes as affected by high-O_2 controlled atmospheres. J. Food Sci., 70(6): M286 - M291.

Robinson J (1996). Bananas and Plantains. CAB-International, University Press. Cambridge. UK.

Salveit ME (2004). Respiratory metabolism. The Commercial Storage of Fruits, Vegetables, and Florist and Nursery Stocks. Agriculture Handbook Number 66.

Seymour GB (1993). Banana. In: Seymour GB, Taylor JE, and Tucker GA (Eds.). Biochemistry of Fruit Ripening. 1: 83-101.

Shin DH, Cheigh HS, Lee DS (2002). The use of Na_2CO_3- based CO_2 absorbent systems to alleviate pressure build up and volume expansion of kimchi packages. J. Food. Eng., 53: 229 - 235.

Shirazi A, Cameron AC (1992). Controlling relative humidity in modified atmosphere packages of tomato fruit. Hortscience. 27: 336 - 339.

Stewart O J, Raghavan GSV, Gariépy Y (2005). MA storage of Cavendish bananas using silicone membrane and diffusion channel systems. Postharvest Biol. Technol. 35(3): 309 - 317.

Theologis A (1993). One rotten apple spoils the whole bushel: the role of ethylene in fruit ripening. Cell, 70(2): 181 - 184.

Uma S (2008). Indigenous Varieties for Export market. International Conference on Banana. October 24-26. NRCB. Tamilnadu. India.

USDA National Nutrient Database (2010). Release23. http://www.ars.usda.gov/SP2UserFiles/Place/12354500/Data/SR24/reports/sr24fg09.pdf as viewed on 19 December, 2011.

Vermeiren L, Heirlings L, Devlieghere F, Devevere J (2003). Oxygen, ethylene and other scavengers. In: Ahvenainen R (Ed). Novel Food Packaging Techniques. Cambridge. UK. Woodhead Publishing. 22 - 49.

Yahia EM (2009). Introduction. In: Yahia EM (Ed). Modified and Controlled Atmospheres for the Storage, Transportation, and Packaging of Horticultural Commodities. CRC Press, Taylor & Francis Group, 1: 1 - 17.

Yahia EM, Singh SP (2009). Tropical Fruits. In Yahia EM (Ed). Modified and Controlled Atmospheres for the Storage, Transportation, and Packaging of Horticultural Commodities. CRC Press, Taylor & Francis Group, 1: 397 - 444.

Yañez L, Armenta M, Mercado E, Yahia EM, Guttierrez P (2004). Integral Handling of Banana. In: Dris R and Jain SM (Eds). Production Practices and Quality Assessment of Food Crops. Quality Handling and Evaluation, 3: 129 - 168.

Yueming J, Joyce DC, Macnish AJ (1999). Extension of the shelf life of banana fruit by1-methylcyclopropene in combination with polyethylene bags. Postharvest Biol. Technol., 16(2): 187 - 193.

Zagory D (1995). Ethylene-removing packaging. In: Rooney ML (Ed). Active Food Packaging. Blackie Academic & Professional. London. pp. 38 - 54.

27

Safe storage guidelines for black gram under different storage conditions

M. Esther Magdalene Sharon, C. V. Kavitha Abirami , K. Alagusundaram and J. Alice Sujeetha

Indian Institute of Crop Processing Technology (IICPT), Thanjavur, Tamil Nadu, India.

India is the leading producer and importer of pulse in the world. Post-harvest loss is very high in India with losses during storage around 5 to 10%. This situation demands the development of storage guidelines for pulses to provide information to farmers on how long storage is possible without deterioration. Black gram is selected for this study as it is an important pulse used in many traditional specialty products in our country. The major storage conditions that affect any grain are temperature and moisture content. Quality parameters of black gram stored at different initial moisture contents (9, 12, 15 and 18% wet basis) at 20, 30 and 40°C were determined. The storage variables (moisture content of the sample, storage temperature and time of storage) had a negative correlation with germination and a positive correlation with fatty acid value (FAV). The maximum storability of 42 weeks with good seed viability and appreciable microbial stability was found in 9% initial moisture content black gram stored at 20°C. The 15 and 18% black gram stored at 30°C was safe up to 10 weeks hence the post harvest treatment like drying to safe moisture content can be recommended. High risk is involved in storing Black gram at higher moisture contents (15 and 18%) at high temperature of 40°C, beyond 2 to 4 weeks because of loss in seed viability and increase in FAV and the early infection with visible and invisible moulds. The safe storage guidelines chart and safe storage time model developed can be used to predict allowable safe storage time of black gram between moisture content and storage temperature ranges of 9 to 18% w.b. and 20 to 40°C, respectively.

Key words: Black gram, moisture content, storage temperature, safe storage guidelines.

INTRODUCTION

Pulses are a gift of nature; they nourish mankind with highly nutritive food and are major sources of protein especially for vegetarians (Savadatti, 2007). India is the largest producer and consumer of pulses in the world contributing around 25 to 28% of the total global production. The production of total pulses in India is presently about 15 million tons covering an area of about 22 to 23 million hectare (GoI, 2013). In India, variety of pulses are grown like chickpea, pigeon pea, lentil, black gram, green gram, lablab bean, moth bean, horse gram, pea, and cowpea (Nene, 2006). Among these black gram is one of the important pulses used in everyday diet of south Indians. Due to its fermentation capacity it is used in the preparation of various foods like idli and dosa (Campbell-Platt, 1994). The crop is of great importance as about 70% of world's black gram production comes from India (Singh and Singh, 1992) and 10 to 12% national share among the total Indian pulse production

(Basu, 2011). India is the largest producer of black gram contributing around 70% of the world's total Black gram production.

Pulses can remain in edible condition for a long time, if properly stored in bags or silos. However, pulses are more difficult to store than cereals and suffer much greater damage from insects and microorganisms resulting in deterioration of quality (Mills, 1994). Most important factors of grain deterioration are the interaction of temperature, humidity and moisture, which are the determining factors in accelerating or delaying the complex degradation reactions (Kreyger, 1972). In general, high temperature and high moisture grain allows a very short time for post harvest operations (Hall, 1980). It becomes essential to determine the allowable time before spoilage for wide range of moisture contents and temperatures. This would help farmers by informing them the number of days before which the grain has to undergo post harvest treatments. This ensures the quality of grain to be maintained throughout the storage period (Nithya et al., 2011). Hence, only good quality grain can be sent for further processing to provide safe and nutritive products for the consumers.

The safe storage guidelines will help farmers to determine the number of days in which the grain has to be dried for a particular temperature moisture combination (Karunakaran, 1999). The time period for which the grain can be stored safely without deterioration or without any significant loss in its quality and quantity is known as the safe storage time (Schroth, 1996). The quality assessment factors of stored grain which could be monitored are the seed germination, fat acidity value, appearance of mould growth, percentage rise in batter volume, and protein changes (Pomeranz, 1992; Mills, 1992). Changes in quality of stored grain have been studied for various grains, oilseeds and pulses (Karunakaran et al., 2001; White et al., 1999). Although safe storage guidelines with respect to moisture content and temperature are available for a pulses like chickpea (Cassells and Caddick, 2000) and red gram (Sravanti et al., 2013), nothing had been reported so far for Black gram. As there are no standard storage guidelines for black gram, an attempt has been made to study the rate of deterioration of black gram by monitoring the quality changes when stored at different moisture and temperature conditions and further developing guidelines for safe storage.

MATERIALS AND METHODS

Sample preparation

The pulse chosen for this study is black gram (ADT-5), procured from local market with initial moisture content of 12.5 ± 0.5% w.b. The black gram was cleaned and sorted using a pulse cleaner cum grader developed at Indian Institute of Crop Processing Technology (IICPT), Thanjavur. Black gram were conditioned to the required moisture contents in the range of 9.0 to 20% w.b. To obtain lower

moisture content, samples were dried in thin layer in the open for natural air drying for about 8 h. The higher moisture content, samples were conditioned by adding a calculated quantity of distilled water and mixed by passing through a screw conveyor. The conditioned grain samples were then stored in sealed polythene bags in a deep freezer at -5 ± 2°C (Singh and Goswami, 1996).

Selection of storage conditions

Temperature and moisture content are the two important factors determining the storage life of grains after harvest. In general, pulses are harvested after the darkening of the pods at in the moisture range of 14 to 18% w.b. Depending on the weather conditions the moisture will be lower in case of a sunny day and higher in case of a rainy day. In general for safe storage pulses are dried to around 11 to 12% moisture before storage. The different levels of moisture contents for the present study, that is, 9 to 18% w.b. were selected on the basis of storage conditions of pulses throughout the world. During the period of harvest and storage, the grains are subjected to a wide range of temperature (20 to 40°C) depending upon the harvesting season. To simulate temperatures that black gram could undergo during harvest and storage in India, temperatures of 20, 30 and 40°C were selected.

Experimental setup

All experiments were performed in three environmental chambers (Industrial, Chennai) maintained at 20, 30 and 40°C with 60 ± 5% relative humidity. The schematic diagram of the experimental setup is shown in Figure 1. For each set of experiments the conditioned black grams were taken in three different mesh bags. Two bags were of 2.5 kg capacity and the third one was of 5 kg capacity. The 5 kg bag was sandwiched between the two 2.5 kg bags which acts as buffer by preventing any loss of moisture from 5 kg sample bag. About 2 L of KOH of known specific gravity was taken in the plastic pails to maintain the required equilibrium relative humidity (Solomon, 1951). The mesh bag were placed in the plastic pail over a support system to hold the grain sample in the mesh bags (Rajarammanna et al., 2010). A lid was loosely placed on the top of each pail. All experiments were performed in environmental chambers maintained at 20, 30 and 40°C with 60% relative humidity. Three replicates were done for each temperature and moisture combination.

Determination of quality assessment parameters of black gram under storage

Sampling was done every week and grain samples were collected from the middle bag, after mixing thoroughly, for quality analysis.

Germination

The germinating ability of the grain is the index used to assess the viability of the stored product (Pomeranz, 1992). The change in seed germination over time was tested every week according to the method of Wallace and Sinha (1962). About 10 g of sub sample was used for the germination test. In a 9 cm diameter Petri dish, 25 seeds were placed on a Whattman no. 3 filter paper and saturated using 5.5 ml of distilled water. The Petri dishes were vertically stacked in a stand; and to prevent desiccation of the filter paper the stacked dishes were covered with a polythene bag. This set up was incubated at 25°C for 4 days. The polythene bag was then removed and incubation was continued for another 3 days. The number of seeds germinated after 7 days of incubation was counted and the

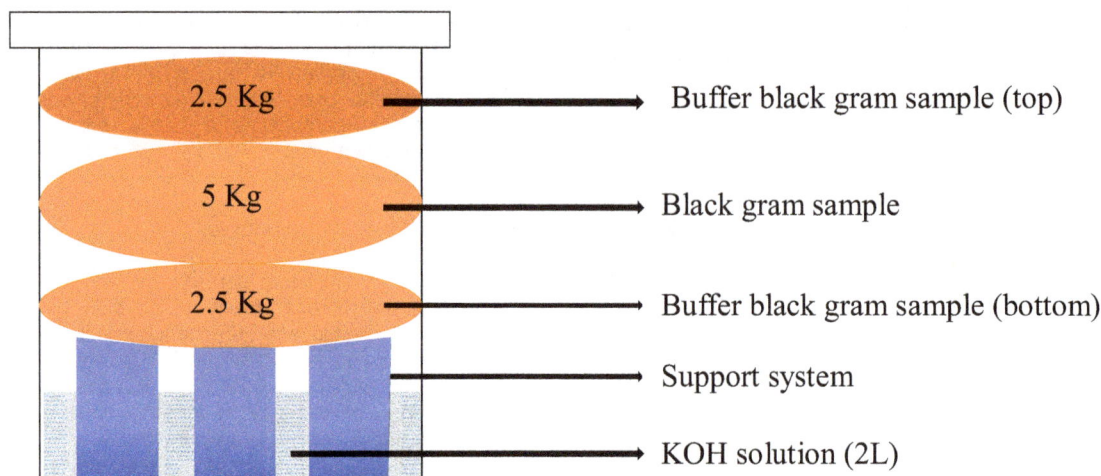

Figure 1. Experimental setup for maintaining equilibrium relative humidity with respect to the moisture content of black gram used for storage studies.

germination (%) was calculated.

Moisture content

The moisture content of the samples was expected to remain constant but needed to be monitored during the study. Therefore, every week, the moisture content was determined by placing approximately 10 g of unground grain in a hot air for 72 h at 130°C (ASAE, 2008). The moisture content of the sample was calculated and expressed in percentage wet basis.

Visible and invisible mould

The deterioration of the grain samples was checked by visual inspection of the samples every week. The presence of invisible mould and the microfloral identification was done every two weeks of storage (Mills et al., 1978). For identification of the invisible microfloral species, 25 seeds were placed in a 9 cm Petri dish with Whattman no. 3 filter paper saturated with 5.5 ml of 7 .5% aqueous sodium chloride (NaCl) solution. The Petri dishes were stacked vertically in stands; covered with plastic bags; and incubated at 25°C for 4 days. On the fourth day the polythene bag was removed and incubation was continued for another 3 days. On the seventh day the microfloral species on the grain were identified using a dissection microscope (Leico Microsystems, India). The percentage of infected seed was calculated and microfloral species were identified based on the appearance of microorganisms.

Fatty acid value (FAV)

The fatty acid values were measured at two week intervals. The analysis was carried out according to the American Association of cereal chemist's procedure (AACC, 1962) with some modifications (Schroth et al., 1998; Karunakaran, 1999). The FAV values were determined by using a fat extractor followed by titration with a KOH solution. The whole grains were dried in hot air oven at 130°C for 19 h. About 5 g of grounded dry grain powder was folded in a whattman no. 5 filter paper. This was placed inside a glass thimble/cylinder and was attached to the fat extractor (Pelican, SOCS PLUS Automatic Solvent Extraction System) with 80 ml of Hexane solvent in beakers. The solvent was heated and allowed to condense, and pass through the same sample for 2 h continuously. The oil was separated from the solvent by heating it again. TAP solution (50% toluene and 50%o ethanol with phenolphthalein indicator) of 25 ml was added to the oil. A KOH solution of known normality was used for titration until the appearance of a pale pink color. The calculated FAV was then expressed as mg KOH/ 100 g of dry grain.

Protein content

Protein content of stored beans were tested prior to storage and after storage to study the effect of storage conditions. Control samples and samples at the end of storage period with different initial moisture contents stored at 20, 30 and 40°C were used for protein analysis using AOAC method 990.03 (AOAC, 1999).

Batter volume rise

The most important quality parameter of Black gram is its fermentation ability. So, batter volume rise was calculated to determine the effect of drying temperature and time on fermentation. Black gram (100g) was soaked for 3 h with 200 ml of distilled water. The soaked black gram was ground in a wet grinder for a constant time of 8 min and the batter was collected. For testing the batter volume rise, 100 ml of the batter was kept for fermentation in the measuring cylinder for 24 h at room temperature (33 ± 2) and the rise in volume was noted to calculate the percentage batter volume rise (Tiwari et al., 2008).

Statistical analysis

Statistical analysis was done to check the effect of moisture content, temperature and storage period on germination rate. The analysis of variance (ANOVA) of a three factorial model (4 moisture contents × 3 temperatures × 43 weeks) was used to study the effects of temperature, storage period and moisture content on the various dependent variables (germination, FAV, protein and percentage batter volume rise). Least significant difference (LSD) was used for pair wise comparisons between quantitative variables

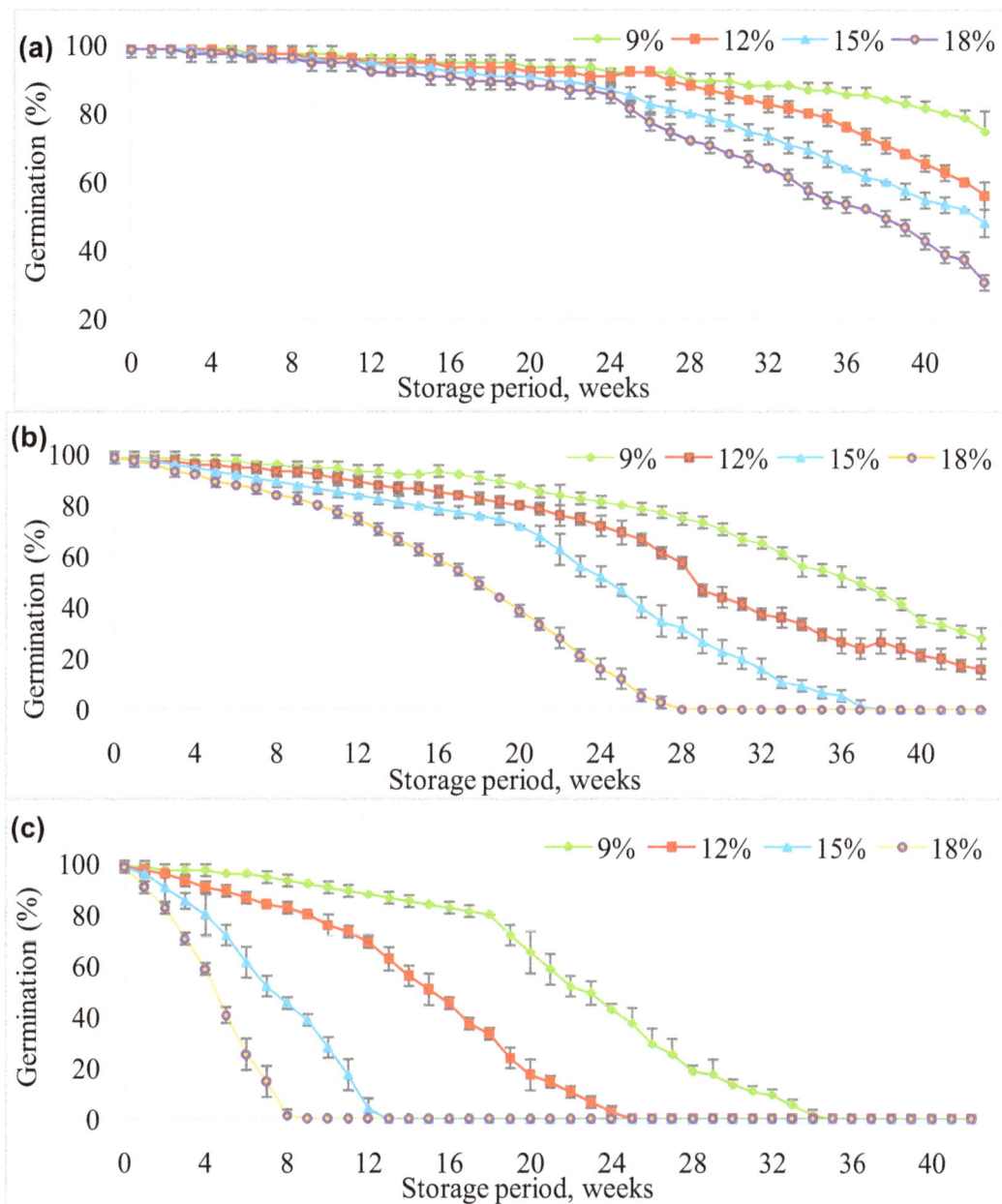

Figure 2. Changes in germination of black gram with different initial moisture contents stored at (a) 20°C, (b) 30°C and (c) 40°C with respect to storage period.

during storage trials. The differences within each level under each variable were tested at a 95% confident interval. General linear models (GLM) procedure in Statistical Analysis System software (SAS version 9.1, Statistical Analysis Systems Institute, Inc., Cary, NC) was used for all the analysis.

RESULTS AND DISCUSSION

Germination

The germination of black gram was used as an indicator

for deterioration during storage as germination is more sensitive to quality changes. Germination is the first factor that gets affected due to improper storage conditions. Figure 2 shows the changes in germination of black gram stored at three different temperatures of 20, 30 and 40°C, with 9, 12, 15 and 18% w.b. initial moisture contents combinations.

The initial germination was around 98.6 % for all the initial moisture contents of black gram stored at different temperature. The grain sample with initial moisture content of 9, 12, 15 and 18% w.b. stored at 20°C did not lose their viability till 25 weeks of storage; however at 40°C, germination reached zero in the 3rd week of storage. The changes in germination at 30°C were in between these two conditions.

At 40°C, the 12, 15 and 18% initial moisture content samples had a significant decrease in germination after a week of storage. Germination trends decreased and reached 0% germination after 35 weeks of storage for the 9, 12, 15 and 18% initial moisture content samples. The highest moisture content (18%) black gram reached 82.6% germination after 2 weeks of storage. Germination of 80% was reached in the initial moisture content black gram of 9, 12 and 15% after 18, 9 and 4 weeks of storage, respectively. The high moisture content samples 15 and 18% reached 0% germination after 13 and 9 weeks of storage whereas, lowest moisture content samples of 9 and   12% samples reached 0 % germination after 35 and 25 weeks of storage (Figure 2a).

In black gram samples stored at 30°C, all the initial moisture content samples were viable up to 10 weeks of storage. The germination of low moisture content samples of 9 and 12% was 78.6% after 26 and 21weeks, respectively. The 15 and 18% initial moisture content samples germination was 78.6 and 80.0% after 16 and 10 weeks of storage, respectively. The 15% initial moisture content black gram lost all germination after 38 weeks of storage. The 18% moisture content black gram followed similar trends and reached 0% germination after the end of the 28 week of storage (Figure 2b).

At 20°C, the entire moisture content black gram was viable up to 25 weeks of storage. The germination rate of 15 and 18% initial moisture content sample was 81.3 and 78.6% after 29 and 25 weeks of storage, respectively. For 12 % initial moisture content samples, germination rate was 78.6% after 35 weeks. The 9% initial moisture content samples germination was 78.6% after 42 weeks of storage (Figure 2c).

According to Pomeranz (1992), germination is the most important factor to assess the quality of grain during storage. The effect of storage parameters like moisture content, temperature and storage time had significant effects (α = 0.05) on the germination. The germination decreased with increasing time, temperature and moisture content. This correlates with the results of Christensen and Kauffmann (1969) who reported that increased storage temperatures cause injury or death to most type of grain. The grain samples with low moisture content were susceptible to spoilage at higher temperature of 40°C. Wallace and Sinha (1962) reported that there exists a negative correlation between germination and storage temperature.

Moisture content

The minimum variation in the moisture content of the samples over time was observed in all the initial moisture content samples. The changes in the initial moisture content of the black gram samples stored at 20, 30 and 40°C with respect to time are shown in Figure 3.

At 20°C, moisture contents of all the black gram samples remained almost constant. The moisture contents of 9, 12, 15 and 18% initial moisture content black gram after 43 weeks of storage increased to 9.9, 12.8, 16.3 and 19.0%, respectively (Figure 3a). At 30°C, the high initial moisture content (15 and 18%) black gram gained moisture over time and increased to 16.2 and 18.1%, respectively after 24 weeks of storage. The buffer samples were replaced to maintain the initial moisture content of the test sample. In 12% initial moisture content samples, there was a gain in moisture content to 13.1% by the end of 43 weeks. The lowest moisture samples (9%) remained almost constant with time and increased to 10.3% moisture content at the end of 43 weeks storage period (Figure 3b).

The grain samples with initial moisture contents of 9, 12, 15 and 18% stored at 40°C, lost moisture over storage time due to drying. The lower initial moisture content of 9 and 12% samples moisture content decreased to 8.3 and 11.4% at the end of 34 and 23 weeks of storage, respectively. The moisture content of higher initial moisture samples (15 and 18%) also decreased with storage time and reached 14.8 and 17.7%, respectively at the end of 12 and 8 weeks (Figure 3c) respectively.

According to Solomon (1951), controlling the relative humidity in biological experiments using chemical solutions like potassium hydroxide has been in practice for a long time. Errors in the humidity control arise if the graded humidity solutions lose too much water through absorption of water vapor by materials enclosed with them or if they absorb water vapor from damp materials. Generally, solutions tend to give too low humidity at elevated temperatures. Furthermore, the equilibrium humidity will deviate from the expected value if the solution is at different temperature than the ambient air temperature above it. This might be the reason for the change in relative humidity inside the storage containers and hence there is a change in moisture contents of the black gram samples over time (Sathya et al., 2008). However by replacement of buffer bags the required initial moisture content of black gram samples was maintained like in the previous studies by Sravanthi et al. (2013) and Rani et al. (2013).

Fatty acid value (FAV)

The fatty acid value (FAV) increases with increase in moisture content of stored black gram. Fatty acid is an intermediate products of hydrolytic reaction caused by the enzymatic secretions of micro-organisms in stored grain. These fatty acids produced in grains have characteristic off-odors. Hence, the FAV has been used

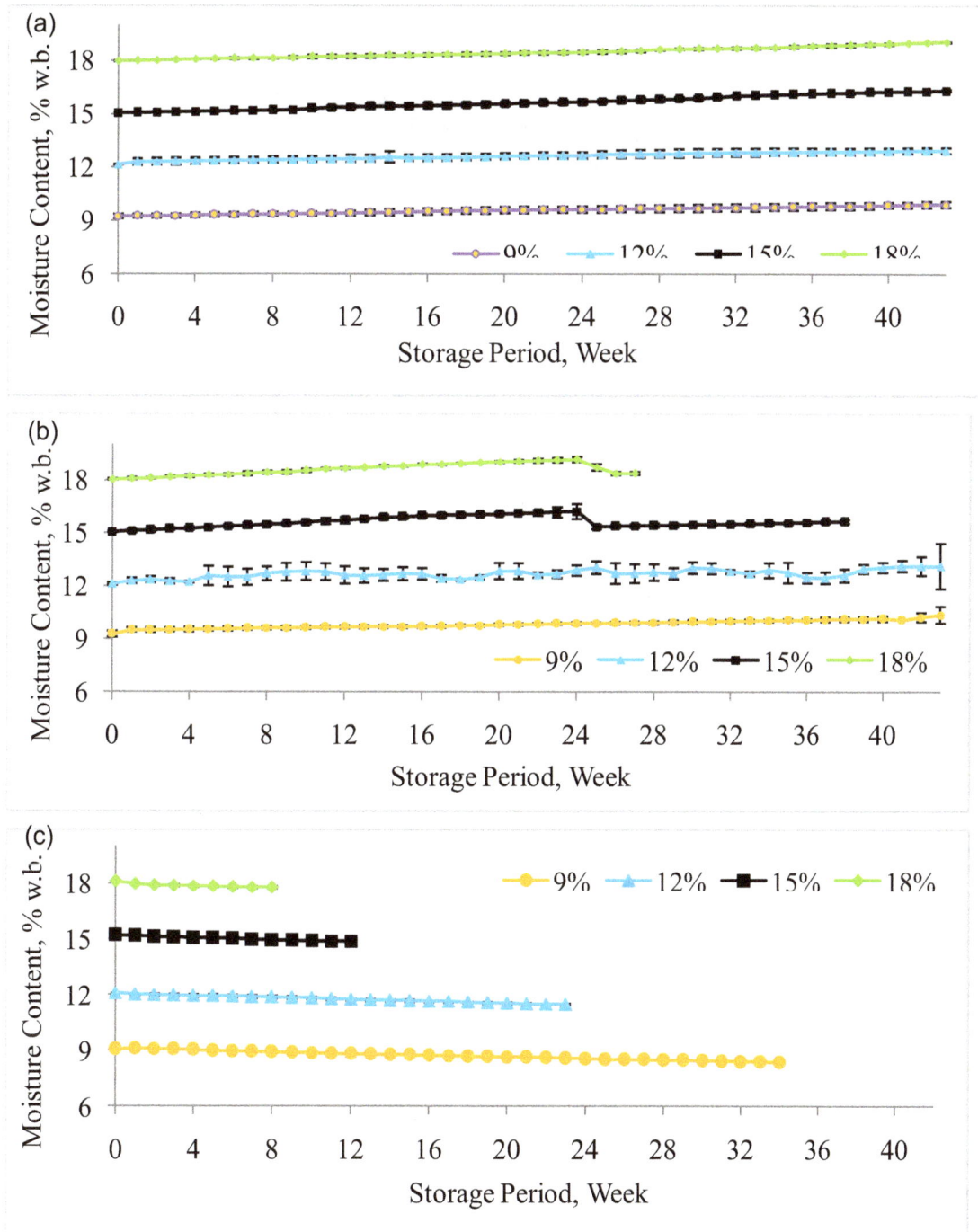

Figure 3. Changes in moisture content of black gram with different initial moisture content stored at (a) 20°C (b) 30°C (c) 40°C with respect to storage period.

as a measure of deterioration of the stored grain (White et al., 1999). The changes in FAV of black gram at three different temperatures of 20, 30 and 40°C, for different initial moisture contents are shown in Figure 4.

The initial free Fatty Acid Value (FAV) of black gram was found to be 2.69 mg KOH/100 g. A positive correlation exists between moisture content, storage temperature, and storage period with the FAV of stored beans as found in previous studies (Karunakaran et al., 2001; Rajarammanna et al., 2010; Nithya et al., 2011; Rani et al., 2013; Sravanthi et al., 2013). The storage period and moisture content of the grain were found to have a significant effect ($\alpha=0.05$) on the FAV values and there was an increase in the FAV values with increase in

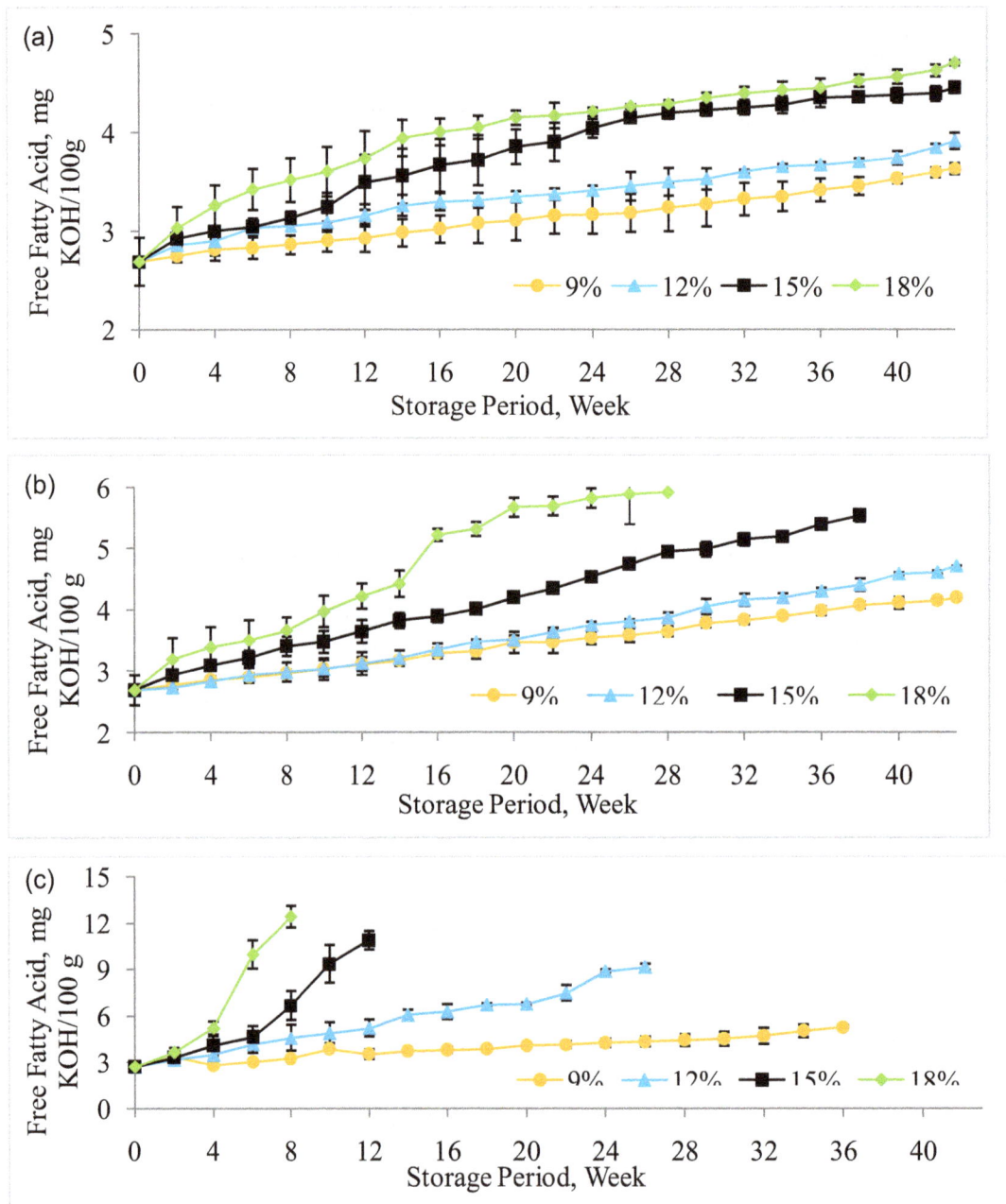

Figure 4. Changes in FAV of different initial moisture content black gram stored at (a) 20°C (b) 30°C (c) 40°C with respect to storage period.

moisture content and storage time. The samples of 15 and 18% initial moisture content stored at 40°C, the FAV values were 11.1 and 12.7 mg KOH/100 g respectively. At 20°C, FAV varied from 3.6 to 4.2 mg KOH/100g for 9 to 18% initial moisture content samples respectively. For the samples stored at 30°C, the FAV of 9 to 18% initial moisture content samples ranged from 4.5 to 5.9 mg KOH/100 g, respectively.

Fatty acids are formed by the hydrolytic reaction caused by the enzymatic secretions of micro-organisms on stored grain. At higher moisture levels grain

undergoes drastic changes due to the proliferation of moulds resulting in production of more free fatty acids at higher moisture and temperature conditions (Nithya et al., 2011).

Visible and invisible mould

The presence of visible mould in grain during storage is always used as an index of deterioration. Increasing the temperature and moisture content favors the growth of

Table 1. Time of first appearance of visible mould on stored red lentil (week) and respective germination (%) of black gram.

Initial moisture content (% wb)	Weeks	Storage temperature (°C)		
	germination	40	30	20
9	Week	23		
	g (%)	49.3 ± 4.6	-	-
12	Week	16	-	-
	g (%)	52 ± 13.8	-	-
15	Week	9	25	-
	g (%)	38.6 ± 2.3	46 ± 2.3	-
18	Week	5	18	33
	g (%)	40 ± 3.0	49.3 ± 2.3	61 ± 2.3

fungi and presence of visible mould. The visible mould appeared in all high moisture samples irrespective of the temperature. The mould growth was first noticed after the germination dropped well below 80% in all the conditions. The appearance of visible mould can be observed mainly around the hilum of the beans and on the cracks in seed coat. The infected seeds were observed under microscope to ensure the presence of mould. There was no mould growth detected for 9% initial moisture content samples stored at 20 and 30°C. Visible moulds appeared in 18% initial moisture content samples stored at 30°C after 18 weeks of storage. At 40°C moulds growth were visible in 5, 9 and 16 weeks in 18, 15 and 12% initial moisture content black gram samples respectively. The time of appearance of visible mould and the respective germination of the black gram samples at that time is given in Table 1.

Different groups and species of moulds was found in the stored black gram at different periods of storage and at different storage temperatures. The commonly found species in black gram were *Aspergillus niger, Aspergillus flavus* and *Rhizopus*. Other microorganisms presented in stored black gram samples were *Penicillium, Chaetomium, Fusarium* and Bacteria. Some of these mould groups might have infected the black gram in the field and their growth and proliferation can be controlled by lowering the storage temperature. Moisture content of the sample is very critical for the development of visible mould. Even at high temperatures, the low moisture samples were devoid of visible mould.

Invisible mould growth was observed after or along with the visible moulds. The results are in accordance with Christensen and Kaufmann (1969) who reported that the invasion of grains by storage fungi is a direct cause of germination loss and some kinds of grains can survive a long time at rather high moisture contents and moderate temperatures, if kept free of storage fungi. Field fungi may present in the freshly harvested grain and storage fungi develop on the stored grain if the storage

conditions are poor (Muir and White, 2001).

Protein content

The protein content of the black gram samples stored at different temperatures was analyzed by the AOAC - 990.03 method (AOAC, 1999). The storage ended for 9, 12, 15 and 18% initial moisture content samples stored at 40°C at 35, 25, 13 and 9 weeks when germination became zero. Similarly 38 and 28 weeks for 15 and 18% initial moisture content samples stored at 30°C. The initial level of protein for 9 and 18% moisture content samples were 24.3 and 22.0%, respectively. The protein content decreased with increase in storage temperature for all the initial moisture contents of stored Black gram. There was a significant difference in protein content of black gram with respect to storage period for all storage temperatures. The protein content of black gram also varied with respect to moisture content ($\alpha=0.05$). Similar results were reported for pinto beans (Rani et al., 2013) and red gram (Sravanthi et al., 2013).

Batter volume rise

Batter volume rise is an important quality parameter as black gram is used in making many fermented products. The decrease in percentage batter volume rise was statistically significant with respect to temperature and moisture content ($\alpha=0.05$). The initial batter volume rise varied from 82 to 86% with increase in moisture from 9 to 18% respectively. As temperature (20 to 40°C) increased, the batter volume rise decreased with storage period. The 20°C stored samples showed a comparatively lesser reduction in batter volume rise. The results can be related to the research by Tiwari et al. (2008), reporting a decrease in fermentation on heat treatment of black gram due to the heat labile muco-protein and inactivation of

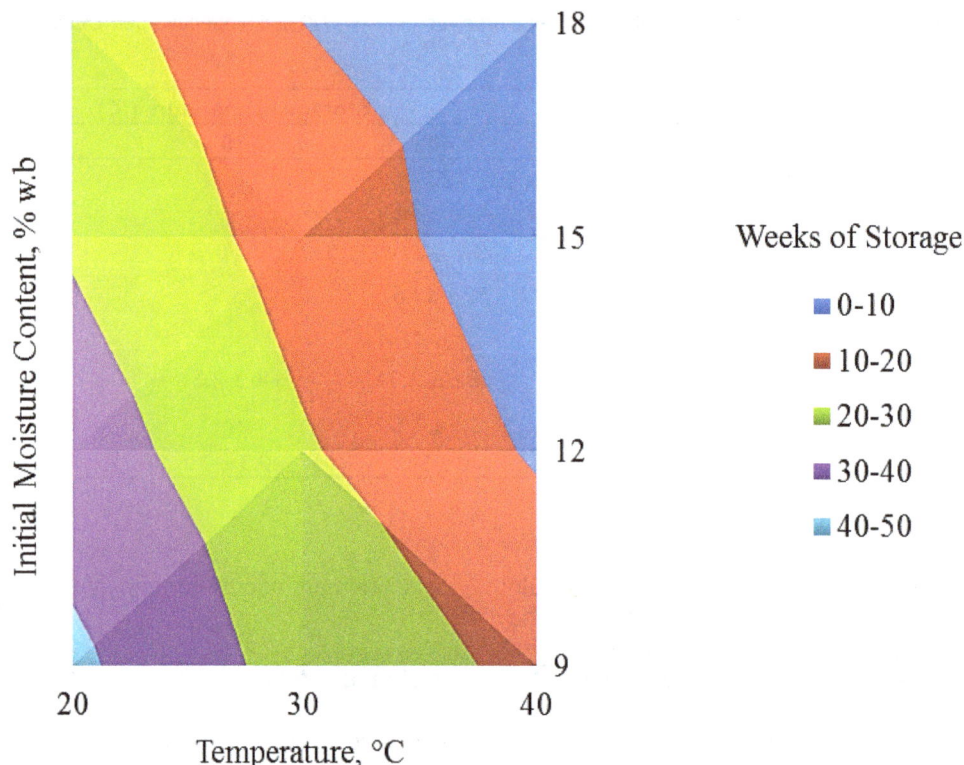

Figure 5. Safe storage guidelines chart for different initial moisture content of black gram stored at different temperature.

enzymes, responsible for fermentation.

Estimated safe storage life of black gram

The estimated safe storage guidelines for black gram were developed based on the decrease in germination rate and appearance of visible mould. When germination was more than 80% of the initial germination (98.8%) and visible mould was absent the black gram is in the safe zone. The number of weeks the samples remained above 80% initial germination without any visible mould was plotted against the initial moisture content and storage temperature to get the estimated safe storage-life guideline (Figure 5). The rate of deterioration increased with increasing initial moisture content and temperature. If the moisture content of the stored grain can be maintained at a sufficiently low level, then that grain can be stored for many years without any significant loss in quality (Tipples, 1995). The time (number of weeks) available for the post-harvest drying treatments before the start of deterioration of stored grain, at a particular moisture and temperature for black gram can be easily determined with the safe storage guidelines chart. The available safe storage time was shown to decrease with increase in moisture content and temperature of the grain during storage.

Conclusion

The storage variables (moisture content of the sample, storage temperature and time of storage) have a negative correlation with germination and a positive correlation with FAV. Germination of the Black gram at 9 and 12% moisture content were maintained above 78.66% at storage temperatures of 20°C for 42 and 34 weeks respectively. But there was significant decrease in germination for higher moisture content (15 and 18%) samples stored at 20, 30 and 40°C with an increase in storage period along with an increase in FAV. The black gram at higher moisture contents (15 and 18%) stored at higher temperature (40°C) lost their viability completely by the end of the 4th and 2nd week of storage with a significant increase in FAV.

The invasion of visible mould around the hilum of the seeds was common in the higher moisture content samples of 15 and 18% irrespective of storage temperature. The lower moisture content samples (9, 12 and 15%) stored at 20°C were free of mould growth for the entire period of storage. *Aspergillus* and *Rhizopus* spp. was the predominant microflora found in the grain. *Pencillium* and *Fusarium* spp. was also common along at lower storage temperature of 20°C. Moisture content, temperature and the storage periods significantly (α=0.05) affected the protein content and batter volume rise of black gram.

Samples with high moisture stored at high temperatures showed visible mould from the beginning of storage. The predominant microfloral species (% of seeds infected) found were *Aspergillus, Rhizopus and Penicillium*. The percentage of infested grains increased with storage time. The maximum storability of 42 weeks with good seed viability and appreciable microbial stability was found in 9% initial moisture content black gram stored at 20°C. The 15 and 18% black gram stored at 30°C, were safe up to 10 weeks of storage period. The higher moisture contents 15 and 18% at high temperature of 40°C, were safe up to 2 to 4 weeks, respectively. The safe storage guidelines chart and safe storage time model developed can be used to predict allowable safe storage time of black gram between moisture content and storage temperature ranges of 9 to 18% w.b. and 20 to 40°C, respectively.

Conflict of Interest

The authors have not declared any conflict of interest.

REFERENCES

AACC (1962). Fat acidity-general method. Method 02-01A. In: Cereal Laboratory Methods, seventh ed. American Association of Cereal Chemists, Inc., St. Paul, MN.
AOAC (1999). Official Method 990.03. Lynch & Barbano: J.AOAC Int. 82(6):1389.
ASAE (2008). Moisture measurement - unground ASAE standards Grain and Seeds. ASAE S352.2. In ASAE standards 2003,555. Sr. Joseph, MI: ASAE.
Basu PS (2011). Vision 2030. Indian Institute of Pulses Research, Kanpur, India.
Campbell-Platt G (1994). Fermented foods-a world perspective. Food Res. Int. 27(3):253-257.
Cassells J, Caddick LP (2000). Storage of desi type chickpeas. In Stored grain in Australia: Proceedings of the Australian Postharvest Technical Conference, pp. 1-4.
Christensen CM, Kaufmann HH (1969). Grain storage; the role of fungi in quality loss. University of Minnesota press, Minneapolis, MN.
Goi (2013). Agricultural Statistics at a glance 2013, Directorate of Economics and Statistics, Department of Agriculture and Cooperation, Ministry of Agriculture.
Hall CW (1980). Drying and storage of agricultural crops. AVI Publishing Company Inc.
Karunakaran C (1999). Modelling safe storage time of high (17 and 19%) moisture content wheat, Masters thesis, University of Manitoba.
Karunakaran C, Muir WE, Jayas DS, White NDG, Abramson D (2001). Safe storage time for high moisture wheat. J. Stored Prod. Res. 37: 303-312.
Kreyger J (1972). Drying and storing grains, seeds and pulses in temperate climates. Institute for storage and processing of agricultural products. Wageningen, Holland.

Mills JT (1992). Safe storage guidelines for grains and their Products. Postharvest News and Information. 3(6):111-115.
Mills JT, Woods SM (1994). Factors affecting storage life of farm-stored field peas (Pisium sativum L.) and white beans (Phaseolus vulgaris L.). J. Stored Products Res. 30:215- 226.
Mills JT, Sinha RN, Wallace HAH (1978). Multivariate evaluation and isolation techniques for fungi associated with stored rapeseed. Phytopathology 68:1520-1525.
Muir WE, White NDG (2001). Microorganisms in stored grain. In Grain Preservation Biosystems, ed. W.E. Muir. Winnipeg, MB: Department of Biosystems Engineering, University of Manitoba, pp. 28-42.
Nithya U, Chelladurai V, Jayas DS, White NDG (2011). Safe storage guidelines for durum wheat. J. Stored Products Res. 47(4):328-333.
Pomeranz Y (1992). Biochemical, functional and nutritive changes during storage. In Storage of Cereal Grains and Their Products, ed. J.A. Anderson and A.W. Alcock, 55-141. St. Paul, MN: American Association of Cereal Chemists.
Rajarammanna R, Jayas DS, White NDG (2010). Comparison of deterioration of rye under two different storage regimes. J. Stored Products Res. 46:87-92.
Rani PR, Chelladurai V, Jayas DS, White NDG, Kavitha-Abirami CV (2013). Storage studies on pinto beans under different moisture contents and temperature regimes. J. Stored Products Res. 52:78-85.
Sathya G, Jayasl DS, White NDG (2008). Safe storage guidelines for rye. Canadian Biosystems Engineering, P 50.
Savadatti PM (2007). An Econometric Analysis of Demand and Supply Response of Pulses in India. Karnataka. J. Agric. Sci. 20(3):545-550.
Schroth E (1996). Modeling allowable storage time of 17% moisture content wheat. Unpublished M. Sc thesis. Winnipeg, MB: Department of Biosystems Engineering, University of Manitoba.
Singh KK, Goswami TK (1996). Physical properties of cumin seed. J. Agric. Eng. Res. (64):93-98.
Singh U, Singh B (1992). Tropical grain legumes as important human foods. Econ. Bot. 46(3):310-321.
Solomon ME (1951). Control of humidity with potassium hydroxide, sulphuric acid, or other solutions. Bull. Entomol. Res. 42:543-554.
Sravanthi B, Jayas DS, Alagusundaram K, Chelladurai V, White NDG (2013). Effect of storage conditions on red lentils. J. Stored Products Res. 53:48–53.
Tiwari BK, Mohan RJ, Vasan BS (2008). Effect of different premilling treatments on dehulling of black gram (Phaseolus mungo L.). J.Of Food Process. Preserv. 32(4):610-620.
Wallace HAH, Sinha RN (1962). Fungi associated with hot spots in farm stored grain. Can. J. Plant Sci. 42:130-141.
White NDG, Hulasare RB, Jayas DS (1999). Effects of storage conditions on quality loss of hull-less and hulled oats and barley. Can. J. Plant Sci. 79(4):475-482.
Schroth E, Muir WE, Jayas DS, White NDG, Abramson D (1998). Storage limit of wheat at 17% moisture content. Can. Agric. Eng. 40(3): 201-205.
Tipples KH (1995). Quality and nutritional changes in stored grain. In Stored-grain Ecosystems, eds. D.S. Jayas, N.D.G. White and W.E. Muir, 325-251. New York, NY: Marcel Decker Inc.

28

Storability of soybean flour and its hazard analysis in Nigeria

Ogundele B. A.[1], Arowora K. A.[2], Abiodun A. A.[3], Afolayan S. S.[3], Ajani A. O.[4] and Adetunji C. O.[3]

[1]Nigerian Stored Products Research Institute, c/o Lake Chad Research Institute, P. M. B. 1293, Maiduguri, Borno State, Nigeria.
[2]Nigerian Stored Products Research Institute, P. M. B.3032, Kano, Kano State, Nigeria.
[3]Nigerian Stored Products Research Institute, P. M. B. 1489, Ilorin, Kwara State, Nigeria.
[4]Nigerian Stored Products Research Institute, P. M. B. 5044, Ibadan, Oyo State, Nigeria.

Soybean was procured from a local market in Ilorin, washed, dried, milled, packaged and stored under hermetic conditions using transparent plastic container. Proximate composition was carried out on the samples 4-monthly (once-in -4 months) for a period of one year. Moisture content was determined by air-oven method while proximate composition was carried out by Standard Methods of Association of Official Analytical Chemists (AOAC). Moisture contents ranged between 8.0 and 8.9% within one year of storage, while protein levels reduced to 34.9 from 40.4%. Whereas increasing trend was recorded in ash content during the course of storage with ranging values between 4.9 and 5.4%. No definite trend was observed in fibre composition, however, the final value at the end of 12-month storage was found to be 5.8%. Hazard analysis and critical control points (HACCP) procedures were developed and applied for production of high quality as well as safe soybean flour for both local consumption and export.

Key words: Storage, soybean, soy flour, hazard analysis and critical control points (HACCP), hazard, multipurpose dryer.

INTRODUCTION

Oilseeds are one of the most important underutilized raw materials. According to Pyke (1964), about 32% of all edible fats and oils in the world market are largely derived from vegetable sources like cotton seed, groundnut, coconut and soybean. World soybean production in the 2009/2010 harvest was roughly 260 million tons, and the major producers were the United States, Brazil and Argentina, producing 91.4, 69.0 and 57.0 million tons, respectively (USDA, 2011). Given the significant world production of soybeans, quality is essential for the sectors involved in production and/or processing of this commodity. Quality is an important parameter for commercialization and processing of the grains and can affect the value of the product and its derivatives. Soybean (Glycine max) is a legume crop classified as an oilseed and a good source of high quality protein because it contains significant amount of essential amino acids. Its cultivation is becoming more popular with farmers in the derived savannah zone of Nigeria and production levels are increasing every year especially now when people are aware of its uses. Good storage management can greatly influence the storability of soybean and subsequent germination when planted in the field as well as other products developed from it. High moisture content and temperature has been reported to increase deterioration and reduce seed viability in storage. Soybean should be stored at a moisture content of 10% or less. At harvest, soybean grains usually contain about 14% moisture. It has been found that soybean can be stored for 6 to 12 months when dried to 13% moisture content, while it can also be stored for longer period when dried to between 10 and 11% moisture content. Open-air drying is the most practical way to protect soybean in storage.

Nutritionally, soybean contains 40.00% crude protein, 19.10% ether extract, 5.71% crude fibre, 5.06% mineral content and 26.05 nitrogen free extract (Oyenuga, 1968). Bates et al. (1977) found that the chemical composition of soybeans changed with the development of the seeds and also reported different chemical composition in vitamins of mature, immature and sprouted vitamins. They found that ascorbic acid and B-carotene decreased with maturity and revealed that the level of B-complex vitamins increased four days after germination, suggesting that sprout could be another nutritious way to consume soybean. The importance of soybean flour in feed formulation and livestock production cannot be overemphasized; this was buttressed by investigations carried out by several authors. Arowora et al. (2004) investigated the utilization of weaner pigs fed soybean and other feed ingredients including biodegraded cassava peel and observed satisfactory growth performance at the end of 8 weeks. Also, Mitaru and Blair (1985) experimented on the comparative effects of dark and yellow rapeseed hulls, soybean hulls and a purified fibre source on growth, feed consumption and digestibility of dietary components in weanling pigs and found that the feed efficiency (gain: feed) values were similar for all dietary treatments with values ranging from 0.53 to 0.57. The market for soybean in Nigeria is growing very fast with opportunities for improving the income of farmers. Currently, SALMA Oil Mills in Kano, Grand Cereals in Jos, ECWA Feeds in Jos, AFCOT Oil Seed Processors, Ngurore, Adamawa State, and PS Mandrides in Kano, all these companies process soybean (Dugje et al., 2009).

In the light of aforementioned, therefore, the objective of this study was to investigate the storability of high quality soybean flour produced in Nigeria with the view of making soybean flour available for its various utilization throughout the year at reasonable prices.

MATERIALS AND METHODS

Soybean was procured from a local market in Ilorin, Kwara State, Nigeria. One batch, divided into three samples was used in this research investigation. The samples were washed and dried using multipurpose dryer (MPD) developed by researchers at Nigerian Stored Products Research Institute (NSPRI). The dried samples were milled and analyzed for proximate composition before packaging in transparent polythene bags with gauge of 0.04 mm and stored under hermetic conditions for one year. Samples were taken for analysis at the end of 4th, 8th and 12th month storage respectively. The three samples were taken as three replicates for analysis. The mean values of 3 replicates with their corresponding standard errors were recorded for analysis. The treatment means of the samples analyzed were subjected to t-test at 5% level of significance using two-tail. Proximate determination of samples was carried out using the standard methods of AOAC (2000). Protein content of *soybean flour* samples was estimated using Microkjeldahl distillation apparatus as per the standard method of AOAC (2000). Crude fat content (triglycerides of fatty acid) of *soyflour* samples was estimated as per the standard method of AOAC (2010) using fat extraction tube of soxhlet apparatus. Ash content was carried out by igniting the sample until only the inorganic residue was left:

$$\text{Ash content} = \frac{\text{Weight of ash}}{\text{Weight of sample}} \times 100$$

Crude fibre was determined as that fraction remaining after digestion with standard solutions of sulphuric acid and sodium hydroxide under carefully controlled conditions.

$$\% \text{ crude fibre in sample} = \frac{Mr - Ma}{Ms}$$

Mr = Mass (g) of crucible + dried residue; Ma = Mass (g) of crucible + ash; Ms = Mass (g) of sample taken
Hazard analysis was developed and applied in the production of safe soybean flour.

RESULTS AND DISCUSSION

Table 1 shows the proximate composition of soybean flour. The results indicated that moisture reduction of samples were still within safe moisture level. Moisture content (MC) represents the amount of extrinsic water in the samples analyzed. There was a decreasing trend in MC values obtained during the course of storage. This decrease was significantly lower (P<0.05) when the final value of treatment mean was compared with the initial of 8.9±1.0%. The decreasing trend observed is similar to the work of Mejule and Lameke (1982) who reported moisture reduction of 0.31% during the course of storage of cocoa beans. Similar result in moisture reduction of 0.30% was obtained by Opadokun and sowunmi (1985) who worked on storability and quality of maize and sorghum stored in metal silos for four years. Gradual reduction was observed in protein levels during the course of storage. The crude protein of freshly dried soybean flour was found to be 40.4±0.4%, while reduction at the end of 12-month storage was found to be 34.9±0.1%. This decreasing trend was found to be significant (P<0.05). This observation is similar to the results obtained by Opadokun and Sowunmi (1985) who

Table 1. Proximate composition of soybean flour in storage.

Parameters	Mean values (before storage)	Mean values (4-month storage)	Mean values (8-month storage)	Mean values (12-month storage)
Crude protein (%)	40.4±0.4	38.3±0.3	37.3±0.1	34.9±0.1
Crude fibre (%)	5.3±0.1	6.1±0.2	5.6±0.1	5.8±0.1
Ash (%)	4.9±0.1	5.1±0.1	5.3±0.1	5.4±0.1
Ether extract (%)	21.6±1.3	14.6±0.4	13.4±0.8	13.8±0.1
N.F.E. (%)	27.3±0.2	35.8±1.4	38.8±1.6	40.5±1.7
M.C. (%)	8.9±1.0	8.3±0.9	8.2±0.9	8.0±1.0

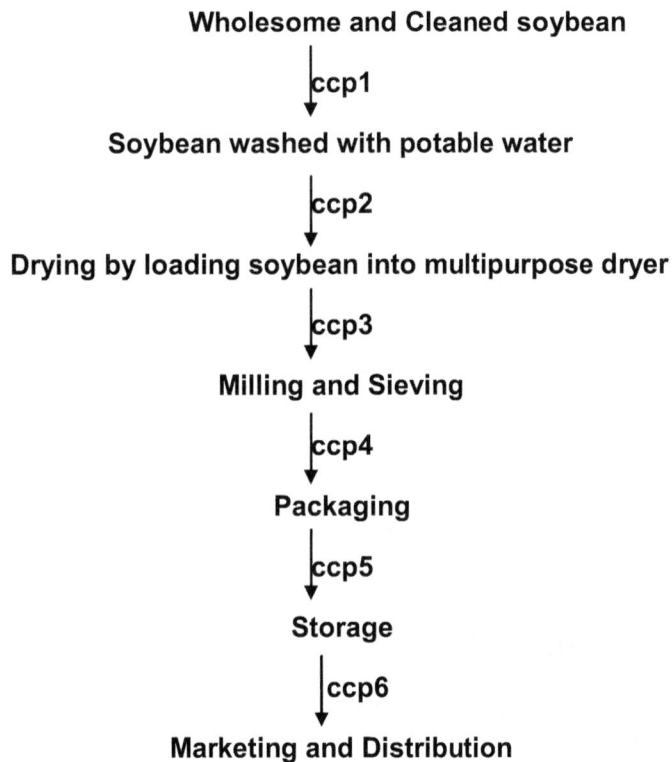

Wholesome and Cleaned soybean

| ccp1

Soybean washed with potable water

| ccp2

Drying by loading soybean into multipurpose dryer

| ccp3

Milling and Sieving

| ccp4

Packaging

| ccp5

Storage

| ccp6

Marketing and Distribution

Figure 1. Flow chart for production of soybean flour.

found reduction of 0.3% in the crude protein of maize stored in silos at the end of 12 months. These authors also found reduction of 0.9% in crude protein of sorghum stored in silos at the end of three years.

There was no definite trend in fibre composition in storage. Fibre was observed to range between 5.3 ± 0.1 and 6.1±0.2% during the course of one year storage. This trend was not significant (P>0.0.05) when the treatment means of the initial and final fibre content were compared. It was observed that the crude fibre composition of soybean flour sample in this study was higher than that of yellow maize and guinea corn which were found to be between 1.32 and 2.94% respectively (Oyenuga, 1968). The total minerals composition (ash) in this study was found to increase from 4.9 to 5.4% in

storage. However, this increase was not significant (P>0.05) between the initial and final values of minerals content. This was in agreement with the results obtained by Opadokun and Sowunmi (1985) that had range between 2.1 and 3.1% during the course of storage of sorghum in metal silos for one year. Ether extract was found to be decreasing generally during the course of storage. Although the lowest value was obtained from samples stored for 8 months with the value of 13.4 ± 0.1% , it was found that this decrease was significantly lower (P<0.05) when compared the initial and final values of ether extract obtained.

Nitrogen free extract (NFE), this simply refers to the carbohydrate composition of the samples analyzed. There was increase in the trend of NFE values obtained during the course of storage. This increase was significant (P<0.05) when compared the treatment means of initial and final NFE values obtained in storage.

Hazard analysis and critical control points (HACCP) was applied to control and reduce the hazards to acceptable levels thereby producing safe soybean flour for human consumption as follows: Wholesome and clean soybean was procured from a local market in Ilorin and transported to the laboratory using transparent, cleaned and covered plastic, (CCP_1) was uncontaminated soybean (Figure 1). Cleaned soybean samples were washed with potable water in order to eliminate microbial hazard and extraneous materials (CCP_2). Washed soybean samples were loaded into the multipurpose dryer (MPD). At this stage, good handling practices were employed in order to eliminate body contamination and pathogens (CCP_3). The milling of dehydrated soybean was carried out using stainless steel hammer mill and stainless sieves which were sterilized (CCP_4). The soybean flour samples were packaged in food grade polythene which conformed with standard specification with gauge of 0.04 mm. This would eliminate leakage and microbial contamination (CCP_5). The packaged soybean flour were stored under hermetic conditions until sold. This would eliminate moisture migration, caking, mouldiness and mycotoxin contamination during the course of storage (CCP_6). The soybean flour sold to the market were monitored until sold completely. The vehicle used for its transportation was thoroughly cleaned to avoid contamination.

Conclusion

The results of this study showed that soybean flour produced was highly nutritious and can be stored without adverse effects on its qualities for one year in transparent polythene bags under hermetic conditions. The final protein level of 34.9% is high enough to justify its consumption at the end of one year. Hence it is still safe for local consumption and export within one year of production.

Conflict of Interest

The authors have not declared any conflict of interest.

REFERENCES

AOAC (2000). Association of Official Analytical Chemists. Official Methods of Analysis. 17th Edition, Washington, D.C.

Arowora KA, Onilude AA, Tewe OO (2004). Performance and nutrient utilization weaner pigs fed different levels of biodegraded cassava-peel based diets, Trop. Anim. Prod. Invest. 1:37-44.

Bates RP, Knapp, FW, Araujo PE (1977). Protein quality of green mature, dry mature and sprouted soybeans. J. Food Sci. 42:271-272. http://dx.doi.org/10.1111/j.1365-2621.1977.tb01269.x

Dugje IY, Omoigui LO, Ekeleme F, Bandyopadhyay R, Lava KP Kamara AY (2009). Farmers' guide to soybean production in Northern Nigeria. International Institute of Tropical Agriculture, Ibadan, Nigeria, P. 21.

Mejule FO, Lameke R (1982). Changes in weight of cocoa during storage. Nigerian Stored Products Research Institute, 19th Annual Report, pp. 137-142.

Mitaru BN, Blair R (1985). Composition of the effect of dark and yellow rapeseed hulls and a purified fibre source on growth, feed consumption and digestibility of dietary components in starter pigs. Can. J. Anim Sci. 65:231-237. http://dx.doi.org/10.4141/cjas85-025

Opadokun JS, Sowunmi O (1985). Chemical Quality of maize and sorghum stored in nitrogen in metal silos for four years. Nigerian Stored Products Research Institute, 22nd Annual Report, pp. 67-68.

Oyenuga VA (1968). Nigerian Foods and Feeding stuffs. Their chemistry and nutritive values. Ibadan University Press, 3rd Edition.

Pyke M (1964). In: Principles of food Science Wikipedia Encyclopedia, http://www.en.wikipedia.org/wiki/soybean

USDA (2011). U.S. soybean inspection March, 14.03.2011, Available from http://www.usda.gov/oce/commodity/wasde/latest.pdf

Permissions

All chapters in this book were first published in JSPPR, by Academic Journals; hereby published with permission under the Creative Commons Attribution License or equivalent. Every chapter published in this book has been scrutinized by our experts. Their significance has been extensively debated. The topics covered herein carry significant findings which will fuel the growth of the discipline. They may even be implemented as practical applications or may be referred to as a beginning point for another development.

The contributors of this book come from diverse backgrounds, making this book a truly international effort. This book will bring forth new frontiers with its revolutionizing research information and detailed analysis of the nascent developments around the world.

We would like to thank all the contributing authors for lending their expertise to make the book truly unique. They have played a crucial role in the development of this book. Without their invaluable contributions this book wouldn't have been possible. They have made vital efforts to compile up to date information on the varied aspects of this subject to make this book a valuable addition to the collection of many professionals and students.

This book was conceptualized with the vision of imparting up-to-date information and advanced data in this field. To ensure the same, a matchless editorial board was set up. Every individual on the board went through rigorous rounds of assessment to prove their worth. After which they invested a large part of their time researching and compiling the most relevant data for our readers.

The editorial board has been involved in producing this book since its inception. They have spent rigorous hours researching and exploring the diverse topics which have resulted in the successful publishing of this book. They have passed on their knowledge of decades through this book. To expedite this challenging task, the publisher supported the team at every step. A small team of assistant editors was also appointed to further simplify the editing procedure and attain best results for the readers.

Apart from the editorial board, the designing team has also invested a significant amount of their time in understanding the subject and creating the most relevant covers. They scrutinized every image to scout for the most suitable representation of the subject and create an appropriate cover for the book.

The publishing team has been an ardent support to the editorial, designing and production team. Their endless efforts to recruit the best for this project, has resulted in the accomplishment of this book. They are a veteran in the field of academics and their pool of knowledge is as vast as their experience in printing. Their expertise and guidance has proved useful at every step. Their uncompromising quality standards have made this book an exceptional effort. Their encouragement from time to time has been an inspiration for everyone.

The publisher and the editorial board hope that this book will prove to be a valuable piece of knowledge for researchers, students, practitioners and scholars across the globe.

List of Contributors

A. K. Musa
Department of Crop Protection, University of Ilorin, P. M. B. 1515, Ilorin, Nigeria

D. M Kalejaiye
Department of Crop Protection, University of Ilorin, P. M. B. 1515, Ilorin, Nigeria

L. E. Ismaila
Department of Crop Protection, University of Ilorin, P. M. B. 1515, Ilorin, Nigeria

A. A. Oyerinde
Department of Biology, University of Abuja, Abuja, Nigeria

Z. D. Osunde
Department of Agricultural Engineering, Federal University of Technology, Minna, Niger State, Nigeria

B. A. Orhevba
Department of Agricultural Engineering, Federal University of Technology, Minna, Niger State, Nigeria

F Appiah
Department of Horticulture, Kwame Nkrumah University of Science and Technology, Kumasi, Ghana

R Guisse
Department of Horticulture, Kwame Nkrumah University of Science and Technology, Kumasi, Ghana

P. K. A Dartey
Crops Research Institute of the Centre for Scientific and Industrial Research, Fumesua, Kumasi, Ghana

M. N. Muchui
Kenya Agricultural Research Institute-Thika, P.O. Box 220-01000, Thika, Kenya

F. M. Mathooko
South Eastern University College (A constituent College of the University of Nairobi), P.O.Box 170-90200, Kitui, Kenya

C. K. Njoroge
Jomo Kenyatta University of Agriculture and Technology, P. O. Box 62000-00200, Nairobi, Kenya

E. M. Kahangi
Jomo Kenyatta University of Agriculture and Technology, P. O. Box 62000-00200, Nairobi, Kenya

C. A. Onyango
Jomo Kenyatta University of Agriculture and Technology, P. O. Box 62000-00200, Nairobi, Kenya

E. M. Kimani
Kenya Agricultural Research Institute-Thika, P.O. Box 220-01000, Thika, Kenya

S. B. Ohen
Department of Agricultural Economics and Extension, University of Calabar, Calabar, Cross River State, Nigeria

S. O. Abang
Department of Agricultural Economics and Extension, University of Calabar, Calabar, Cross River State, Nigeria

Mir Javad Chailoo
Department of Horticulture, Faculty of Agriculture, Urmia University, Iran

Mohammad Reza Asghari
Department of Horticulture, Faculty of Agriculture, Urmia University, Iran

D. Baskaran
Department. of Dairy Science, Madras Veterinary College, Chennai- 600 007, India

K. Muthupandian
Department. of Dairy Science, Madras Veterinary College, Chennai- 600 007, India

K. S. Gnanalakshmi
Department. of Dairy Science, Madras Veterinary College, Chennai- 600 007, India

T. R. Pugazenthi
Department. of Dairy Science, Madras Veterinary College, Chennai- 600 007, India

S. Jothylingam
Department. of Dairy Science, Madras Veterinary College, Chennai- 600 007, India

K. Ayyadurai
Department of Veterinary Biochemistry, MVC, Chennai- 600 007, India

Sree S. Lekha
Central Tuber Crops Research Institute, Sreekariyam, Trivandrum 695017, Kerala, India

Jaime A. Teixeira da Silva
Faculty of Agriculture and Graduate School of Agriculture, Kagawa University, Miki-cho, Ikenobe 2393, Kagawa-ken, 761-0795, Japan

Santha V. Pillai
Central Tuber Crops Research Institute, Sreekariyam, Trivandrum 695017, Kerala, India

Gurmesa Umeta
Adami Tulu Agricultural Research Center, P. O. Box 35, Zeway, Ethiopia

Feyisa Hundesa
Adami Tulu Agricultural Research Center, P. O. Box 35, Zeway, Ethiopia

Misgana Duguma
Adami Tulu Agricultural Research Center, P. O. Box 35, Zeway, Ethiopia

Merga Muleta
Adami Tulu Agricultural Research Center, P. O. Box 35, Zeway, Ethiopia

FF Ilesanmi
Nigerian Stored Products Research Institute, PMB 5044, Ibadan, Nigeria

OA Oyebanji
Nigerian Stored Products Research Institute, PMB 5044, Ibadan, Nigeria

AR Olagbaju
Nigerian Stored Products Research Institute, PMB 5044, Ibadan, Nigeria

MO Oyelakin
Nigerian Stored Products Research Institute, PMB 5044, Ibadan, Nigeria

KO Zaka
Nigerian Stored Products Research Institute, PMB 5044, Ibadan, Nigeria

AO Ajani
Nigerian Stored Products Research Institute, PMB 5044, Ibadan, Nigeria

MF Olorunfemi
Nigerian Stored Products Research Institute, PMB 5044, Ibadan, Nigeria

TM Awoite
Nigerian Stored Products Research Institute, PMB 5044, Ibadan, Nigeria

IO Ikotun
Nigerian Stored Products Research Institute, PMB 5044, Ibadan, Nigeria

AO Lawal
Nigerian Stored Products Research Institute, PMB 5044, Ibadan, Nigeria

JP Alimi
Nigerian Stored Products Research Institute, PMB 5044, Ibadan, Nigeria

Ngando Ebongue Georges Frank
Centre spécialisé de Recherche sur le palmier à huile de La Dibamba, BP 243 Douala, Cameroun

Mpondo Mpondo Emmanuel Albert
University of Douala, Faculty of science, Department of Biochemistry, P. O. Box 24 157 Douala, Cameroon

Dikotto Ekwe Emmanuel Laverdure
University of Douala, Faculty of science, Department of Biochemistry, P. O. Box 24 157 Douala, Cameroon

Koona Paul
Centre spécialisé de Recherche sur le palmier à huile de La Dibamba, BP 243 Douala, Cameroun

Ozumba Isaac C
Processing and Storage Engineering Department, National Centre for Agricultural Mechanization (NCAM), P. M. B. 1525, Ilorin, Nigeria

Obiakor Sylvester I
Processing and Storage Engineering Department, National Centre for Agricultural Mechanization (NCAM), P. M. B. 1525, Ilorin, Nigeria

M. A. Basunia
Department of Soils, Water and Agricultural Engineering, Sultan Qaboos University, P. O. Box 34, Al-Khod 123, Muscat, Sultanate of Oman

M. A. Rabbani
Department of Farm Power and Machinery, Bangladesh Agricultural University, Mymensingh 2202, Bangladesh

Sushil Kumar Shahi
Bioresource Technology Laboratory, Department of Microbiology, CCS University, Meerut- 250005, India

Mamta Patra Shahi
Department of Biotechnology, Meerut Institute of Engineering and Technology, Meerut-250005, India

Kinfe Asayehegn
Institute of Environment, Gender and Development Studies, Awassa College of Agriculture, Hawassa University, Ethiopia

Chilot Yirga
Ethiopian Institute of Agricultural Research (EIAR), Ethiopia

Sundara Rajan
Department of Rural Development and Agricultural Extension, Harapmaya University, Ethiopia

P. K. Baidoo
Department of Theoretical and Applied Biology, Kwame Nkrumah University of Science and Technology, Kumasi, Ghana

M. B. Mochiah
Entomology Section, Crop Research Institute, P. O. Box 3785, Kumasi Ghana

M. Owusu -Akyaw
Entomology Section, Crop Research Institute, P. O. Box 3785, Kumasi Ghana

C. M. Ngatia
KARI – Kabete, Post harvest research, NARL. P. O. Box 14733-00800 Nairobi, Kenya

M. Kimondo
KARI – Kabete, Post harvest research, NARL. P. O. Box 14733-00800 Nairobi, Kenya

B. J. Opara
Michael Okpara University of Agriculture, Umudike, Abia State, Nigeria

N. Uchechukwu
Michael Okpara University of Agriculture, Umudike, Abia State, Nigeria

R. M. Omodamiro
National Root Crops Research Institute, Umudike, Abia State, Nigeria

N. Paul
National Root Crops Research Institute, Umudike, Abia State, Nigeria

S. Ajijola
Institute of Agricultural Research and Training, PMB 5029, Moor Plantation, Ibadan, Nigeria

J. M. Usman
Federal College of Forestry, P.M.B.5087, Ibadan, Nigeria

O. A Egbetokun
Institute of Agricultural Research and Training, PMB 5029, Moor Plantation, Ibadan, Nigeria

J. Akoun
Federal College of Forestry, P.M.B.5087, Ibadan, Nigeria

C. S. Osalusi
Federal College of Forestry, P.M.B.5087, Ibadan, Nigeria

E. Y. Nyamah
Department of Horticulture, Kwame Nkrumah University of Science and Technology, Kumasi, Ghana

B. K. Maalekuu
Department of Horticulture, Kwame Nkrumah University of Science and Technology, Kumasi, Ghana

D. Oppong-skyere
Department of Horticulture, Kwame Nkrumah University of Science and Technology, Kumasi, Ghana

K. Kalidas
Vanavarayar Institute of Agriculture, Pollachi -642103, Tamil Nadu State, India

K. Akila
Bank of Baroda, Senjerimalai, Coimbatore – 642002, Tamil Nadu State, India

Kietsuda Luengwilai
Department of Plant Sciences, University of California, Mail Stop 3, One Shields Avenue, Davis CA 95616, USA

Diane M. Beckles
Department of Plant Sciences, University of California, Mail Stop 3, One Shields Avenue, Davis CA 95616, USA

AMRITESH C. SHUKLA
Department of Horticulture, Aromatic and Medicinal Plants, Mizoram University, Aizawl-796 004, India

J. M. Usman
Federal College of Forestry, PMB 5087, Jericho, Ibadan, Nigeria

J. A. Adeboye
Federal College of Forestry, PMB 5087, Jericho, Ibadan, Nigeria

K. A. Oluyole
Cocoa Research Institute of Nigeria, PMB 5244, Idi-Ayunre, Ibadan, Nigeria

S. Ajijola
Institute of Agricultural Research and Training, PMB 5029, Moor Plantation, Ibadan, Nigeria

Abimbola O. Adepoju
Department of Agricultural Economics, University of Ibadan, Oyo State, Nigeria

Marcos Ribeiro da Silva Vieira
Universidade Federal Rural de Pernambuco, Unidade Acadêmica de Serra Talhada, CEP: 59909-460, Serra Talhada, PE, Brasil

Adriano do Nascimento Simões
Universidade Federal Rural de Pernambuco, Unidade Acadêmica de Serra Talhada, CEP: 59909-460, Serra Talhada, PE, Brasil

Glauber Henrique Sousa Nunes
Departamento de Ciências Vegetais/UFERSA, Caixa Postal 137, CEP: 59625-900, Mossoró, RN, Brasil

Pahlevi Augusto de Souza
Instituto Federal de Educação, Ciência e Tecnologia do Ceará, CEP: 62930-000, Limoeiro do Norte, CE, Brasil

C. Sen
Agricultural and Food Engineering Department, Indian Institute of Technology Kharagpur, Kharagpur-721302, West Bengal, India

H. N. Mishra
Agricultural and Food Engineering Department, Indian Institute of Technology Kharagpur, Kharagpur-721302, West Bengal, India

P. P. Srivastav
Agricultural and Food Engineering Department, Indian Institute of Technology Kharagpur, Kharagpur-721302, West Bengal, India

M. Esther Magdalene Sharon
Indian Institute of Crop Processing Technology (IICPT), Thanjavur, Tamil Nadu, India

C. V. Kavitha Abirami
Indian Institute of Crop Processing Technology (IICPT), Thanjavur, Tamil Nadu, India

K. Alagusundaram
Indian Institute of Crop Processing Technology (IICPT), Thanjavur, Tamil Nadu, India

J. Alice Sujeetha
Indian Institute of Crop Processing Technology (IICPT), Thanjavur, Tamil Nadu, India

B. A. Ogundele
Nigerian Stored Products Research Institute, c/o Lake Chad Research Institute, P. M. B. 1293, Maiduguri, Borno State, Nigeria

K. A. Arowora
Nigerian Stored Products Research Institute, P. M. B.3032, Kano, Kano State, Nigeria

A. A. Abiodun
Nigerian Stored Products Research Institute, P. M. B. 1489, Ilorin, Kwara State, Nigeria

S. S. Afolayan
Nigerian Stored Products Research Institute, P. M. B. 1489, Ilorin, Kwara State, Nigeria

A. O. Ajani
Nigerian Stored Products Research Institute, P. M. B. 5044, Ibadan, Oyo State, Nigeria

C. O. Adetunji
Nigerian Stored Products Research Institute, P. M. B. 1489, Ilorin, Kwara State, Nigeria

www.ingramcontent.com/pod-product-compliance
Lightning Source LLC
Chambersburg PA
CBHW050443200326

41458CB00014B/5052